# WEIGHT
# IN AMERICA
## OBESITY, EATING DISORDERS, AND OTHER HEALTH RISKS

ISSN 1551-2118

# WEIGHT IN AMERICA

## OBESITY, EATING DISORDERS, AND OTHER HEALTH RISKS

Barbara Wexler

**INFORMATION PLUS® REFERENCE SERIES**
Formerly published by Information Plus, Wylie, Texas

GALE
CENGAGE Learning

Detroit • New York • San Francisco • New Haven, Conn • Waterville, Maine • London

**Weight in America: Obesity, Eating Disorders, and Other Health Risks**

**Barbara Wexler**
**Paula Kepos, Series Editor**

Project Editors: Kathleen J. Edgar, Elizabeth Manar

Rights Acquisition and Management: Jackie Jones, Barb McNeil, Tim Sisler

Composition: Evi Abou-El-Seoud, Mary Beth Trimper

Manufacturing: Cynde Bishop

Product Management: Carol Nagel

For product information and technology assistance, contact us at
**Gale Customer Support, 1-800-877-4253.**
For permission to use material from this text or product,
submit all requests online at **www.cengage.com/permissions.**
Further permissions questions can be e-mailed to
**permissionrequest@cengage.com**

Cover photograph: Image copyright Ljupco Smokovski, 2008. Used under license from Shutterstock.com.

Gale
27500 Drake Rd.
Farmington Hills, MI 48331-3535

ISBN-13 978-0-7876-5103-9 (set)          ISBN-10 0-7876-5103-6 (set)
ISBN-13 978-1-4144-0782-1                     ISBN-10 1-4144-0782-3

ISSN 1551-2118

This title is also available as an e-book.
ISBN-13: 978-1-4144-3819-1 (set)
ISBN-10: 1-4144-3819-2 (set)
Contact your Gale sales representative for ordering information.

Printed in the United States of America
1 2 3 4 5 6 7 12 11 10 09 08

# TABLE OF CONTENTS

# PREFACE

*Weight in America: Obesity, Eating Disorders, and Other Health Risks* is part of the *Information Plus Reference Series*. The purpose of each volume of the series is to present the latest facts on a topic of pressing concern in modern American life. These topics include today's most controversial and studied social issues: abortion, capital punishment, care for the elderly, crime, health care, the environment, immigration, minorities, social welfare, women, youth, and many more. Even though this series is written especially for high school and undergraduate students, it is an excellent resource for anyone in need of factual information on current affairs.

By presenting the facts, it is the intention of Gale, a part of Cengage Learning, to provide its readers with everything they need to reach an informed opinion on current issues. To that end, there is a particular emphasis in this series on the presentation of scientific studies, surveys, and statistics. These data are generally presented in the form of tables, charts, and other graphics placed within the text of each book. Every graphic is directly referred to and carefully explained in the text. The source of each graphic is presented within the graphic itself. The data used in these graphics are drawn from the most reputable and reliable sources, in particular from the various branches of the U.S. government and from major independent polling organizations. Every effort has been made to secure the most recent information available. Readers should bear in mind that many major studies take years to conduct, and that additional years often pass before the data from these studies are made available to the public. Therefore, in many cases the most recent information available in 2008 is dated from 2005 or 2006. Older statistics are sometimes presented as well, if they are of particular interest and no more-recent information exists.

Even though statistics are a major focus of the *Information Plus Reference Series*, they are by no means its only content. Each book also presents the widely held positions and important ideas that shape how the book's subject is discussed in the United States. These positions are explained in detail and, where possible, in the words of their proponents. Some of the other material to be found in these books includes historical background; descriptions of major events related to the subject; relevant laws and court cases; and examples of how these issues play out in American life. Some books also feature primary documents, or have pro and con debate sections giving the words and opinions of prominent Americans on both sides of a controversial topic. All material is presented in an even-handed and unbiased manner; readers will never be encouraged to accept one view of an issue over another.

## HOW TO USE THIS BOOK

The United States has a serious weight problem. The majority of Americans weigh more than they should, and roughly one-third of them are considered obese. Overweight and obesity have serious health consequences, and their epidemic levels in the United States have had a major impact on society. Yet overweight and obesity are not the only problems that Americans face when it comes to food. Some suffer from eating disorders, such as anorexia nervosa and bulimia, that can have a devastating effect on their health. This book brings together information from academic and governmental sources on every aspect of overweight, obesity, and eating disorders, including their prevalence in the United States, their consequences, public opinion about them, and methods of combating them.

*Weight in America: Obesity, Eating Disorders, and Other Health Risks* consists of eleven chapters and three appendixes. Each of the chapters is devoted to a particular aspect of weight in the United States. For a summary of the information covered in each chapter, please see the synopses provided in the Table of Contents at the front of the book. Chapters generally begin with an overview of the basic facts and background information on the chapter's topic, then proceed to examine subtopics of particular interest. For example, Chapter 2: Weight and Physical Health begins by examining trends in mortality and disease in the United States during the twentieth century and the prevalence of weight-related diseases, such as heart disease and cancer. This is followed by a detailed discussion of whether obesity is a disease. The next section describes the role that genetics plays in determining body weight and obesity. The chapter concludes with a section examining the health risks and consequences of overweight and obesity. Readers can find their way through a chapter by looking for the section and subsection headings, which are clearly set off from the text. Or, they can refer to the book's extensive index if they already know what they are looking for.

## Statistical Information

The tables and figures featured throughout *Weight in America: Obesity, Eating Disorders, and Other Health Risks* will be of particular use to readers in learning about this issue. These tables and figures represent an extensive collection of the most recent and important statistics on weight and related issues. For example, graphics in the book cover dietary intake of Americans by age and sex, the relationship of a high body mass index and cholesterol levels, names for added sugars found on ingredients labels, dubious diet claims, and the percentage of Americans who say they are trying to lose weight, by weight status. Gale, a part of Cengage Learning, believes that making this information available to readers is the most important way to fulfill the goal of this book: to help readers understand the issues and controversies surrounding weight in the United States and reach their own conclusions.

Each table or figure has a unique identifier appearing above it, for ease of identification and reference. Titles for the tables and figures explain their purpose. At the end of each table or figure, the original source of the data is provided.

To help readers understand these often complicated statistics, all tables and figures are explained in the text. References in the text direct readers to the relevant statistics. Furthermore, the contents of all tables and figures are fully indexed. Please see the opening section of the index at the back of this volume for a description of how to find tables and figures within it.

## Appendixes

Besides the main body text and images, *Weight in America: Obesity, Eating Disorders, and Other Health Risks* has three appendixes. The first is the Important Names and Addresses directory. Here readers will find contact information for a number of government and private organizations that can provide further information on aspects of weight, eating disorders, and their impact on health. The second appendix is the Resources section, which can also assist readers in conducting their own research. In this section, the author and editors of *Weight in America: Obesity, Eating Disorders, and Other Health Risks* describe some of the sources that were most useful during the compilation of this book. The final appendix is the detailed index, which facilitates reader access to specific topics in this book.

## ADVISORY BOARD CONTRIBUTIONS

The staff of Information Plus would like to extend its heartfelt appreciation to the Information Plus Advisory Board. This dedicated group of media professionals provides feedback on the series on an ongoing basis. Their comments allow the editorial staff who work on the project to continually make the series better and more user-friendly. Our top priorities are to produce the highest-quality and most useful books possible, and the Advisory Board's contributions to this process are invaluable.

The members of the Information Plus Advisory Board are:

- Kathleen R. Bonn, Librarian, Newbury Park High School, Newbury Park, California

- Madelyn Garner, Librarian, San Jacinto College–North Campus, Houston, Texas

- Anne Oxenrider, Media Specialist, Dundee High School, Dundee, Michigan

- Charles R. Rodgers, Director of Libraries, Pasco-Hernando Community College, Dade City, Florida

- James N. Zitzelsberger, Library Media Department Chairman, Oshkosh West High School, Oshkosh, Wisconsin

## COMMENTS AND SUGGESTIONS

The editors of the *Information Plus Reference Series* welcome your feedback on *Weight in America: Obesity, Eating Disorders, and Other Health Risks*. Please direct all correspondence to:

Editors
*Information Plus Reference Series*
27500 Drake Rd.
Farmington Hills, MI 48331-3535

# CHAPTER 1

# AMERICANS WEIGH IN OVER TIME

More die in the United States of too much food than of too little.

—John Kenneth Galbraith, *The Affluent Society* (1958)

In 2007 more Americans were fatter than ever before—in fact, they were the heaviest since the U.S. government started tracking patterns of body weight of the U.S. adult population in the first half of the twentieth century. The American Obesity Association reports in the fact sheet "Obesity in the U.S." (May 2, 2005, http://obesity1.temp domainname.com/subs/fastfacts/obesity_US.shtml) that an estimated 127 million adults weigh more than what is considered healthy, and of this group, more than 60 million are considered obese. In *Health, United States, 2006, with Chartbook on Trends in the Health of Americans* (November 2006, http://www.cdc.gov/nchs/data/hus/hus06.pdf), the National Center for Health Statistics (NCHS) notes that in 2003–04, 67% of American adults were overweight, including 34% who were classified as obese. Despite billions of dollars spent on diet programs, overweight and obesity are widespread and increasingly prevalent throughout the United States.

Even though Americans' body weight had been increasing incrementally during the last century, overweight and obesity skyrocketed between 1985 and 2006. The Centers for Disease Control and Prevention (CDC) indicates in *Overweight and Obesity: Obesity Trends: U.S. Obesity Trends 1985–2006* (July 27, 2007, http://www.cdc.gov/nccdphp/dnpa/obesity/trend/maps/) that during this period obesity among adults more than doubled, and obesity among adolescents tripled. Normal-weight adults are now a minority in the United States; nearly one-third of the adult population is obese, and childhood obesity is at an all-time high. In 1990 ten states had obesity prevalence rates of less than 10% and no states had rates at or above 15%. By 2006 just four states had obesity prevalence rates of less than 20%, twenty-two states had rates of 25% or higher, and two of these states (Mississippi and

West Virginia) reported rates of 30%. (The prevalence rate is the number of cases of a disease or condition present during a specified interval of time, usually a year, divided by the population.) Figure 1.1 maps the geographic distribution of obesity throughout the United States in 1990, 1998, and 2006.

The prevalence of obesity varies by state. Jeffrey Levi, Laura M. Segal, and Emily Gadola indicate in *F As in Fat: How Obesity Policies Are Failing in America, 2007* (August 2007, http://healthyamericans.org/reports/obesity2007/Obesity2007Report.pdf), an analysis of 2004 data from the CDC's Behavioral Risk Factor Surveillance System, that Colorado reported the lowest percentage of obesity (17.6%) in 2006, followed by Massachusetts (19.8%), Vermont (20%), Connecticut (20.1%), Rhode Island (20.5%), and Montana (20.7%). Mississippi reported the highest rate of obesity (30.6%), followed by West Virginia (29.8%), Alabama (29.4%), Louisiana (28.2%), and Tennessee (27.8%).

An analysis of data from the Behavioral Risk Factor Surveillance System reveals that the obesity epidemic affects men and women of all ages, races, ethnic origin, smoking status, and educational attainment. Even though the prevalence of obesity among U.S. adults disproportionately affects older age groups, African-Americans, and Hispanics, and declines with increasing educational attainment, no group remains untouched by this epidemic.

In "The Obesity Epidemic in the United States—Gender, Age, Socioeconomic, Racial/Ethnic, and Geographic Characteristics: A Systematic Review and Meta-regression Analysis" (*Epidemiologic Reviews*, vol. 29, May 2007), Youfa Wang and May A. Beydoun of the Johns Hopkins Bloomberg School of Public Health Center for Human Nutrition examine various sociodemographic groups to assess whether overweight and obesity trends varied in response to education, income, race/ethnicity,

**FIGURE 1.1**

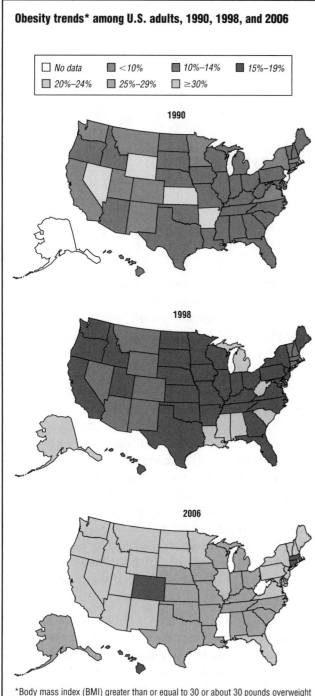

**Obesity trends* among U.S. adults, 1990, 1998, and 2006**

| ☐ No data | ▨ <10% | ▨ 10%–14% | ■ 15%–19% |
| ▨ 20%–24% | ▨ 25%–29% | ▨ ≥30% | |

**1990**

**1998**

**2006**

*Body mass index (BMI) greater than or equal to 30 or about 30 pounds overweight for a 5'4" person.

SOURCE: "Obesity Trends among U.S. Adults, BRFSS, 1990, 1998, 2006," in *Overweight and Obesity: Obesity Trends: U.S. Obesity Trends 1985–2006*, Centers for Disease Control and Prevention, Division of Nutrition, Physical Activity, and Obesity, National Center for Chronic Disease Prevention and Health Promotion, July 27, 2007, http://www.cdc.gov/nccdphp/dnpa/obesity/trend/maps/ (accessed October 6, 2007)

Specific groups—such as non-Hispanic African-American women and children; Mexican-American women and children; low socioeconomic status African-American men and white women and children; Native Americans; and Pacific Islanders—are disproportionately affected, and in general less educated people have a higher prevalence of obesity than their more educated peers. Wang and Beydoun project that by 2015, 75% of adults will be overweight or obese and 41% will be obese.

In the United States, many researchers believe that obesity is the second-leading cause of preventable death after smoking. There is conclusive scientific evidence that mortality (death) risk increases with increasing weight and that even slightly overweight adults—people of average height who are ten to twenty pounds above their ideal weight—are at increased risk of premature death. The rising prevalence of overweight and obesity not only foretells increasing adverse effects on health and longevity but also guarantees increased costs for medical care. Overweight and obesity increase the risk of developing a range of ailments including heart disease, stroke, selected cancers, sleep apnea (interrupted breathing while sleeping), respiratory problems, osteoarthritis (loss of joint bone and cartilage), gallbladder disease, fatty liver disease, and Type 2 diabetes. (Insulin is necessary for the body to be able to use sugar, the basic fuel for the cells in the body. People with diabetes do not produce enough insulin or their cells are resistant to the effects of the insulin.) Kenneth E. Thorpe et al. estimate in "The Impact of Obesity on Rising Medical Spending" (October 20, 2004, http://content.healthaffairs.org/cgi/content/full/hlthaff.w4.480/DC1) that the annual medical costs of an obese person are nearly 37% higher than those incurred by a person of normal weight.

Overweight and obesity also exact a personal toll, with affected individuals at increased risk for emotional, psychological, and social problems. Overweight children, teens, and adults suffer from depression, low self-esteem, and other mental health and emotional problems more than their normal-weight counterparts. Along with a physical inability to participate in many activities, people who are overweight or obese may encounter weight-based stigmatization, bias, and discrimination in school and at the workplace and may be excluded from social opportunities.

## TRENDS IN U.S. BIRTH WEIGHTS

Americans are not born overweight. In fact, Joyce A. Martin et al. indicate in "Births: Final Data for 2004" (*National Vital Statistics Reports*, vol. 55, no. 1, September 29, 2006) that the mean birth weight of infants born as singletons (births of one infant as opposed to twins or other multiple births) has steadily declined since 1990. In 2004 the mean birth weight of all singletons was approximately 7 pounds, 5 ounces (3,316 grams), and the average

and gender. Their analysis of data from the Behavioral Risk Factor Surveillance System finds that overall, each sociodemographic group experienced increases in overweight and obesity, although women aged twenty to thirty-four had the fastest increase in rates of overweight and obesity.

**TABLE 1.1**

**Percentage of live births, very low and low birthweight, by race and Hispanic origin of mother, 1981–2004**

| Year | Very low birthweight[d] All races[a] | Non-Hispanic White[b] | Non-Hispanic Black[b] | Hispanic[c] | Low birthweight[e] All races[a] | Non-Hispanic White[b] | Non-Hispanic Black[b] | Hispanic[c] |
|------|------|------|------|------|------|------|------|------|
| 2004 | 1.48 | 1.20 | 3.15 | 1.20 | 8.1 | 7.2 | 13.7 | 6.8 |
| 2003 | 1.45 | 1.18 | 3.12 | 1.16 | 7.9 | 7.0 | 13.6 | 6.7 |
| 2002 | 1.46 | 1.17 | 3.15 | 1.17 | 7.8 | 6.9 | 13.4 | 6.5 |
| 2001 | 1.44 | 1.17 | 3.08 | 1.14 | 7.7 | 6.8 | 13.1 | 6.5 |
| 2000 | 1.43 | 1.14 | 3.10 | 1.14 | 7.6 | 6.6 | 13.1 | 6.4 |
| 1999 | 1.45 | 1.15 | 3.18 | 1.14 | 7.6 | 6.6 | 13.2 | 6.4 |
| 1998 | 1.45 | 1.15 | 3.11 | 1.15 | 7.6 | 6.6 | 13.2 | 6.4 |
| 1997 | 1.42 | 1.12 | 3.05 | 1.13 | 7.5 | 6.5 | 13.1 | 6.4 |
| 1996 | 1.37 | 1.08 | 3.02 | 1.12 | 7.4 | 6.4 | 13.1 | 6.3 |
| 1995 | 1.35 | 1.04 | 2.98 | 1.11 | 7.3 | 6.2 | 13.2 | 6.3 |
| 1994 | 1.33 | 1.01 | 2.99 | 1.08 | 7.3 | 6.1 | 13.3 | 6.2 |
| 1993 | 1.33 | 1.00 | 2.99 | 1.06 | 7.2 | 5.9 | 13.4 | 6.2 |
| 1992[f] | 1.29 | 0.94 | 2.97 | 1.04 | 7.1 | 5.7 | 13.4 | 6.1 |
| 1991[f] | 1.29 | 0.94 | 2.97 | 1.02 | 7.1 | 5.7 | 13.6 | 6.1 |
| 1990[g] | 1.27 | 0.93 | 2.93 | 1.03 | 7.0 | 5.6 | 13.3 | 6.1 |
| 1989[h] | 1.28 | 0.93 | 2.97 | 1.05 | 7.0 | 5.6 | 13.6 | 6.2 |
| 1988 | 1.24 | — | — | — | 6.9 | — | — | — |
| 1987 | 1.24 | — | — | — | 6.9 | — | — | — |
| 1986 | 1.21 | — | — | — | 6.8 | — | — | — |
| 1985 | 1.21 | — | — | — | 6.8 | — | — | — |
| 1984 | 1.19 | — | — | — | 6.7 | — | — | — |
| 1983 | 1.19 | — | — | — | 6.8 | — | — | — |
| 1982 | 1.18 | — | — | — | 6.8 | — | — | — |
| 1981 | 1.16 | — | — | — | 6.8 | — | — | — |

—Data not available.
[a]Includes races other than white and black and origin not stated.
[b]Race and Hispanic origin are reported separately on birth certificates. Persons of Hispanic origin may be of any race. Race categories are consistent with the 1977 Office of Management and Budget (OMB) standards. Fifteen states reported multiple-race data for 2004. Multiple-race data for these states were bridged to the single-race categories of the 1977 OMB standards for comparability with other states.
[c]Includes all persons of Hispanic origin of any race.
[d]Less than 1,500 grams (3 lb 4 oz).
[e]Less than 2,500 grams (5 lb 8 oz).
[f]Data by Hispanic origin exclude New Hampshire, which did not report Hispanic origin.
[g]Data by Hispanic origin exclude New Hampshire and Oklahoma, which did not report Hispanic origin.
[h]Data by Hispanic origin exclude New Hampshire, Oklahoma, and Louisiana, which did not report Hispanic origin.

SOURCE: Joyce A. Martin et al., "Table 32. Percentage of Live Births Very Preterm and Preterm and Percentage of Very Low Birthweight and Low Birthweight, by Race and Hispanic Origin of Mother: United States, 1981–2004," in "Births: Final Data for 2004," *National Vital Statistics Reports*, vol. 55, no. 1, September 29, 2006, http://www.cdc.gov/nchs/data/nvsr/nvsr55/nvsr55_01.pdf (accessed October 6, 2007)

white non-Hispanic singleton (7 pounds, 7 ounces, or 3,375 grams) weighed 9.2 ounces more than the average non-Hispanic African-American singleton (6 pounds, 13 ounces, or 3,115 grams). The percent of infants born with higher-than-average birth weight (8 pounds, 14 ounces or more, or at least 4,000 grams) has declined by 20% since 1990 and was down by 4% from 2003 to 2004.

Even though ideal birth weight varies based on the expectant mother's ethnicity, for women in the United States, the average ideal birth weight is approximately 7.5 pounds, close to the average weight of singletons born in 2004. In the United States, the percent of babies born with low birth weight (LBW)—less than 5 pounds, 8 ounces (2,500 grams)—has risen steadily since the 1980s. (See Table 1.1.) According to Martin et al., the LBW rate rose from 7.6% in 2000 to 8.1% in 2004, the highest level reported in more than three decades. The percent of infants with very low birth weight (VLBW)—less than 3 pounds, 5 ounces (1,500 grams)—remained nearly steady between 2000 (1.43%) and 2004 (1.48%).

LBW and VLBW are major predictors of infant morbidity (illness or disease) and mortality. For LBW infants, the risk of dying during the first year of life is more than five times that of infants born at normal weight; the risk for VLBW infants is nearly one hundred times higher. The risk of delivering an LBW infant is greatest among the youngest and oldest mothers; however, many of the LBW births among older mothers are attributable to their higher rates of multiple births.

**Birth Weight Influences Risk of Disease**

Even though the relationship between birth weight and development of disease in adulthood is an emerging field of research, and scientists cannot yet fully explain how and why birth weight is a predictor of health and illness in later life, mounting evidence indicates that both LBW and higher-than-average birth weight are linked to future health problems. Research reveals that LBW infants are more likely than normal-weight infants to develop disease in later life. Male infants with LBW who gain weight rapidly before their first birthday appear to be at the highest risk. Investigators hypothesize that LBW infants have fewer muscle cells at birth and that rapid weight gain during the first year of life may lead to disproportionate amounts of fat to muscle and above average body mass. Infants with LBW who later develop above average body mass are at an increased risk for developing diseases such as Type 2 diabetes, hypertension (high blood pressure), and cardiovascular disease (heart disease and stroke).

Thiemo Pfab et al. find in "Low Birth Weight, a Risk Factor for Cardiovascular Diseases in Later Life, Is Already Associated with Elevated Fetal Glycosylated Hemoglobin at Birth" (*Circulation*, vol. 114, October 2006) an inverse relationship between birth weight and cardiovascular disease. In general, rates of cardiovascular disease decreased with increasing birth weight. The association was strong, did not depend on adjustment for size in later childhood, and was independent of social class and other maternal and pregnancy characteristics.

LBW was also linked to childhood asthma in Ann-Marie Brooks et al.'s study, "Impact of Low Birth Weight on Early Childhood Asthma in the United States" (*Archives of Pediatrics and Adolescent Medicine*, vol. 155, no. 3, March 2001), which found that babies born at 5.5 pounds or less faced the greatest risk of respiratory complications such as asthma. Research has also demonstrated that both LBW and abnormally high birth weight are associated with a risk of developing diabetes later in life.

Evidence also indicates that birth weight is related to a risk of developing breast cancer. Valerie A. McCormack et al. investigated whether the size at birth and the rate of fetal growth influenced the risk of developing breast cancer in adulthood. The results of the study were published in "Fetal Growth and Subsequent Risk of Breast Cancer: Results from Long Term Follow up of Swedish Cohort" (*British Medical Journal*, vol. 326, no. 7383, February 2003). By examining birth and medical records of 5,358 singleton females born between 1915 and 1929, the investigators determined that size at birth was associated with breast cancer in perimenopausal (the stage of reproductive life immediately before the onset of menopause) women aged fifty or younger—the larger and longer the baby, the greater the risk.

Athanasios Michos, Fei Xue, and Karin B. Michels find in "Birth Weight and the Risk of Testicular Cancer: A Meta-Analysis" (*International Journal of Cancer*, vol. 121, no. 5, September 1, 2007) that both low and high birth weight increase the risk of testicular cancer in men. Men with low birth weight were 18% more likely to develop testicular cancer, and men with high birth weight were 12% more likely to develop the cancer than men of average birth weight.

In "Aerobic Capacity, Strength, Flexibility, and Activity Level in Unimpaired Extremely Low Birth Weight (≤800 g) Survivors at Seventeen Years of Age Compared with Term-Born Control Subjects" (*Pediatrics*, vol. 116, no. 1, July 2005), Marilyn Rogers et al. note that infants born either prematurely or with an extremely low birth weight (ELBW)—1 pound, 12 ounces (800 grams)—were significantly more likely to suffer a lower level of fitness later in life, including less strength, endurance, and flexibility, and a greater risk of health problems as adults. When compared to teens born at normal weight, the ELBW teens had lower aerobic capacity, grip strength, leg power, and vertical jump. They were unable to perform as many push-ups, had less abdominal strength as measured by curl-ups, showed less flexibility in their lower backs, and had tighter hamstrings. The ELBW teens reported less previous and current sports participation, lower physical activity level, and poorer coordination compared to term-born control subjects. ELBW teens also had more trouble maintaining rhythm and tempo than their peers who were born at normal weight.

The only action able to alter the birth weight of an infant is to modify weight gain during pregnancy. In 2007 health professionals concurred that for normal-weight women the ideal weight gain during pregnancy ranges from twenty-five to thirty-five pounds of fat and lean mass. Furthermore, research published in 2003 revealed that a newborn's birth weight and mother's postpregnancy weight are influenced not only by how much weight is gained during pregnancy but also by the source of the excess weight. In "Composition of Gestational Weight Gain Impacts Maternal Fat Retention and Infant Birth Weight" (*American Journal of Obstetrics and Gynecology*, vol. 189, no. 5, November 2003), Nancy F. Butte et al. conducted body scans of sixty-three women before, during, and after their pregnancies and recorded changes in women's weight from water, protein, fat, and potassium—a marker for changes in muscle tissue, which is one component of lean mass. They found that only increases in lean mass, and not fat mass, appeared to influence infant size. Independent of how much fat the women gained during pregnancy, only lean body mass increased the birth weight of the infant, with women who gained more lean body mass giving birth to larger infants.

**FIRST WEEK OF LIFE MAY DETERMINE ADULT OBESITY.** Research demonstrates that low birth weight and low weight gain during infancy are associated with coronary heart disease. Similarly, rapid weight gain in infancy has been shown to predict obesity in childhood. In 2004 research funded by the National Institutes of Health and conducted at the Children's Hospital of Philadelphia, University of Pennsylvania School of Medicine, and the Fomon Infant Nutrition Unit, University of Iowa, sought to determine which periods of weight gain in infancy might be associated with adult obesity.

In "Weight Gain in the First Week of Life and Over-weight in Adulthood: A Cohort Study of European American Subjects Fed Infant Formula" (*Circulation*, vol. 111, no. 15, April 2005), Nicolas Stettler et al. reviewed data for 653 subjects who had been weighed on seven occasions during infancy and were contacted when they were young adults, aged twenty to thirty-two, when they again reported their height and weight. The researchers pin-pointed the period between birth and age eight days as potentially critical because weight gain during the first week of life was associated with adulthood overweight status. The formula-fed babies who gained weight rapidly during their first week of life were significantly more likely to be overweight decades later. Stettler et al. con-clude that "in formula-fed infants, weight gain during the first week of life may be a critical determinant for the development of obesity several decades later." The investigators also observe that their findings reinforce the American Academy of Pediatrics recommendation that infants should be exclusively breast-fed for the first six months of life. Among the many health benefits associ-ated with breastfeeding is the fact that breast-fed babies are much less likely than formula-fed babies to become obese adults.

Janis Baird et al. report in "Being Big or Growing Fast: Systematic Review of Size and Growth in Infancy and Later Obesity" (*British Medical Journal*, vol. 331, no. 929, October 22, 2005) that big babies who grow quickly in the first two years of life risk being obese in childhood and adulthood. Baird et al. looked at twenty-four studies that found an association between infant size or growth during the first two years of life and obesity later in life. They note that the heaviest infants and those who gained weight rapidly during the first and second year of life faced a ninefold greater risk of obesity in childhood, adolescence, and adulthood. Their findings suggest that factors in infant growth are probably influ-encing the risk of later obesity. Baird et al. do not know why big and fast-growing babies have a higher risk of obesity, but they believe that some factors related to how an infant grows are important in influencing their later risk of obesity and suggest that the timing of weaning and social circumstances are factors that merit further investigation.

**DEFINING AND ASSESSING IDEAL WEIGHT, OVERWEIGHT, AND OBESITY**

Historically, the determination of desirable, healthy, or ideal weights have been derived from demographic and actuarial statistics (data compiled to assess insurance risk and formulate insurance premiums). The NCHS compiles and analyzes demographic data—the heights and weights of a representative sample of the U.S. pop-ulation to develop standards for desirable weight. In 1943 the Metropolitan Life Insurance Company (MetLife)

introduced standard weight-for-height tables for men and women based on an analysis of actuarial data. The MetLife weight-for-height tables assisted adults in deter-mining if their weight was within an appropriate range for height and frame size. Revised in 1959 and 1983, the tables were based on actuarial data, in which desirable or ideal weight is defined as the weight for height associated with the lowest mortality rate, or longest life span, among the client population of adults (policyholders) insured by MetLife.

Even though the MetLife and other weight-for-height tables remained in use in 2007, many health professionals and medical researchers believe these table have limited utility. Nearly every weight-for-height table shows differ-ent acceptable weight ranges for men and women, and considerable debate continues among health professio-nals over which table to use. The tables lack information about body composition, such as the ratio of fat to lean muscle mass; their data are derived primarily from white populations and do not represent the entire U.S. popula-tion; they generally do not take age into consideration; and it is often unclear how the frame size is determined. Furthermore, it is now known that ideal, healthy, or low-risk weight varies for different populations and varies for the same population at different times and in relation to different causes of morbidity and mortality.

The limitations of weight-for-height tables have prompted health-care practitioners and researchers to adopt other measures that allow comparison of weight independent of height and frame across populations to define desirable or healthy weight as well as overweight and obesity. For example, the *Dietary Guidelines for Americans, 2005* (January 2005, http://www.health.gov/dietaryguidelines/dga2005/document/pdf/DGA2005.pdf), which is published jointly by the U.S. Department of Agriculture (USDA) and U.S. Department of Health and Human Services, and weight-control information pub-lished by the National Institutes of Health's National Institute of Diabetes and Digestive and Kidney Diseases include updated weight-for-height tables for adults that incorporate height, weight, and body mass index (BMI). (See Table 1.2.)

Overweight is generally defined as excess body weight in relation to height, when compared to a prede-termined standard of acceptable, desirable, or ideal weight. One definition characterizes individuals as over-weight if they are between ten and thirty pounds heavier than the desirable weight for height. Overweight does not necessarily result from excessive body fat; people may become overweight as the result of an increase in lean muscle. For example, even though muscular bodybuilders with minimal body fat frequently weigh more than non-athletes of the same height, they are overweight because of their increased muscle mass rather than increased fat.

# TABLE 1.2

## Adult BMI (body mass index) chart

| BMI Height | 19 | 20 | 21 | 22 | 23 | 24 | 25 | 26 | 27 | 28 | 29 | 30 | 31 | 32 | 33 | 34 | 35 |
|---|---|---|---|---|---|---|---|---|---|---|---|---|---|---|---|---|---|
| | Healthy weight | | | | | | Overweight | | | | | Obese | | | | | |
| | | | | | | | Weight in pounds | | | | | | | | | | |
| 4'10" | 91 | 96 | 100 | 105 | 110 | 115 | 119 | 124 | 129 | 134 | 138 | 143 | 148 | 153 | 158 | 162 | 167 |
| 4'11" | 94 | 99 | 104 | 109 | 114 | 119 | 124 | 128 | 133 | 138 | 143 | 148 | 153 | 158 | 163 | 168 | 173 |
| 5' | 97 | 102 | 107 | 112 | 118 | 123 | 128 | 133 | 138 | 143 | 148 | 153 | 158 | 163 | 168 | 174 | 179 |
| 5'1" | 100 | 106 | 111 | 116 | 122 | 127 | 132 | 137 | 143 | 148 | 153 | 158 | 164 | 169 | 174 | 180 | 185 |
| 5'2" | 104 | 109 | 115 | 120 | 126 | 131 | 136 | 142 | 147 | 153 | 158 | 164 | 169 | 175 | 180 | 186 | 191 |
| 5'3" | 107 | 113 | 118 | 124 | 130 | 135 | 141 | 146 | 152 | 158 | 163 | 169 | 175 | 180 | 186 | 191 | 197 |
| 5'4" | 110 | 116 | 122 | 128 | 134 | 140 | 145 | 151 | 157 | 163 | 169 | 174 | 180 | 186 | 192 | 197 | 204 |
| 5'5" | 114 | 120 | 126 | 132 | 138 | 144 | 150 | 156 | 162 | 168 | 174 | 180 | 186 | 192 | 198 | 204 | 210 |
| 5'6" | 118 | 124 | 130 | 136 | 142 | 148 | 155 | 161 | 167 | 173 | 179 | 186 | 192 | 198 | 204 | 210 | 216 |
| 5'7" | 121 | 127 | 134 | 140 | 146 | 153 | 159 | 166 | 172 | 178 | 185 | 191 | 198 | 204 | 211 | 217 | 223 |
| 5'8" | 125 | 131 | 138 | 144 | 151 | 158 | 164 | 171 | 177 | 184 | 190 | 197 | 203 | 210 | 216 | 223 | 230 |
| 5'9" | 128 | 135 | 142 | 149 | 155 | 162 | 169 | 176 | 182 | 189 | 196 | 203 | 209 | 216 | 223 | 230 | 236 |
| 5'10" | 132 | 139 | 146 | 153 | 160 | 167 | 174 | 181 | 188 | 195 | 202 | 209 | 216 | 222 | 229 | 236 | 243 |
| 5'11" | 136 | 143 | 150 | 157 | 165 | 172 | 179 | 186 | 193 | 200 | 208 | 215 | 222 | 229 | 236 | 243 | 250 |
| 6' | 140 | 147 | 154 | 162 | 169 | 177 | 184 | 191 | 199 | 206 | 213 | 221 | 228 | 235 | 242 | 250 | 258 |
| 6'1" | 144 | 151 | 159 | 166 | 174 | 182 | 189 | 197 | 204 | 212 | 219 | 227 | 235 | 242 | 250 | 257 | 265 |
| 6'2" | 148 | 155 | 163 | 171 | 179 | 186 | 194 | 202 | 210 | 218 | 225 | 233 | 241 | 249 | 256 | 264 | 272 |
| 6'3" | 152 | 160 | 168 | 176 | 184 | 192 | 200 | 208 | 216 | 224 | 232 | 240 | 248 | 256 | 264 | 272 | 279 |

Notes: Locate the height of interest in the left-most column and read across the row for that height to the weight of interest. Follow the column of the weight up to the top row that lists the BMI. BMI of 18.5–24.9 is the healthy range, BMI of 25–29.9 is the overweight range, and BMI of 30 and above is the obese range.

source: "Figure 2. Adult BMI Chart," in *Dietary Guidelines for Americans, 2005*, 6th ed., U.S. Department of Health and Human Services and U.S. Department of Agriculture, January 2005, http://www.health.gov/dietaryguidelines/dga2005/document/pdf/DGA2005.pdf (accessed October 6, 2007)

Rather than viewing overweight and obesity as distinct conditions, many researchers prefer to consider weight as a curve or continuum with obesity at the far end of the curve. People who are obese constitute a subset of the overweight population. In this definition, only some overweight people are obese, but all obese people are overweight.

Similarly, there is still no uniform definition of obesity. Some health professionals describe anyone who is more than thirty pounds above his or her desirable weight for height as obese. Others assert that body weight 20% or more above desirable or ideal body weight constitutes obesity. Extreme or clinically severe obesity is often defined as weight twice the desirable weight or one hundred pounds more than the desirable weight. Obesity is also defined as an excessively high amount of adipose tissue (body fat) in relation to lean body mass such as muscle and bone. The amount of body fat (also known as adiposity), the distribution of fat throughout the body, and the size of the adipose tissue deposits are also used to assess obesity because the location and distribution of body fat are important predictors of the health risks associated with obesity. The location and distribution of body fat may be measured by the ratio of waist-to-hip circumference. High ratios are associated with higher risks of morbidity and mortality.

Overweight and obese body types may be characterized as apple- or pear-shaped, depending on the anatomical site where fat is more prominent. In the apple or android type of obesity, fat is mainly located in the trunk (upper body, nape of the neck, shoulder, and abdomen). Gynoid obesity, or the pear-shape, features rounded hips and more fat located in the buttocks, thighs, and lower abdomen. Fat cells around the waist, flank, and abdomen are more active metabolically than those in the thighs, hips, and buttocks. This increased metabolic activity is thought to produce the increased health risks associated with android obesity. In general, women are more likely to have gynoid obesity. However, those women with the android type of obesity are subject to similar health risks as males with android overweight.

There are many ways to measure body fat. Weighing an individual underwater in a laboratory with specialized equipment provides a highly accurate assessment of body fat. By performing hydrostatic or underwater weighing, an examiner obtains an estimate of whole-body density and uses this to calculate the percentage of the body that is fat. First, the subject is weighed on a land scale. The subject puts on a diver's belt with weights to prevent floating during the weighing procedure, sits on a chair suspended from a precision scale, and is completely submerged. When maximum expiration of breath is achieved, the subject remains in this submerged position for about ten seconds while the investigator reads the

scale. This procedure is repeated as many as ten times to obtain reliable, consistent values. The weight of the diver's belt and chair are subtracted from this weight to obtain the true value of the subject's mass in water.

Simpler, but potentially less accurate assessments of body fat include skinfold thickness measurements, which involve measuring subcutaneous (immediately below the skin) fat deposits using an instrument called a caliper in locations such as the upper arm. Skinfold thickness measurements rely on the fact that a certain fraction of total body fat is subcutaneous and by using a representative sample of that fat the overall body fatness (density) may be predicted. Several skinfold measurements are obtained, and the values are used in equations to calculate body density. Using a caliper, the examiner grasps a fold of skin and subcutaneous fat firmly, pulling it away from the underlying muscle tissue following the natural contour of the skin. The caliper jaws exert a relatively constant tension at the point of contact and measure skinfold thickness in millimeters. Most obesity researchers believe there is an acceptable correlation between skinfold thickness and body fat—that it is possible to estimate body fatness from the use of skinfold calipers. Skinfold thickness measurements are considered more subjective than underwater weights because the accuracy of measurements of skinfold thickness depends on the technique and skill of the examiner, and there may be variations in readings from one examiner to another.

Another technique used to evaluate body fat is bioelectric impedance analysis (BIA). BIA offers an indirect estimate of body fat and lean body mass. It entails passing an electrical current through the body and assessing the body's ability to conduct the current. It is based on the principle that resistance is inversely proportional to total body water when an electrical current (seventy megahertz) is applied through several electrodes placed on body extremities. Because greater conductivity occurs when there is a higher percent of body water and because a higher percent of body water indicates larger amounts of muscle and other lean tissue (fat cells contain less water than muscle cells), people with less fat are better able to conduct electrical current. BIA has been shown to correlate well with total body water assessed by other methods.

Other means of estimating the location and distribution of body fat include waist-to-hip circumference ratios and imaging techniques such as ultrasound, computed tomography, or magnetic resonance imaging.

### Waist Circumference and Waist-to-Hip Ratio

Along with height and weight, waist circumference is a common measure used to assess abdominal fat content. An excess of body fat in the abdomen or upper body is considered to increase the risk of developing heart

**TABLE 1.3**

**Classification of overweight and obesity by body mass index (BMI), waist circumference, and associated disease risk**

| | BMI (kg/m$^2$) | Obesity class | Disease risk[a] relative to normal weight and waist circumference | |
|---|---|---|---|---|
| | | | Men ≤ 102 cm (≤ 40 in) Women ≤ 88 cm (≤ 35 in) | > 102 cm (> 40 in) > 88 cm (> 35 in) |
| Underweight | <18.5 | | — | — |
| Normal[b] | 18.5–24.9 | | — | — |
| Overweight | 25.0–29.9 | | Increased | High |
| Obesity | 30.0–34.9 | I | High | Very high |
| | 35.0–39.9 | II | Very high | Very high |
| Extreme obesity | ≥40 | III | Extremely high | Extremely high |

[a]Disease risk for type 2 diabetes, hypertension, and cardiovascular disease.
[b]Increased waist circumference can also be a marker for increased risk even in persons of normal weight.

SOURCE: "Table ES-4. Classification of Overweight and Obesity by BMI, Waist Circumference, and Associated Disease Risk," in *Clinical Guidelines on the Identification, Evaluation, and Treatment of Overweight and Obesity in Adults: The Evidence Report*, National Institutes of Health, National Heart, Lung, and Blood Institute in cooperation with The National Institute of Diabetes and Digestive and Kidney Diseases, September 1998, http://www.nhlbi.nih.gov/guidelines/obesity/ob_gdlns.htm (accessed October 6, 2007)

disease, high blood pressure, diabetes, stroke, and certain cancers. Like body fat, health risks increase as waist circumference increases. For men, waist circumference greater than forty inches is considered to confer increased health risks. Women are considered at increased risk when their waist measurements are thirty-five inches or greater. Waist circumference measures lose their incremental predictive value in people with a BMI greater or equal to 35 because these individuals generally exceed the cutoff points for increased risk. Table 1.3 shows the relationship between BMI, waist circumference, and disease risk for people who are underweight, normal weight, overweight, obese, and extremely obese.

Waist-to-hip ratio is the ratio of waist circumference to hip circumference, which is calculated by dividing waist circumference by hip circumference. For men and women, a waist-to-hip ratio of 1 or more is considered to place them at greater risk. Most people store body fat at the waist and abdomen (android body fat distribution) or at the hips (gynoid body fat distribution). Interestingly, even though overweight and obesity both increase health risks, body fat that is concentrated in the lower body is thought to be less harmful in terms of morbidity and mortality than abdominal fat.

## Body Mass Index

BMI is a single number that evaluates an individual's weight status in relation to height. It does not directly measure the percent of body fat; however, it offers a more accurate assessment of overweight and obesity than weight alone. It is a direct calculation based on height and weight, and it is not gender specific. BMI is the preferred measurement of health-care professionals and obesity researchers to assess body fat and is the most common method of tracking overweight and obesity among adults. BMI, which is calculated by dividing

**TABLE 1.4**

**How to calculate body mass index (BMI)**

You can calculate BMI as follows

$$BMI = \frac{Weight\ (kg)}{Height\ squared\ (m^2)}$$

If pounds and inches are used

$$BMI = \frac{Weight\ (pounds) \times 703}{Height\ squared\ (inches^2)}$$

Calculation directions and sample

Here is a shortcut method for calculating BMI. (Example: for a person who is 5 feet 5 inches tall weighing 180 lbs.)

1. Multiply weight (in pounds) by 703

$$180 \times 703 = 126,540$$

2. Multiply height (in inches) by height (in inches)

$$65 \times 65 = 4,225$$

3. Divide the answer in step 1 by the answer in step 2 to get the BMI.

$$126,540/4,225 \times 29.9$$

$$BMI = 29.9$$

SOURCE: "You Can Calculate BMI as Follows," in *The Practical Guide: Identification, Evaluation, and Treatment of Overweight and Obesity in Adults*, National Institutes of Health, National Heart, Lung, and Blood Institute, North American Association for the Study of Obesity, October 2000, http://www.nhlbi.nih.gov/guidelines/obesity/prctgd_b.pdf (accessed October 6, 2007)

weight in kilograms by the square of height in meters, classifies people as underweight, normal weight, overweight, or obese. Table 1.4 shows the formula used to calculate BMI when height is measured in either inches or centimeters and weight is measured in either pounds or kilograms.

The World Health Organization and the National Institutes of Health consider individuals overweight when their BMI is between 25 and 29.9, and they are classified as obese when their BMI exceeds 30. Table 1.2 shows the relationship between height, weight, and BMI. Table 1.3 shows the classification of overweight and obesity by BMI and distinguishes between three levels of obesity.

Even though BMI is a simple, inexpensive tool for assessing weight, it has several limitations. BMI calculations may deem a muscular athlete overweight, when he or she is extremely fit and excess weight is the result of a larger amount of lean muscle. It may similarly misrepresent the health of older adults who as the result of muscle wasting—loss of muscle mass—may be considered normal or healthy weights when they may actually be nutritionally depleted or overweight in terms of body fat composition. Even though it is an imperfect method for assessing individuals, BMI is extremely useful for tracking weight trends in the population.

### Definitions and Estimates of Prevalence Vary

Historically, varying definitions of, and criteria for, overweight and obesity have affected prevalence statistics and made it difficult to compare data. Some overweight- and obesity-related prevalence rates are crude or unadjusted estimates; others are age-adjusted estimates that offer different values. Early efforts to track overweight and obesity in the U.S. population relied on the 1943, 1959, and 1983 MetLife tables of desirable weight-for-height as the reference standard for overweight. During the last three decades, most government agencies and public health organizations have estimated overweight using data from a series of surveys conducted by the NCHS. These surveys include the National Health Examination Surveys, the National Health and Nutrition Examination Surveys (NHANES), and the Behavioral Risk Factor Surveillance System.

Despite changing definitions of overweight and obesity and various methods to track changes in the U.S. population, there is irrefutable evidence that the prevalence of overweight and obesity has steadily increased among people of both genders, all ages, all racial and ethnic groups, all educational levels, and all smoking levels. The prevalence of obesity in the United States was first reported in the 1960 National Health Examination Survey, and subsequent reports were derived from three NHANES: NHANES I, 1971; NHANES II, 1976–1980; and NHANES III, 1988–1994. Most obesity data referenced in the medical literature in 2007 were drawn from the NHANES study conducted between 2003 and 2004 and the 1997–2007 National Health Interview Studies, along with several other national studies. Data from the National Health Examination Survey, NHANES I, and NHANES II indicated that the prevalence of obesity was relatively constant from 1960 to 1980; however, the results of the NHANES III indicated a sharp increase in the prevalence of obesity.

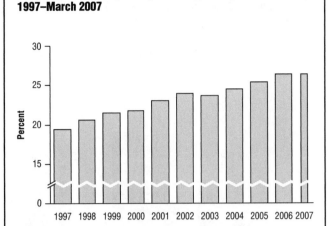

FIGURE 1.2

**Prevalence of obesity among adults aged 20 years and older, 1997–March 2007**

Notes: Obesity is defined as a Body Mass Index (BMI) of 30 kg/m² or more. The measure is based on self-reported height and weight. The analyses excluded people with unknown height or weight (about 4% of respondents each year).

SOURCE: P. Barnes, K. M. Heyman, and J. S. Schiller, "Figure 6.1. Prevalence of Obesity among Adults Aged 20 Years and Over: United States, 1997–March 2007," in *Early Release of Selected Estimates Based on Data from the January–March 2007 National Health Interview Survey*, Centers for Disease Control and Prevention, National Center for Health Statistics, September 26, 2007, http://www.cdc.gov/nchs/data/nhis/earlyrelease/earlyrelease200709.pdf (accessed October 6, 2007)

Overweight and obesity have steadily progressed at an alarming rate over the course of the past three decades. The NCHS reports in *Health, United States, 2006* that from 1960 to 2004 the prevalence of overweight (defined as BMI greater than 25 but less than 30) increased from 31.5% to 33.2% in U.S. adults aged twenty to seventy-four. The prevalence of obesity (BMI of 30 or more) during this same period more than doubled, with most of the rise occurring in the past twenty years. Among adults aged twenty and older, the prevalence of obesity rose 7% in less than a decade—from 19% in 1997 to 26% in 2006. (See Figure 1.2.)

The prevalence of overweight and obesity generally increases with advancing age, then starts to decline among people over sixty. In 2007, for men and women combined, the prevalence of obesity was highest among adults aged forty to fifty-nine (28%) and lowest among adults aged twenty to thirty-nine (24%). (See Figure 1.3.) There was no significant difference in the prevalence of obesity between men and women in all three age groups.

The age-adjusted prevalence of obesity in racial and ethnic minorities, especially minority women, is generally higher than in whites in the United States. P. Barnes, K. M. Heyman, and J. S. Schiller of the NCHS indicate in *Early Release of Selected Estimates Based on Data from the January–March 2007 National Health Interview Survey*

FIGURE 1.3

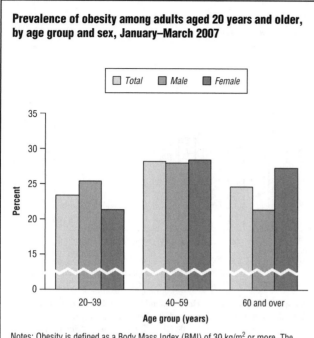

**Prevalence of obesity among adults aged 20 years and older, by age group and sex, January–March 2007**

Notes: Obesity is defined as a Body Mass Index (BMI) of 30 kg/m² or more. The measure is based on self-reported height and weight. The analyses excluded 362 people (6.2%) with unknown height or weight.

SOURCE: P. Barnes, K. M. Heyman, and J. S. Schiller, "Figure 6.2. Prevalence of Obesity among Adults Aged 20 Years and Over, by Age Group and Sex: United States, January–March 2007," in *Early Release of Selected Estimates Based on Data From the January–March 2007 National Health Interview Survey*, Centers for Disease Control and Prevention, National Center for Health Statistics, September 26, 2007, http://www.cdc.gov/nchs/data/nhis/earlyrelease/earlyrelease200709.pdf (accessed October 6, 2007)

(September 26, 2007, http://www.cdc.gov/nchs/data/nhis/earlyrelease/earlyrelease200709.pdf) that in 2007 for both genders, non-Hispanic African-Americans were more likely than Hispanics and non-Hispanic whites to be obese. The age-adjusted prevalence of obesity was highest among non-Hispanic African-American women (39.4%) and lowest among non-Hispanic white women (22.7%). Earlier studies, including the NHANES, reported a high prevalence of overweight and obesity among Hispanics and Native Americans and a lower prevalence of overweight and obesity in Asian-Americans than in the U.S. population as a whole.

## WHY ARE SO MANY AMERICANS OVERWEIGHT?

Historically, overweight and obesity were largely attributed to gluttony—solely the result of inappropriate eating. The scientific study of obesity has identified genetic, biochemical, viral, and metabolic alterations in humans and experimental animals, as well as the complex interactions of psychosocial and cultural factors that create susceptibility to overweight and obesity. Even though obesity is thought to result from multiple causes, for the overwhelming majority of Americans, overweight and obesity result from excessive consumption of calories and inad-

equate physical activity—eating too much and exercising too little.

Some observers maintain that Americans were destined to become overweight when their diets remained unchanged even as products of the industrial revolution such as cars, automation, and a variety of laborsaving devices sharply reduced levels of physical activity. The widespread availability of high-caloric foods and less physically demanding jobs conspired to make Americans fatter. Others contend that the rise in overweight and obesity began during the 1970s, when women entered the workforce in large numbers and had less time to cook, so Americans came to rely instead on processed, convenient, and calorie-dense, saturated-fat-laden fast foods. The CDC reports in "Trends in Intake of Energy and Macronutrients—United States, 1971–2000" (*Morbidity and Mortality Weekly Report*, vol. 53, no. 4, February 6, 2004) that in 2000 women ate 1,877 calories per day, 335 calories more per day than they did in 1971. Men, averaging 2,618 calories per day, consumed 168 calories more per day than their counterparts in 1971.

Recent research even implicates a viral cause of obesity. Alan Mozes of HealthDay.com notes in "Virus Could Help Drive Obesity" (August 21, 2007, http://www.healthday.com/Article.asp?AID=607462) that researchers at the Pennington Biomedical Research Center at Louisiana State University find evidence that adenovirus-36, a common virus responsible for many respiratory and eye infections, may also promote obesity. The researchers find that the virus, which acts to increase both the size and number of fat cells, is much more prevalent among obese people than among those of a healthy weight.

### The American Diet Has Changed

The American diet has changed dramatically since the middle of the twentieth century. According to the USDA in the *Agriculture Fact Book 2001–2002* (March 2003, http://www.usda.gov/factbook/2002factbook.pdf), during the 1950s food production in the United States provided about eight hundred fewer calories per person per day than in 2000. Of the thirty-eight hundred calories produced per person per day in 2000, the USDA estimates that about eleven hundred calories were wasted, either through spoilage, plate waste, or cooking, leaving an average of about twenty-seven hundred calories per person per day. The USDA data reveal that average daily caloric intake increased 24.5%, or about 530 calories, between 1970 and 2000. Of that 24.5% increase, grains (primarily refined grain products) accounted for 9.5%; added fats and oils, 9%; added sugars, 4.7%; fruits and vegetables together, 1.5%; and meats and nuts together, 1%; whereas dairy products and eggs together declined by 1.5%.

The USDA states that Americans consumed an average of four pounds more fish and shellfish, seven pounds more red meat, and forty-six pounds more poultry per person per year in 2000 than they did during the 1950s. Americans consumed more meat—fifty-seven pounds more per year in 2000 than they did during the 1950s. Despite record-high per capita consumption of meat in 2000, the proportion of fat in the U.S. food supply from meat, poultry, and fish declined from one-third (33%) in the 1950s to one-quarter (24%) in 2000. This decline resulted from marketing of lower-fat ground and processed meat products, a shift away from red meat to poultry, and closer trimming of outside fat on meat, which commenced in 1986.

The USDA also reports that in 2000 Americans drank an average of 38% less milk and ate nearly four times as much cheese (excluding cottage, pot, and baker's cheese) as they had in the 1950s. Consumption of milk dropped from an annual average of 36.4 gallons per person in the 1950s to 22.6 gallons in 2000, although consumption of lower-fat milk increased during this period. The USDA posits a link between the trend toward dining out and the reduction in beverage milk consumption. According to the USDA, soft drinks, fruit drinks, and flavored teas appear to be displacing milk as the beverages of choice for Americans.

The average use of added fats and oils in 2000 was two-thirds (67%) higher than the average use in the 1950s. Added fats included those used directly by consumers, such as butter on bread, as well as shortenings and oils used in commercially prepared foods. All fats that naturally occur in foods, such as those in milk and meat, were excluded from the USDA analysis. In 2000 Americans consumed an average of 23% more salad and cooking oil than they did annually in the 1950s, and more than twice as much shortening; however, the use of table spreads (butter and margarine) declined by 25% during this same period. During the 1950s the added fats and oils group contributed the most fat to the food supply (41%), followed by the meat, poultry, and fish group (32%). By 2000 the fats and oils group's contribution to total fat had jumped to 53%, probably because of the higher consumption of fried foods in fast-food outlets, the increase in consumption of high-fat snack foods, and the increased use of salad dressings. Margarine, salad dressings and mayonnaise, cakes and other sweet baked goods, and oils continue to appear in the top ten foods for fat contribution, according to USDA food intake surveys, which examine the ongoing prevalence of discretionary fats in Americans' diets.

According to the USDA, Americans in 2000 consumed 20% more fruit and vegetables than did their counterparts in the 1970s. The USDA attributes some of the increase to the introduction of convenient, ready-to-eat, pre-cut, and packaged fruit and vegetables and to increasing consumer health awareness.

Per capita use of flour and cereal products reached 199.9 pounds in 2000 from an annual average of 155.4 pounds in the 1950s and 138.2 pounds in the 1970s, when grain consumption was at a record low. This increase reflects plentiful grain stocks, robust consumer demand for store-bought bakery items and grain-based snack foods, and increased consumption of fast-food products such as buns, pizza dough, and tortillas. Despite the overall increase in grain consumption, the average American's diet contained mostly refined grain products and fell short of the recommended minimum three daily servings of whole grain products.

The USDA cites a variety of factors that have contributed to the changes in the American diet over the past fifty years, including fluctuations in food prices and availability, increases in real (adjusted for inflation) disposable income, and more food assistance for the poor. New products, particularly the expanding array of convenience foods, also alter patterns of consumption, as do more imports, growth in the away-from-home food market, intensified advertising campaigns, and increases in nutrient-enrichment standards and food fortification. The social and demographic trends driving changes in food choices include smaller households, more two-wage earner households, more single-parent households, an aging population, and increased ethnic diversity.

## Americans Enjoy Eating Out

A variety of societal trends are thought to contribute to Americans' propensity to overeat, including eating outside the home, as well as ready access to and preference for sugar- and fat-laden foods. Table 1.5 shows how expenditures for eating away from home have increased steadily, and more than doubled between 1991 and 2006. Besides less strenuous work, many Americans spend their leisure time in relatively sedentary pursuits—watching television, using computers, or playing videogames—that not only do not expend calories but also, as in the case of television, actually encourage excessive eating.

Many nutritionists and obesity researchers assert that controlling portion size, which is key to controlling calorie consumption, is more difficult in restaurants, where portions are frequently quite large. Increasingly, restaurants have translated consumer demand for value into more food for less money. Because humans are genetically programmed to eat when food is abundant, larger portions trigger the natural impulse to eat more.

Barbara Rolls, Erin L. Morris, and Liane S. Roe of Pennsylvania State University confirm in "Portion Size of Food Affects Energy Intake in Normal-Weight and Overweight Men and Women" (*American Journal of Clinical Nutrition*, vol. 76, no. 6, December 2002) the notion that when presented with larger portions, people will generally consume more. When they offered

**TABLE 1.5**

**Food away from home, total expenditures, 1929–2006**

| Year | Eating and drinking places[a] | Hotels and motels[a] | Retail stores, direct selling[b] | Recreational places[c] | Schools and colleges[d] | All other[e] | Total[f] |
|---|---|---|---|---|---|---|---|
| | | | Million dollars | | | | |
| 1929 | 2,101 | 362 | — | — | 175 | 1,483 | 4,121 |
| 1933 | 1,235 | 250 | — | — | 105 | 869 | 2,459 |
| 1935 | 1,257 | 271 | — | — | 161 | 1,145 | 2,834 |
| 1936 | 1,430 | 320 | — | — | 175 | 1,236 | 3,161 |
| 1937 | 1,696 | 351 | — | — | 194 | 1,375 | 3,616 |
| 1938 | 1,626 | 312 | — | — | 191 | 1,260 | 3,389 |
| 1939 | 1,782 | 321 | — | — | 203 | 1,307 | 3,613 |
| 1940 | 1,938 | 353 | — | — | 219 | 1,385 | 3,895 |
| 1941 | 2,369 | 386 | — | — | 263 | 1,781 | 4,799 |
| 1942 | 2,992 | 453 | — | — | 310 | 2,539 | 6,294 |
| 1943 | 3,837 | 604 | — | — | 332 | 3,572 | 8,345 |
| 1944 | 4,471 | 681 | — | — | 326 | 4,415 | 9,893 |
| 1945 | 5,218 | 736 | — | — | 373 | 4,908 | 11,235 |
| 1946 | 5,859 | 846 | — | — | 525 | 3,802 | 11,032 |
| 1947 | 6,243 | 854 | — | — | 842 | 3,864 | 11,803 |
| 1948 | 6,338 | 846 | — | — | 983 | 4,069 | 12,236 |
| 1949 | 6,294 | 786 | — | — | 979 | 3,943 | 12,002 |
| 1950 | 6,472 | 774 | — | — | 1,051 | 4,172 | 12,469 |
| 1951 | 7,172 | 783 | — | — | 1,124 | 5,167 | 14,246 |
| 1952 | 7,549 | 805 | — | — | 1,138 | 5,435 | 14,927 |
| 1953 | 7,834 | 790 | — | — | 1,215 | 5,392 | 15,231 |
| 1954 | 8,008 | 752 | 1,416 | 274 | 1,311 | 3,676 | 15,437 |
| 1955 | 8,490 | 809 | 1,468 | 313 | 1,390 | 3,539 | 16,009 |
| 1956 | 8,992 | 875 | 1,534 | 354 | 1,530 | 3,506 | 16,791 |
| 1957 | 9,409 | 932 | 1,592 | 342 | 1,661 | 3,609 | 17,545 |
| 1958 | 9,447 | 922 | 1,599 | 356 | 1,809 | 3,756 | 17,889 |
| 1959 | 10,102 | 982 | 1,677 | 385 | 1,949 | 3,739 | 18,834 |
| 1960 | 10,505 | 1,028 | 1,716 | 421 | 2,082 | 3,855 | 19,607 |
| 1961 | 10,907 | 1,061 | 1,740 | 452 | 2,264 | 3,961 | 20,385 |
| 1962 | 11,624 | 1,134 | 1,812 | 472 | 2,463 | 4,090 | 21,595 |
| 1963 | 12,247 | 1,200 | 1,854 | 484 | 2,624 | 4,148 | 22,557 |
| 1964 | 13,156 | 1,289 | 1,988 | 496 | 2,814 | 4,279 | 24,022 |
| 1965 | 14,444 | 1,409 | 2,162 | 522 | 3,062 | 4,598 | 26,197 |
| 1966 | 15,768 | 1,541 | 2,346 | 544 | 3,329 | 5,173 | 28,701 |
| 1967 | 16,595 | 1,623 | 2,436 | 563 | 3,632 | 5,570 | 30,419 |
| 1968 | 18,695 | 1,703 | 2,713 | 616 | 3,903 | 5,830 | 33,460 |
| 1969 | 20,207 | 1,716 | 2,984 | 661 | 4,256 | 6,291 | 36,115 |
| 1970 | 22,617 | 1,894 | 3,325 | 721 | 4,475 | 6,551 | 39,583 |
| 1971 | 24,166 | 2,086 | 3,626 | 762 | 4,990 | 6,621 | 42,251 |
| 1972 | 27,167 | 2,390 | 3,811 | 832 | 5,370 | 7,017 | 46,587 |
| 1973 | 31,265 | 2,639 | 4,218 | 963 | 5,605 | 7,960 | 52,650 |
| 1974 | 34,029 | 2,864 | 4,520 | 1,167 | 6,287 | 9,178 | 58,045 |
| 1975 | 41,384 | 3,199 | 4,952 | 1,369 | 7,060 | 10,145 | 68,109 |
| 1976 | 47,536 | 3,769 | 5,341 | 1,511 | 7,854 | 10,822 | 76,833 |
| 1977 | 52,491 | 4,115 | 5,663 | 2,606 | 8,413 | 11,547 | 84,835 |
| 1978 | 60,042 | 4,863 | 6,323 | 2,810 | 9,034 | 13,012 | 96,084 |
| 1979 | 68,872 | 5,551 | 7,157 | 2,921 | 9,914 | 14,756 | 109,171 |
| 1980 | 75,883 | 5,906 | 8,158 | 3,040 | 11,115 | 16,194 | 120,296 |
| 1981 | 83,358 | 6,639 | 8,830 | 2,979 | 11,357 | 17,751 | 130,914 |
| 1982 | 90,390 | 6,888 | 9,256 | 2,887 | 11,692 | 18,663 | 139,776 |
| 1983 | 98,710 | 7,660 | 9,827 | 3,271 | 12,338 | 19,077 | 150,883 |
| 1984 | 105,836 | 8,409 | 10,315 | 3,489 | 12,950 | 20,047 | 161,046 |
| 1985 | 111,760 | 9,168 | 10,499 | 3,737 | 13,534 | 20,133 | 168,831 |
| 1986 | 121,699 | 9,665 | 11,116 | 4,059 | 14,401 | 20,755 | 181,695 |
| 1987 | 137,190 | 11,117 | 11,860 | 4,396 | 13,470 | 21,122 | 199,155 |
| 1988 | 150,724 | 11,905 | 12,972 | 5,082 | 13,889 | 22,471 | 217,044 |
| 1989 | 160,226 | 12,179 | 14,153 | 6,089 | 14,609 | 24,005 | 231,261 |
| 1990 | 171,616 | 12,508 | 15,763 | 7,206 | 15,299 | 25,744 | 248,136 |
| 1991 | 180,062 | 12,460 | 16,513 | 7,936 | 16,186 | 26,379 | 259,535 |
| 1992 | 184,860 | 13,204 | 13,595 | 8,513 | 17,666 | 27,128 | 264,966 |
| 1993 | 197,987 | 13,362 | 13,704 | 9,365 | 18,330 | 27,216 | 279,964 |
| 1994 | 207,545 | 13,880 | 14,008 | 10,107 | 19,271 | 27,655 | 292,466 |
| 1995 | 216,091 | 14,211 | 14,040 | 11,081 | 20,064 | 28,138 | 303,624 |

research subjects a five-cup portion of macaroni and cheese, the subjects all responded by eating 30% more than they had when they were given portions half that size. Rolls, Morris, and Roe observe that both "restrained and unrestrained eaters" ate more when offered larger portions and assert that Americans have become accustomed to eating too much at one sitting. The problem of portion size is compounded by the observation that Americans are eating larger portions of foods that are high in calories and fat.

**TABLE 1.5**

**Food away from home, total expenditures, 1929–2006** [CONTINUED]

| Year | Eating and drinking places[a] | Hotels and motels[a] | Retail stores, direct selling[b] | Recreational places[c] | Schools and colleges[d] | All other[e] | Total[f] |
|------|-----|-----|-----|-----|-----|-----|-----|
| | | | **Million dollars** | | | | |
| 1996 | 223,546 | 14,553 | 14,056 | 11,515 | 20,867 | 28,602 | 313,139 |
| 1997 | 237,475 | 15,381 | 13,764 | 12,283 | 21,901 | 31,006 | 331,810 |
| 1998 | 250,495 | 16,069 | 14,872 | 13,048 | 23,053 | 32,053 | 349,590 |
| 1999 | 261,527 | 16,710 | 16,492 | 13,839 | 23,920 | 33,461 | 365,951 |
| 2000 | 282,235 | 18,003 | 16,932 | 14,662 | 24,468 | 35,147 | 391,447 |
| 2001 | 289,331 | 20,813 | 18,056 | 15,316 | 25,394 | 35,777 | 404,688 |
| 2002 | 300,753 | 21,812 | 19,753 | 16,235 | 26,735 | 36,185 | 421,472 |
| 2003 | 317,557 | 22,049 | 19,701 | 16,775 | 28,077 | 37,472 | 441,630 |
| 2004 | 338,229 | 22,543 | 20,009 | 17,417 | 29,287 | 39,194 | 466,678 |
| 2005 | 360,027 | 22,923 | 20,507 | 18,174 | 30,271 | 41,584 | 493,485 |
| 2006 | 389,810 | 23,093 | 23,623 | 18,907 | 30,899 | 42,768 | 529,100 |

—=Not available.
[a]Includes tips.
[b]Includes vending machine operators but not vending machines operated by organization.
[c]Motion picture theaters, bowling alleys, pool parlors, sports arenas, camps, amusement parks, golf and country clubs (includes concessions beginning in 1977).
[d]Includes school food subsidies.
[e]Military exchanges and clubs; railroad dining cars; airlines; food service in manufacturing plants, institutions, hospitals, boarding houses, fraternities and sororities, and civic and social organizations; and food supplied to military forces, civilian employees and child day care.
[f]Computed from unrounded data.

SOURCE: "Table 3. Food Away from Home: Total Expenditures," in *Food CPI, Prices, and Expenditures: Food Away from Home*, United States Department of Agriculture, Economic Research Service, July 2, 2007, http://www.ers.usda.gov/briefing/CPIFoodAndExpenditures/Data/table3.htm (accessed October 6, 2007)

**BIGGER PORTIONS IN RESTAURANTS.** Samara Joy Nielsen and Barry M. Popkin of the University of North Carolina at Chapel Hill looked at portion size consumed in the United States to determine whether average portion sizes had increased over time and reported their findings in "Patterns and Trends in Food Portion Sizes, 1977–1998" (*Journal of the American Medical Association*, vol. 289, no. 4, January 22, 2003). They analyzed data collected by national nutrition surveys—the Nationwide Food Consumption Survey and the Continuing Survey of Food Intake by Individuals—conducted in the United States in 1977, 1989, 1994, and 1996, detailing the consumption habits of more than sixty-three thousand people. For each survey year, the researchers analyzed average portion sizes consumed of specific food items (salty snacks, desserts, soft drinks, fruit drinks, French fries, hamburgers, cheeseburgers, pizza, and Mexican food) by eating location—home, restaurant, or fast-food outlet. Nielsen and Popkin report that over the past two decades the average portions of salty snacks such as popcorn and chips have increased by 60%, and soft drinks have grown by 50%. The average dispensed soft drink measured 13.1 ounces in 1977, but by 1996 it was 19.9 ounces. During this same period, an average bag of chips grew from 1 ounce to 1.6 ounces. As a result, the average chips-and-soda snack contains 150 more calories than it did two decades before.

The portion-size changes were observed with many fast-food offerings. During the twenty years studied, the size of the average hamburger grew by 23%, to 7 ounces, and servings of fries grew by 16%, to 3.5 ounces. A regular-sized burger-and-fries meal contained 155 calories more than it did in 1977. Worse still, Nielsen and Popkin find that portion size had also expanded in Americans' homes, indicating widespread ignorance about appropriate portion size. Interestingly, portion sizes were smallest in restaurants, although they, too, had increased during the study period. For example, the average restaurant portion of spaghetti with tomato sauce and meatballs doubled in size from 500 to 1,025 calories.

In "Increased Portion Size Leads to Increased Energy Intake in a Restaurant Meal" (*Obesity Research*, vol. 12, no. 3, March 2004), Nicole Diliberti et al. of Pennsylvania State University find that larger portions served in restaurants resulted in patrons consuming more calories. The investigators covertly recorded the food intake of patrons who selected a pasta entrée over a ten-day period of a cafeteria-style restaurant on a university campus. On five days, the portion size of the entrée was the standard portion, and on five different days, the size was increased to 150% of the standard portion. Subjects were also asked to complete a survey to determine perceptions of the portion size of the entrée and of the amount that they ate. The subjects who completed a survey were unaware that their intake was being monitored.

Diliberti et al. posited that when the portion size of an entrée was increased by 150%, the subjects would consume significantly more than when the standard portion was offered. They also sought to determine whether the subjects would compensate for the increased intake from the entrée by reducing their consumption of other foods at the meal and whether they could identify any characteristics of subjects that would predict how they would respond to the increased portion size.

When the larger portion size was offered, subjects who purchased it consumed 43% more of the entrée than those who purchased the standard portion size. Subjects given the larger portion also ate significantly more of the entrée accompaniments (tomato, roll, and butter) than those who purchased the standard portion, even though the portion size of the accompaniments was the same for all subjects.

Overall, ratings of the appropriateness of the portion size of the entrée did not differ between subjects given the 150% portion and those who received the standard portion. There was, however, an effect of subject body size on this rating. Underweight and normal-weight subjects who purchased the 150% portion rated it as closer to the "too large" end of the seven-point scale than those who purchased the standard portion. In contrast, overweight and obese subjects did not rate the portion size as "too large." Diliberti et al. conclude that subjects ate significantly more when the portion size was increased, and their responses to the survey indicate that they were unaware that the portion was larger than normal or that they had consumed more food.

AND BIGGER PORTIONS AT HOME. Increased portion sizes at home are reflected in recipes and cookbooks. Lisa R. Young notes in *The Portion Teller: Smartsize Your Way to Permanent Weight Loss* (2005) that recipes call for bigger portions using the same ingredients than they did in past decades. For example, a brownie recipe from Irma S. Rombauer and Marion Rombauer Becker's *Joy of Cooking* (1964) recommended dividing it into thirty servings, whereas the same recipe in the 1997 edition of the book is divided into only sixteen servings. Similarly, a 1984 recipe for Toll House cookies yielded one hundred servings, whereas today the same recipe yields only sixty. Other popular food items have increased in size and caloric content. When the National Heart, Lung, and Blood Institute (http://hp2010.nhlbihin.net/portion/) compared portion sizes and the corresponding calories of several popular foods from 1983 and 2003, researchers found that two decades earlier a bagel measured three inches in diameter and contained 140 calories. In the early twenty-first century, six-inch bagels contain 350 calories.

Nielsen and Popkin also note other changes in eating behavior. For example, they find that Americans obtain 19% of their total calories from snacks—double the amount of 1977—and 81% from meals. They conclude that "control of portion size must be systematically addressed both in general and as it relates to fast food pricing and marketing. The best way to encourage people to eat smaller portions is if food portions served inside and outside the home are smaller."

## Technology Satisfies the Hunger for Quick, Inexpensive Food

In "Why Have Americans Become More Obese?" (*Journal of Economic Perspectives*, vol. 17, no. 3, summer 2003), David M. Cutler, Edward L. Glaeser, and Jesse M. Shapiro refute the notion that increased portion sizes, increasingly sedentary lifestyles, or restaurant dining are responsible for Americans' widening waistlines. After examining nearly one hundred years of nutritional data, they determine that technological advances have increased the efficiency of food production and made food more varied, convenient, tastier, and cheaper.

Cutler, Glaeser, and Shapiro illustrate how efficiencies in food preparation have revolutionized Americans' eating habits. They compare the speed and ease of preparation of commercial French fries with the previously time-consuming, labor-intensive process of scrubbing, peeling, paring, and frying required to prepare French fries. They observe that during the 1960s women spent an average of two hours per day on meal preparation—twice as long as the average American nonworking woman devoted to meal preparation in 2003. It takes considerably less time today to prepare food because of advances in food processing and packaging. Furthermore, technology improvements in the home, such as the microwave oven, have made it easier to eat quickly on demand.

According to Cutler, Glaeser, and Shapiro, increased food consumption is the direct "result of technological innovations which made it possible for food to be mass prepared far from the point of consumption, and consumed with lower time costs of preparation and cleaning. Price changes are normally beneficial, but may not be if people have self-control problems." They find that the average number of daily snacks between meals has risen by 60% since the late 1970s. Unable to resist the tempting, affordable variety of foods, Americans engage in more frequent snacking, consuming the excess calories that ultimately result in overweight.

## Is the Food Industry the Culprit?

Kelly D. Brownell, the director of the Yale Center for Eating and Weight Disorders, and Katherine Battle Horgen cite in *Food Fight: The Inside Story of the Food Industry, America's Obesity Crisis, and What We Can Do about It* (2004) a "near-total surrender to a powerful food industry" as one of the main causes of the obesity epidemic in the United States. Brownell and Horgen contend that the obesity epidemic represents more than a failure of Americans to assume personal responsibility and exercise willpower over their appetites. They exhort consumers to agitate against a food industry intent on fattening them and to work to counteract a variety of unhealthy social trends. They lament the supersized meals and sedentary lifestyles, including Americans' "car-centric" culture that actively discourages walking and encourages children to sit in front of television, videogames, and computers while eliminating physical education classes from schools, but they insist that the food industry bears most of the responsibility for the rise in obesity. Brownell and Horgen argue that Americans feed their pets better than their children and that children are induced and manipulated by food industry media advertising to adopt poor eating habits and to consume high-caloric, low-nutrition junk food.

Brownell and Horgen cite toy giveaways, movie tie-ins, and in-school promotions as evidence of effective strategies employed by the politically powerful food industry to promote fast-food consumption. They feel that the battle against these pervasive influences is one that parents cannot win because even children receiving consistent, sound nutritional counseling from parents are not immune to the effects of multiple, powerful exposures to media advertising. Brownell and Horgen call for a nationwide, grassroots movement to reverse these trends and advocate specific measures such as junk-food taxes and banning advertisements that target children.

Greg Critser also indicts the food industry in *Fat Land: How Americans Became the Fattest People in the World* (2003). He presents a critical analysis of the many social and economic factors that make Americans among the most overweight people in the world. Critser believes that chief among these factors is high-fructose corn syrup, a low-cost sweetener that was developed by Japanese scientists in response to an overabundance of cheap corn. Corn syrup does more than sweeten, it also acts as a preservative, giving sweet foods a longer shelf life. Since the 1970s high-fructose corn syrup has been used to sweeten nearly every product on supermarket shelves, from cereal to soda. Some researchers feel that because it is so ubiquitous, many Americans are unknowingly consuming excessive amounts of fructose. Table 1.6 shows per capita consumption of high-fructose corn syrup, which peaked in 1999 and has since slowly declined.

Unfortunately, fructose also appears to trigger fat storage more efficiently than other sugars do. New studies show that the body does not metabolize high-fructose corn syrup well. Even though all sugars are stored in the body as fat, some researchers think that fructose is more readily converted into fat than other sugars. The fructose encourages the liver to promote fat by activating enzymes that create higher levels of cholesterol and triglycerides and make muscles more insulin resistant. Elevated levels of cholesterol and triglycerides, fatty substances normally present in the bloodstream and all cells of the body, increase the risk of coronary heart disease. Insulin resistance can lead to diabetes.

Critser also explains that once the staples used to produce fast foods became cheaper, the industry intensified marketing efforts to induce consumers to buy and eat more. Table 1.7 shows that food expenditures have consistently decreased as a percent of disposable personal income, declining from 23.4% of personal disposable income in 1929 to just 9.9% in 2006. Critser observes that a serving of McDonald's French fries "ballooned from 200 calories (1960) ... to the present 610 calories" and that Americans' appetites grew to expect and demand the bigger servings. He notes that changing values and lifestyles conspired to fatten Americans. Furthermore, he describes the rise of a "new boundary-free culture" that promoted consumption of sugar- and fat-laden foods. Traditionally, families convened for home-cooked dinners, but Critser describes the rushed parents of the 1980s as preferring to eat out or take in prepared foods. Childcare experts popularized the theory that children instinctively knew when they were sated and encouraged busy parents to relinquish control over their children's food consumption. In some parts of the country, budget cuts prompted schools to allow fast-food franchises to sell lunches and snacks to students on the school campuses. Finally, Critser observes that to accommodate—or even camouflage—Americans' expanding bodies, clothing manufacturers marketed large, loose-fitting clothing.

**TABLE 1.6**

**Estimated number of per capita calories of high fructose corn syrup consumed daily, 1970–2006**

| Year | Primary weight (market level)[a] Pounds/year | Loss from primary to retail weight Percent | Weight at retail level Pounds/year | Loss from retail/ institutional to consumer level Percent | Weight at consumer level Pounds/year | Loss at consumer level Nonedible share Percent | Loss at consumer level other (uneaten food, spoilage, etc.) Percent | Per capita consumption (adjusted for loss) Pounds/year | Per capita consumption (adjusted for loss) Ounces/daily | Grams/daily | Calories per serving Number | Serving weight Grams | Calories consumed daily[b] Number | Servings (teaspoons) consumed daily[c] Teaspoons |
|---|---|---|---|---|---|---|---|---|---|---|---|---|---|---|
| 1970 | 0.5 | 0.0 | 0.5 | 11.0 | 0.5 | 0.0 | 20.0 | 0.4 | 0.0 | 0.5 | 16.0 | 4.2 | 2 | 0.1 |
| 1971 | 0.8 | 0.0 | 0.8 | 11.0 | 0.7 | 0.0 | 20.0 | 0.6 | 0.0 | 0.7 | 16.0 | 4.2 | 3 | 0.2 |
| 1972 | 1.2 | 0.0 | 1.2 | 11.0 | 1.0 | 0.0 | 20.0 | 0.8 | 0.0 | 1.0 | 16.0 | 4.2 | 4 | 0.2 |
| 1973 | 2.1 | 0.0 | 2.1 | 11.0 | 1.8 | 0.0 | 20.0 | 1.5 | 0.1 | 1.8 | 16.0 | 4.2 | 7 | 0.4 |
| 1974 | 2.8 | 0.0 | 2.8 | 11.0 | 2.5 | 0.0 | 20.0 | 2.0 | 0.1 | 2.4 | 16.0 | 4.2 | 9 | 0.6 |
| 1975 | 4.9 | 0.0 | 4.9 | 11.0 | 4.3 | 0.0 | 20.0 | 3.5 | 0.2 | 4.3 | 16.0 | 4.2 | 16 | 1.0 |
| 1976 | 7.2 | 0.0 | 7.2 | 11.0 | 6.4 | 0.0 | 20.0 | 5.1 | 0.2 | 6.3 | 16.0 | 4.2 | 24 | 1.5 |
| 1977 | 9.6 | 0.0 | 9.6 | 11.0 | 8.5 | 0.0 | 20.0 | 6.8 | 0.3 | 8.5 | 16.0 | 4.2 | 32 | 2.0 |
| 1978 | 10.8 | 0.0 | 10.8 | 11.0 | 9.6 | 0.0 | 20.0 | 7.7 | 0.3 | 9.5 | 16.0 | 4.2 | 36 | 2.3 |
| 1979 | 14.8 | 0.0 | 14.8 | 11.0 | 13.1 | 0.0 | 20.0 | 10.5 | 0.5 | 13.1 | 16.0 | 4.2 | 50 | 3.1 |
| 1980 | 19.0 | 0.0 | 19.0 | 11.0 | 16.9 | 0.0 | 20.0 | 13.5 | 0.6 | 16.8 | 16.0 | 4.2 | 64 | 4.0 |
| 1981 | 22.8 | 0.0 | 22.8 | 11.0 | 20.3 | 0.0 | 20.0 | 16.3 | 0.7 | 20.2 | 16.0 | 4.2 | 77 | 4.8 |
| 1982 | 26.6 | 0.0 | 26.6 | 11.0 | 23.7 | 0.0 | 20.0 | 19.0 | 0.8 | 23.6 | 16.0 | 4.2 | 90 | 5.6 |
| 1983 | 31.2 | 0.0 | 31.2 | 11.0 | 27.8 | 0.0 | 20.0 | 22.2 | 1.0 | 27.6 | 16.0 | 4.2 | 105 | 6.6 |
| 1984 | 37.2 | 0.0 | 37.2 | 11.0 | 33.1 | 0.0 | 20.0 | 26.5 | 1.2 | 32.9 | 16.0 | 4.2 | 125 | 7.8 |
| 1985 | 45.2 | 0.0 | 45.2 | 11.0 | 40.2 | 0.0 | 20.0 | 32.2 | 1.4 | 40.0 | 16.0 | 4.2 | 152 | 9.5 |
| 1986 | 45.7 | 0.0 | 45.7 | 11.0 | 40.7 | 0.0 | 20.0 | 32.5 | 1.4 | 40.4 | 16.0 | 4.2 | 154 | 9.6 |
| 1987 | 47.7 | 0.0 | 47.7 | 11.0 | 42.5 | 0.0 | 20.0 | 34.0 | 1.5 | 42.2 | 16.0 | 4.2 | 161 | 10.1 |
| 1988 | 49.0 | 0.0 | 49.0 | 11.0 | 43.6 | 0.0 | 20.0 | 34.9 | 1.5 | 43.3 | 16.0 | 4.2 | 165 | 10.3 |
| 1989 | 48.2 | 0.0 | 48.2 | 11.0 | 42.9 | 0.0 | 20.0 | 34.3 | 1.5 | 42.6 | 16.0 | 4.2 | 162 | 10.2 |
| 1990 | 49.6 | 0.0 | 49.6 | 11.0 | 44.1 | 0.0 | 20.0 | 35.3 | 1.5 | 43.9 | 16.0 | 4.2 | 167 | 10.4 |
| 1991 | 50.3 | 0.0 | 50.3 | 11.0 | 44.8 | 0.0 | 20.0 | 35.8 | 1.6 | 44.5 | 16.0 | 4.2 | 170 | 10.6 |
| 1992 | 51.8 | 0.0 | 51.8 | 11.0 | 46.1 | 0.0 | 20.0 | 36.9 | 1.6 | 45.8 | 16.0 | 4.2 | 175 | 10.9 |
| 1993 | 54.5 | 0.0 | 54.5 | 11.0 | 48.5 | 0.0 | 20.0 | 38.8 | 1.7 | 48.2 | 16.0 | 4.2 | 184 | 11.5 |
| 1994 | 56.2 | 0.0 | 56.2 | 11.0 | 50.0 | 0.0 | 20.0 | 40.0 | 1.8 | 49.7 | 16.0 | 4.2 | 189 | 11.8 |
| 1995 | 57.6 | 0.0 | 57.6 | 11.0 | 51.3 | 0.0 | 20.0 | 41.0 | 1.8 | 51.0 | 16.0 | 4.2 | 194 | 12.1 |
| 1996 | 57.8 | 0.0 | 57.8 | 11.0 | 51.4 | 0.0 | 20.0 | 41.1 | 1.8 | 51.1 | 16.0 | 4.2 | 195 | 12.2 |
| 1997 | 60.4 | 0.0 | 60.4 | 11.0 | 53.7 | 0.0 | 20.0 | 43.0 | 1.9 | 53.4 | 16.0 | 4.2 | 204 | 12.7 |
| 1998 | 61.9 | 0.0 | 61.9 | 11.0 | 55.1 | 0.0 | 20.0 | 44.1 | 1.9 | 54.8 | 16.0 | 4.2 | 209 | 13.0 |
| 1999 | 63.7 | 0.0 | 63.7 | 11.0 | 56.7 | 0.0 | 20.0 | 45.4 | 2.0 | 56.4 | 16.0 | 4.2 | 215 | 13.4 |
| 2000 | 62.7 | 0.0 | 62.7 | 11.0 | 55.8 | 0.0 | 20.0 | 44.6 | 2.0 | 55.4 | 16.0 | 4.2 | 211 | 13.2 |
| 2001 | 62.6 | 0.0 | 62.6 | 11.0 | 55.7 | 0.0 | 20.0 | 44.6 | 2.0 | 55.4 | 16.0 | 4.2 | 211 | 13.2 |
| 2002 | 62.9 | 0.0 | 62.9 | 11.0 | 56.0 | 0.0 | 20.0 | 44.8 | 2.0 | 55.6 | 16.0 | 4.2 | 212 | 13.2 |
| 2003 | 61.0 | 0.0 | 61.0 | 11.0 | 54.2 | 0.0 | 20.0 | 43.4 | 1.9 | 53.9 | 16.0 | 4.2 | 205 | 12.8 |
| 2004 | 59.9 | 0.0 | 59.9 | 11.0 | 53.3 | 0.0 | 20.0 | 42.7 | 1.9 | 53.0 | 16.0 | 4.2 | 202 | 12.6 |
| 2005 | 59.2 | 0.0 | 59.2 | 11.0 | 52.7 | 0.0 | 20.0 | 42.2 | 1.8 | 52.4 | 16.0 | 4.2 | 200 | 12.5 |
| 2006 | 58.3 | 0.0 | 58.3 | 11.0 | 51.9 | 0.0 | 20.0 | 41.5 | 1.8 | 51.6 | 16.0 | 4.2 | 197 | 12.3 |

Note: Estimated number of daily per capita calories calculated by adjusting high fructose corn syrup deliveries for domestic food and beverage use for food losses.

[a]U.S. per capita high fructose corn syrup estimated deliveries for domestic food and beverage use, calendar year.

[b]Number of daily teaspoons multiplied by calories per serving.

[c]Grams per day divided by serving weight.

SOURCE: "Table 52. High Fructose Corn Syrup: Estimated Number of Per Capita Calories Consumed Daily, by Calendar Year," in *Sugar and Sweeteners: Data Tables*, United States Department of Agriculture, Economic Research Service, April 11, 2007, http://www.ers.usda.gov/Briefing/Sugar/Data/Table52.xls (accessed October 8, 2007)

TABLE 1.7

**Food expenditures by families and individuals as a share of disposable personal income, 1929–2006**

| Year | Disposable personal income Billion dollars | Expenditures for food | | | | | |
|---|---|---|---|---|---|---|---|
| | | At home[a] | | Away from home[b] | | Total[c] | |
| | | Billion dollars | Percent | Billion dollars | Percent | Billion dollars | Percent |
| 1929 | 83.4 | 16.9 | 20.3 | 2.6 | 3.1 | 19.5 | 23.4 |
| 1930 | 74.7 | 15.8 | 21.2 | 2.3 | 3.1 | 18.1 | 24.2 |
| 1931 | 64.3 | 12.7 | 19.8 | 2.1 | 3.3 | 14.8 | 23.0 |
| 1932 | 49.2 | 9.6 | 19.5 | 1.7 | 3.5 | 11.3 | 23.0 |
| 1933 | 46.1 | 10.1 | 21.9 | 1.5 | 3.3 | 11.6 | 25.2 |
| 1934 | 52.8 | 11.1 | 21.0 | 1.7 | 3.2 | 12.8 | 24.2 |
| 1935 | 59.3 | 12.1 | 20.4 | 1.8 | 3.0 | 13.9 | 23.4 |
| 1936 | 67.4 | 12.7 | 18.8 | 2.0 | 3.0 | 14.7 | 21.8 |
| 1937 | 72.2 | 13.3 | 18.4 | 2.2 | 3.0 | 15.5 | 21.5 |
| 1938 | 66.6 | 12.6 | 18.9 | 2.1 | 3.2 | 14.7 | 22.1 |
| 1939 | 71.4 | 13.0 | 18.1 | 2.3 | 3.2 | 15.2 | 21.3 |
| 1940 | 76.8 | 13.5 | 17.6 | 2.4 | 3.1 | 15.9 | 20.7 |
| 1941 | 93.8 | 15.3 | 16.3 | 2.9 | 3.1 | 18.2 | 19.4 |
| 1942 | 118.6 | 18.5 | 15.6 | 3.6 | 3.0 | 22.1 | 18.6 |
| 1943 | 135.4 | 20.7 | 15.3 | 4.5 | 3.3 | 25.2 | 18.6 |
| 1944 | 148.3 | 22.1 | 14.9 | 5.1 | 3.4 | 27.2 | 18.4 |
| 1945 | 152.2 | 23.6 | 15.5 | 5.7 | 3.7 | 29.3 | 19.2 |
| 1946 | 161.4 | 28.4 | 17.6 | 6.5 | 4.0 | 34.9 | 21.6 |
| 1947 | 171.2 | 32.8 | 19.2 | 7.4 | 4.3 | 40.2 | 23.5 |
| 1948 | 190.6 | 34.9 | 18.3 | 7.5 | 3.9 | 42.4 | 22.3 |
| 1949 | 190.4 | 34.3 | 18.0 | 7.8 | 4.1 | 42.0 | 22.1 |
| 1950 | 210.1 | 35.7 | 17.0 | 7.6 | 3.6 | 43.3 | 20.6 |
| 1951 | 231.0 | 40.0 | 17.3 | 8.4 | 3.6 | 48.4 | 20.9 |
| 1952 | 243.4 | 41.8 | 17.2 | 8.8 | 3.6 | 50.6 | 20.8 |
| 1953 | 258.6 | 42.3 | 16.4 | 9.0 | 3.5 | 51.3 | 19.9 |
| 1954 | 264.3 | 42.4 | 16.0 | 9.3 | 3.5 | 51.7 | 19.6 |
| 1955 | 283.3 | 42.9 | 15.1 | 9.8 | 3.5 | 52.7 | 18.6 |
| 1956 | 303.0 | 44.4 | 14.7 | 10.4 | 3.4 | 54.8 | 18.1 |
| 1957 | 319.8 | 48.1 | 15.0 | 10.9 | 3.4 | 59.0 | 18.4 |
| 1958 | 330.5 | 49.8 | 15.1 | 11.1 | 3.4 | 60.9 | 18.4 |
| 1959 | 350.5 | 50.1 | 14.3 | 12.1 | 3.5 | 62.3 | 17.8 |
| 1960 | 365.4 | 51.5 | 14.1 | 12.6 | 3.4 | 64.0 | 17.5 |
| 1961 | 381.8 | 52.0 | 13.6 | 13.1 | 3.4 | 65.1 | 17.1 |
| 1962 | 405.1 | 52.9 | 13.1 | 13.9 | 3.4 | 66.8 | 16.5 |
| 1963 | 425.1 | 53.3 | 12.5 | 14.5 | 3.4 | 67.9 | 16.0 |
| 1964 | 462.5 | 55.5 | 12.0 | 15.7 | 3.4 | 71.2 | 15.4 |
| 1965 | 498.1 | 58.4 | 11.7 | 16.9 | 3.4 | 75.4 | 15.1 |
| 1966 | 537.5 | 61.0 | 11.3 | 18.6 | 3.5 | 79.6 | 14.8 |
| 1967 | 575.3 | 61.4 | 10.7 | 19.8 | 3.4 | 81.1 | 14.1 |
| 1968 | 625.0 | 64.5 | 10.3 | 21.7 | 3.5 | 86.2 | 13.8 |
| 1969 | 674.0 | 69.0 | 10.2 | 23.4 | 3.5 | 92.3 | 13.7 |
| 1970 | 735.7 | 75.5 | 10.3 | 26.4 | 3.6 | 102.0 | 13.9 |
| 1971 | 801.8 | 79.5 | 9.9 | 28.1 | 3.5 | 107.6 | 13.4 |
| 1972 | 869.1 | 86.0 | 9.9 | 31.3 | 3.6 | 117.3 | 13.5 |
| 1973 | 978.3 | 94.9 | 9.7 | 34.9 | 3.6 | 129.8 | 13.3 |
| 1974 | 1,071.6 | 107.3 | 10.0 | 38.5 | 3.6 | 145.8 | 13.6 |
| 1975 | 1,187.4 | 117.4 | 9.9 | 45.9 | 3.9 | 163.3 | 13.8 |
| 1976 | 1,302.5 | 125.1 | 9.6 | 52.6 | 4.0 | 177.7 | 13.6 |
| 1977 | 1,435.7 | 133.8 | 9.3 | 58.5 | 4.1 | 192.3 | 13.4 |
| 1978 | 1,608.3 | 147.3 | 9.2 | 67.5 | 4.2 | 214.8 | 13.4 |
| 1979 | 1,793.5 | 164.0 | 9.1 | 76.9 | 4.3 | 240.9 | 13.4 |
| 1980 | 2,009.0 | 180.8 | 9.0 | 85.2 | 4.2 | 266.0 | 13.2 |
| 1981 | 2,246.1 | 195.5 | 8.7 | 95.8 | 4.3 | 291.3 | 13.0 |
| 1982 | 2,421.2 | 201.0 | 8.3 | 104.5 | 4.3 | 305.5 | 12.6 |
| 1983 | 2,608.4 | 211.4 | 8.1 | 113.7 | 4.4 | 325.1 | 12.5 |
| 1984 | 2,912.0 | 224.0 | 7.7 | 121.9 | 4.2 | 345.8 | 11.9 |
| 1985 | 3,109.3 | 234.0 | 7.5 | 128.6 | 4.1 | 362.6 | 11.7 |
| 1986 | 3,285.1 | 242.7 | 7.4 | 137.9 | 4.2 | 380.6 | 11.6 |
| 1987 | 3,458.1 | 252.7 | 7.3 | 146.4 | 4.2 | 399.0 | 11.5 |
| 1988 | 3,748.7 | 255.5 | 6.8 | 157.5 | 4.2 | 413.0 | 11.0 |
| 1989 | 4,021.7 | 274.2 | 6.8 | 165.4 | 4.1 | 439.6 | 10.9 |
| 1990 | 4,285.8 | 298.9 | 7.0 | 177.4 | 4.1 | 476.3 | 11.1 |

**TABLE 1.7**

**Food expenditures by families and individuals as a share of disposable personal income, 1929–2006** [CONTINUED]

| Year | Disposable personal income Billion dollars | At home[a] Billion dollars | At home[a] Percent | Away from home[b] Billion dollars | Away from home[b] Percent | Total[c] Billion dollars | Total[c] Percent |
|------|------|------|------|------|------|------|------|
| 1991 | 4,464.3 | 312.8 | 7.0 | 186.3 | 4.2 | 499.1 | 11.2 |
| 1992 | 4,751.4 | 313.1 | 6.6 | 191.9 | 4.0 | 505.0 | 10.6 |
| 1993 | 4,911.9 | 322.5 | 6.6 | 205.9 | 4.2 | 528.4 | 10.8 |
| 1994 | 5,151.8 | 336.4 | 6.5 | 216.5 | 4.2 | 552.9 | 10.7 |
| 1995 | 5,408.2 | 345.0 | 6.4 | 226.2 | 4.2 | 571.2 | 10.6 |
| 1996 | 5,688.5 | 360.3 | 6.3 | 233.2 | 4.1 | 593.5 | 10.4 |
| 1997 | 5,988.8 | 376.9 | 6.3 | 246.3 | 4.1 | 623.2 | 10.4 |
| 1998 | 6,395.9 | 386.8 | 6.0 | 259.7 | 4.1 | 646.6 | 10.1 |
| 1999 | 6,695.0 | 408.5 | 6.1 | 272.0 | 4.1 | 680.5 | 10.2 |
| 2000 | 7,194.0 | 419.4 | 5.8 | 291.3 | 4.0 | 710.8 | 9.9 |
| 2001 | 7,486.8 | 441.4 | 5.9 | 301.3 | 4.0 | 742.7 | 9.9 |
| 2002 | 7,830.1 | 453.7 | 5.8 | 314.1 | 4.0 | 767.8 | 9.8 |
| 2003 | 8,162.5 | 472.4 | 5.8 | 329.5 | 4.0 | 801.9 | 9.8 |
| 2004 | 8,681.6 | 491.5 | 5.7 | 348.2 | 4.0 | 839.7 | 9.7 |
| 2005 | 9,036.1 | 519.2 | 5.7 | 368.5 | 4.1 | 887.8 | 9.8 |
| 2006 | 9,534.8 | 551.1 | 5.8 | 396.1 | 4.2 | 947.1 | 9.9 |

[a]Food purchases from grocery stores and other retail outlets, including purchases with food stamps and WIC (Women, Infants and Children) vouchers and food produced and consumed on farms (valued at farm prices) because the value of these foods is included in personal income. Excludes government-donated foods.
[b]Purchases of meals and snacks by families and individuals, and food furnished to employees since it is included in personal income. Excludes food paid for by government and business, such as donated foods to schools, meals in prisons and other institutions, and expense-account meals.
[c]Total may not add due to rounding.

SOURCE: "Table 7. Food Expenditures by Families and Individuals as a Share of Disposable Personal Income," in *Food CPI, Prices, and Expenditures: Food Expenditures by Families and Individuals as a Share of Disposable Income*, United States Department of Agriculture, Economic Research Service, July 2, 2007, http://www.ers.usda.gov/Briefing/CPIFoodAndExpenditures/Data/table7.htm (accessed October 8, 2007)

# CHAPTER 2
# WEIGHT AND PHYSICAL HEALTH

*If we could give every individual the right amount of nourishment and exercise, not too little and not too much, we would have found the safest way to health.*

—Hippocrates

During the twentieth century, advances in public health and medical care helped Americans lead longer, healthier lives. Two important measures of the health of the population are infant mortality (death) rates and life expectancy at birth rates. By the end of the last century, infant mortality rates had significantly decreased and life expectancy had increased by thirty years. Table 2.1 shows the long-term upward trend in life expectancy as well as recent gains. In 2004 life expectancy at birth for the total population reached a record high of 77.8 years, up from 75.4 years in 1990. The Central Intelligence Agency estimates in the *World Factbook* (December 13, 2007, https://www.cia.gov/library/publications/the-world-factbook/geos/us.html) that in 2007 life expectancy at birth had increased to seventy-eight years.

As deaths from infectious diseases declined during the second half of the twentieth century, mortality from chronic diseases, such as heart disease and cancer, increased. Table 2.2 displays the ten leading causes of death in the United States in 1980 and 2004. Overweight and obesity are considered contributing factors to at least four of the ten leading causes of death in 2004: diseases of the heart, malignant neoplasms (tumors), cerebrovascular diseases (diseases affecting the supply of blood to the brain), and diabetes mellitus. Obesity may also be implicated in some deaths attributable to another leading cause of death: nephritis, nephrotic syndrome, and nephrosis (kidney disease or chronic renal failure). Table 2.2 also reveals the rise of diabetes as a cause of death. In 1980 it was the seventh-leading cause of death, claiming 34,851 lives. By 2004 it rose to being the sixth-leading cause of death and was the underlying cause of 73,138 deaths. Epidemiologists (scientists who study the occur-

rence and distribution of diseases and the factors that govern their spread) and medical researchers believe that the increasing prevalence of diabetes in the U.S. population and the resultant rise in deaths attributable to diabetes are direct consequences of the obesity epidemic in the United States.

Figure 2.1 reveals little change in the prevalence of overweight and obesity between the 1960s and 1980, and increasing prevalence of overweight and obesity between 1980 and 2000. As Figure 2.1 shows, 67% of American adults were overweight during 2003–04, including the 34% who were classified as obese. The prevalence of obesity varies somewhat by gender, race, and ethnicity. In 2001–04 slightly more women (34%) than men (30.2%) and half (51.6%) of non-Hispanic African-American women were obese. (See Table 2.3.)

Overweight and obesity increase not only the risk of morbidity (illness or disease) and mortality but also the severity of diseases such as hypertension (high blood pressure), arthritis, and other musculoskeletal problems. Table 2.4 lists the health consequences that may result from overweight and obesity among adults and children. It also estimates the likelihood of these health consequences. For example, adults who are obese are twice as likely to suffer from high blood pressure as adults who are at a healthy weight.

In "A Potential Decline in Life Expectancy in the United States in the 21st Century" (*New England Journal of Medicine*, vol. 352, no. 11, March 17, 2005), S. Jay Olshansky et al. suggest that the steady rise in life expectancy the United States enjoyed during the past two centuries might soon come to an end. The investigators use obesity prevalence data and previously published estimates of years of life lost from obesity to project life expectancy. Instead of using historical trends to forecast life expectancy, they calculated in reverse, assessing the fall in death rates that would occur if all obese Americans

TABLE 2.1

**Life expectancy at birth, at 65 years of age, and at 75 years of age, according to race and sex, selected years 1900–2004**

[Data are based on death certificates]

| Specified age and year | All races | | | White | | | Black or African American[a] | | |
|---|---|---|---|---|---|---|---|---|---|
| | Both sexes | Male | Female | Both sexes | Male | Female | Both sexes | Male | Female |
| **At birth** | | | | | Remaining life expectancy in years | | | | |
| 1900[b, c] | 47.3 | 46.3 | 48.3 | 47.6 | 46.6 | 48.7 | 33.0 | 32.5 | 33.5 |
| 1950[c] | 68.2 | 65.6 | 71.1 | 69.1 | 66.5 | 72.2 | 60.8 | 59.1 | 62.9 |
| 1960[c] | 69.7 | 66.6 | 73.1 | 70.6 | 67.4 | 74.1 | 63.6 | 61.1 | 66.3 |
| 1970 | 70.8 | 67.1 | 74.7 | 71.7 | 68.0 | 75.6 | 64.1 | 60.0 | 68.3 |
| 1980 | 73.7 | 70.0 | 77.4 | 74.4 | 70.7 | 78.1 | 68.1 | 63.8 | 72.5 |
| 1990 | 75.4 | 71.8 | 78.8 | 76.1 | 72.7 | 79.4 | 69.1 | 64.5 | 73.6 |
| 1995 | 75.8 | 72.5 | 78.9 | 76.5 | 73.4 | 79.6 | 69.6 | 65.2 | 73.9 |
| 1996 | 76.1 | 73.1 | 79.1 | 76.8 | 73.9 | 79.7 | 70.2 | 66.1 | 74.2 |
| 1997 | 76.5 | 73.6 | 79.4 | 77.1 | 74.3 | 79.9 | 71.1 | 67.2 | 74.7 |
| 1998 | 76.7 | 73.8 | 79.5 | 77.3 | 74.5 | 80.0 | 71.3 | 67.6 | 74.8 |
| 1999 | 76.7 | 73.9 | 79.4 | 77.3 | 74.6 | 79.9 | 71.4 | 67.8 | 74.7 |
| 2000 | 77.0 | 74.3 | 79.7 | 77.6 | 74.9 | 80.1 | 71.9 | 68.3 | 75.2 |
| 2001 | 77.2 | 74.4 | 79.8 | 77.7 | 75.0 | 80.2 | 72.2 | 68.6 | 75.5 |
| 2002 | 77.3 | 74.5 | 79.9 | 77.7 | 75.1 | 80.3 | 72.3 | 68.8 | 75.6 |
| 2003 | 77.5 | 74.8 | 80.1 | 78.0 | 75.3 | 80.5 | 72.7 | 69.0 | 76.1 |
| 2004 | 77.8 | 75.2 | 80.4 | 78.3 | 75.7 | 80.8 | 73.1 | 69.5 | 76.3 |
| **At 65 years** | | | | | | | | | |
| 1950[c] | 13.9 | 12.8 | 15.0 | — | 12.8 | 15.1 | 13.9 | 12.9 | 14.9 |
| 1960[c] | 14.3 | 12.8 | 15.8 | 14.4 | 12.9 | 15.9 | 13.9 | 12.7 | 15.1 |
| 1970 | 15.2 | 13.1 | 17.0 | 15.2 | 13.1 | 17.1 | 14.2 | 12.5 | 15.7 |
| 1980 | 16.4 | 14.1 | 18.3 | 16.5 | 14.2 | 18.4 | 15.1 | 13.0 | 16.8 |
| 1990 | 17.2 | 15.1 | 18.9 | 17.3 | 15.2 | 19.1 | 15.4 | 13.2 | 17.2 |
| 1995 | 17.4 | 15.6 | 18.9 | 17.6 | 15.7 | 19.1 | 15.6 | 13.6 | 17.1 |
| 1996 | 17.5 | 15.7 | 19.0 | 17.6 | 15.8 | 19.1 | 15.8 | 13.9 | 17.2 |
| 1997 | 17.7 | 15.9 | 19.2 | 17.8 | 16.0 | 19.3 | 16.1 | 14.2 | 17.6 |
| 1998 | 17.8 | 16.0 | 19.2 | 17.8 | 16.1 | 19.3 | 16.1 | 14.3 | 17.4 |
| 1999 | 17.7 | 16.1 | 19.1 | 17.8 | 16.1 | 19.2 | 16.0 | 14.3 | 17.3 |
| 2000 | 18.0 | 16.2 | 19.3 | 18.0 | 16.3 | 19.4 | 16.2 | 14.2 | 17.7 |
| 2001 | 18.1 | 16.4 | 19.4 | 18.2 | 16.5 | 19.5 | 16.4 | 14.4 | 17.9 |
| 2002 | 18.2 | 16.6 | 19.5 | 18.2 | 16.6 | 19.5 | 16.6 | 14.6 | 18.0 |
| 2003 | 18.4 | 16.8 | 19.8 | 18.5 | 16.9 | 19.8 | 17.0 | 14.9 | 18.5 |
| 2004 | 18.7 | 17.1 | 20.0 | 18.7 | 17.2 | 20.0 | 17.1 | 15.2 | 18.6 |
| **At 75 years** | | | | | | | | | |
| 1980 | 10.4 | 8.8 | 11.5 | 10.4 | 8.8 | 11.5 | 9.7 | 8.3 | 10.7 |
| 1990 | 10.9 | 9.4 | 12.0 | 11.0 | 9.4 | 12.0 | 10.2 | 8.6 | 11.2 |
| 1995 | 11.0 | 9.7 | 11.9 | 11.1 | 9.7 | 12.0 | 10.2 | 8.8 | 11.1 |
| 1996 | 11.1 | 9.8 | 12.0 | 11.1 | 9.8 | 12.0 | 10.3 | 9.0 | 11.2 |
| 1997 | 11.2 | 9.9 | 12.1 | 11.2 | 9.9 | 12.1 | 10.7 | 9.3 | 11.5 |
| 1998 | 11.3 | 10.0 | 12.2 | 11.3 | 10.0 | 12.2 | 10.5 | 9.2 | 11.3 |
| 1999 | 11.2 | 10.0 | 12.1 | 11.2 | 10.0 | 12.1 | 10.4 | 9.2 | 11.1 |
| 2000 | 11.4 | 10.1 | 12.3 | 11.4 | 10.1 | 12.3 | 10.7 | 9.2 | 11.6 |
| 2001 | 11.5 | 10.2 | 12.4 | 11.5 | 10.2 | 12.3 | 10.8 | 9.3 | 11.7 |
| 2002 | 11.5 | 10.3 | 12.4 | 11.5 | 10.3 | 12.3 | 10.9 | 9.5 | 11.7 |
| 2003 | 11.8 | 10.5 | 12.6 | 11.7 | 10.5 | 12.6 | 11.4 | 9.8 | 12.4 |
| 2004 | 11.9 | 10.7 | 12.8 | 11.9 | 10.7 | 12.8 | 11.4 | 9.9 | 12.2 |

—Data not available.

[a]Data shown for 1900–1960 are for the nonwhite population.

[b]Death registration area only. The death registration area increased from 10 states and the District of Columbia in 1900 to the coterminous United States in 1933.

[c]Includes deaths of persons who were not residents of the 50 states and the District of Columbia.

Notes: Populations for computing life expectancy for 1991–1999 are 1990-based postcensal estimates of U.S. resident population. In 1997, life table methodology was revised to construct complete life tables by single years of age that extend to age 100 (Anderson RN. Method for constructing complete annual U.S. life tables. National Center for Health Statistics. Vital Health Stat 2(129). 1999). Previously, abridged life tables were constructed for 5-year age groups ending with 85 years and over. Life table values for 2000 and later years were computed using a slight modification of the new life table method due to a change in the age detail of populations received from the U.S. Census Bureau. Starting with 2003 data, some states reported multiple-race data. The multiple-race data for these states were bridged to the single-race categories of the 1977 Office of Management and Budget standards for comparability with other states. In 2003, California, Hawaii, Idaho, Maine, Montana, New York, and Wisconsin reported multiple-race data. In 2004, 15 states reported multiple-race data. In addition to the seven states listed above, Michigan, Minnesota, New Hampshire, New Jersey, Oklahoma, South Dakota, Washington, and Wyoming reported multiple-race data.

SOURCE: "Table 27. Life Expectancy at Birth, at 65 Years of Age, and at 75 Years of Age, by Race and Sex: United States, Selected Years 1900–2004," in *Health, United States, 2006, with Chartbook on Trends in the Health of Americans*, Centers for Disease Control and Prevention, National Center for Health Statistics, February 2007, http://www.cdc.gov/nchs/data/hus/hus06.pdf#summary (accessed October 8, 2007)

had a normal weight. Their projections reveal that within fifty years obesity is likely to reduce the average life expectancy in the United States by at least two to five years. The impact of obesity and its health consequences on life expectancy was considered larger than cancer or heart disease.

TABLE 2.2

**Leading causes of death and numbers of deaths, according to sex and race, 1980 and 2004**

[Data are based on death certificates]

| Sex, race, and rank order | 1980 | | 2004 | |
|---|---|---|---|---|
| | Cause of death | Deaths | Cause of death | Deaths |
| **All persons** | | | | |
| — | All causes | 1,989,841 | All causes | 2,397,615 |
| 1 | Diseases of heart | 761,085 | Diseases of heart | 652,486 |
| 2 | Malignant neoplasms | 416,509 | Malignant neoplasms | 553,888 |
| 3 | Cerebrovascular diseases | 170,225 | Cerebrovascular diseases | 150,074 |
| 4 | Unintentional injuries | 105,718 | Chronic lower respiratory diseases | 121,987 |
| 5 | Chronic obstructive pulmonary diseases | 56,050 | Unintentional injuries | 112,012 |
| 6 | Pneumonia and influenza | 54,619 | Diabetes mellitus | 73,138 |
| 7 | Diabetes mellitus | 34,851 | Alzheimer's disease | 65,965 |
| 8 | Chronic liver disease and cirrhosis | 30,583 | Influenza and pneumonia | 59,664 |
| 9 | Atherosclerosis | 29,449 | Nephritis, nephrotic syndrome and nephrosis | 42,480 |
| 10 | Suicide | 26,869 | Septicemia | 33,373 |
| **Male** | | | | |
| — | All causes | 1,075,078 | All causes | 1,181,668 |
| 1 | Diseases of heart | 405,661 | Diseases of heart | 321,973 |
| 2 | Malignant neoplasms | 225,948 | Malignant neoplasms | 286,830 |
| 3 | Unintentional injuries | 74,180 | Unintentional injuries | 72,050 |
| 4 | Cerebrovascular diseases | 69,973 | Cerebrovascular diseases | 58,800 |
| 5 | Chronic obstructive pulmonary diseases | 38,625 | Chronic lower respiratory diseases | 58,646 |
| 6 | Pneumonia and influenza | 27,574 | Diabetes mellitus | 35,267 |
| 7 | Suicide | 20,505 | Influenza and pneumonia | 26,861 |
| 8 | Chronic liver disease and cirrhosis | 19,768 | Suicide | 25,566 |
| 9 | Homicide | 18,779 | Nephritis, nephrotic syndrome and nephrosis | 20,370 |
| 10 | Diabetes mellitus | 14,325 | Alzheimer's disease | 18,974 |
| **Female** | | | | |
| — | All causes | 914,763 | All causes | 1,215,947 |
| 1 | Diseases of heart | 355,424 | Diseases of heart | 330,513 |
| 2 | Malignant neoplasms | 190,561 | Malignant neoplasms | 267,058 |
| 3 | Cerebrovascular diseases | 100,252 | Cerebrovascular diseases | 91,274 |
| 4 | Unintentional injuries | 31,538 | Chronic lower respiratory diseases | 63,341 |
| 5 | Pneumonia and influenza | 27,045 | Alzheimer's disease | 46,991 |
| 6 | Diabetes mellitus | 20,526 | Unintentional injuries | 39,962 |
| 7 | Atherosclerosis | 17,848 | Diabetes mellitus | 37,871 |
| 8 | Chronic obstructive pulmonary diseases | 17,425 | Influenza and pneumonia | 32,803 |
| 9 | Chronic liver disease and cirrhosis | 10,815 | Nephritis, nephrotic syndrome and nephrosis | 22,110 |
| 10 | Certain conditions originating in the perinatal period | 9,815 | Septicemia | 18,362 |
| **White** | | | | |
| — | All causes | 1,738,607 | All causes | 2,056,643 |
| 1 | Diseases of heart | 683,347 | Diseases of heart | 565,703 |
| 2 | Malignant neoplasms | 368,162 | Malignant neoplasms | 478,134 |
| 3 | Cerebrovascular diseases | 148,734 | Cerebrovascular diseases | 127,868 |
| 4 | Unintentional injuries | 90,122 | Chronic lower respiratory diseases | 112,914 |
| 5 | Chronic obstructive pulmonary diseases | 52,375 | Unintentional injuries | 95,890 |
| 6 | Pneumonia and influenza | 48,369 | Alzheimer's disease | 61,087 |
| 7 | Diabetes mellitus | 28,868 | Diabetes mellitus | 58,087 |
| 8 | Atherosclerosis | 27,069 | Influenza and pneumonia | 52,430 |
| 9 | Chronic liver disease and cirrhosis | 25,240 | Nephritis, nephrotic syndrome and nephrosis | 33,691 |
| 10 | Suicide | 24,829 | Suicide | 29,251 |

## IS OBESITY A DISEASE?

Researchers now recognize that obesity does not simply result from willful overeating and laziness, but from a complex combination of genetic, metabolic, behavioral, and environmental factors. Rather than viewing it as a lifestyle choice or personal failing, many groups favor declaring obesity a disease. Proponents assert that many public health benefits would result from designating obesity as a disease including:

- Reducing the social stigma and prejudice associated with obesity, and promoting attitudinal changes to reduce weight-based discrimination.

- Enabling more people to seek treatment for obesity by providing health insurance coverage for treatment.

- Increasing public awareness of the severity of obesity as a threat to health and longevity.

- Stimulating scientific and medical research about the prevention and treatment of the condition and speeding approval of new antiobesity drugs.

Advocates of classifying obesity as a disease, including the World Health Organization, the National Institutes of Health, the National Academy of Sciences, the Federal Trade Commission, the Maternal and Child Health Bureau,

TABLE 2.2

**Leading causes of death and numbers of deaths, according to sex and race, 1980 and 2004** [CONTINUED]

[Data are based on death certificates]

| Sex, race, and rank order | 1980 | | 2004 | |
|---|---|---|---|---|
| | Cause of death | Deaths | Cause of death | Deaths |
| **Black or African American** | | | | |
| — | All causes | 233,135 | All causes | 287,315 |
| 1 | Diseases of heart | 72,956 | Diseases of heart | 74,225 |
| 2 | Malignant neoplasms | 45,037 | Malignant neoplasms | 62,499 |
| 3 | Cerebrovascular diseases | 20,135 | Cerebrovascular diseases | 18,118 |
| 4 | Unintentional injuries | 13,480 | Diabetes mellitus | 12,834 |
| 5 | Homicide | 10,172 | Unintentional injuries | 12,670 |
| 6 | Certain conditions originating in the perinatal period | 6,961 | Homicide | 8,135 |
| 7 | Pneumonia and influenza | 5,648 | Nephritis, nephrotic syndrome and nephrosis | 7,834 |
| 8 | Diabetes mellitus | 5,544 | Chronic lower respiratory diseases | 7,400 |
| 9 | Chronic liver disease and cirrhosis | 4,790 | Human immunodeficiency virus (HIV) disease | 7,271 |
| 10 | Nephritis, nephrotic syndrome, and nephrosis | 3,416 | Septicemia | 6,010 |

—Category not applicable.
Notes: Starting with 2003 data, some states reported multiple-race data. The multiple-race data for these states were bridged to the single-race categories of the 1977 Office of Management and Budget standards for comparability with other states. In 2003, California, Hawaii, Idaho, Maine, Montana, New York, and Wisconsin reported multiple-race data. In 2004, 15 states reported multiple-race data. In addition to the seven states listed above, Michigan, Minnesota, New Hampshire, New Jersey, Oklahoma, South Dakota, Washington, and Wyoming reported multiple-race data.

SOURCE: Adapted from "Table 31. Leading Causes of Death and Numbers of Deaths, by Sex, Race, and Hispanic Origin: United States, 1980 and 2004," in *Health, United States, 2006, with Chartbook on Trends in the Health of Americans*, Centers for Disease Control and Prevention, National Center for Health Statistics, February 2007, http://www.cdc.gov/nchs/data/hus/hus06.pdf#summary (accessed October 8, 2007)

the American Heart Association, the American Academy of Family Physicians, the American Society for Bariatric Surgery, the American Society of Bariatric Physicians, and the American Obesity Association (AOA), observe that not long ago in U.S. history alcoholism was viewed as a personal choice or moral weakness, whereas today it is considered a disease. They also observe that eating disorders such as anorexia and bulimia are classified as diseases. In view of the size and scope of the obesity epidemic, proponents argue that the social and financial costs of allowing it to go unchecked will far exceed the costs associated with extending health-care coverage for weight-reduction programs.

The AOA argues that obesity meets the criteria for disease because according to *Stedman's Medical Dictionary* (2006) a disease should have at least two of the following three features:

- Recognized etiologic (causes) agents

- Identifiable signs and symptoms

- Consistent anatomical alterations

The AOA describes causative agents for obesity as social, behavioral, cultural, physiological, metabolic, and genetic factors. The identifiable signs and symptoms of obesity include an excess accumulation of adipose tissue (fat), an increase in the size or number of fat cells, insulin resistance, decreased levels of high-density lipoprotein (HDL) and norepinephrine, alterations in the activity of the sympathetic and parasympathetic nervous system, and elevated blood pressure, blood glucose, cholesterol levels, and triglyceride levels. The consistent anatomic alteration of obesity is the increase in body mass.

Opponents contend that even though obesity increases the risk of developing many diseases, it is not an ailment in itself but an unhealthy consequence of poor lifestyle choices. They liken it to cigarette smoking, a risk factor that predisposes people to disease, and they dispute the notion that labeling obesity as a disease will have a beneficial effect on the ability of public health organizations to alter the course of the obesity epidemic. They maintain that the public tends to view diseases as conditions that are contracted or contagious, and that with disease comes a victim mentality, rather than an assumption of personal responsibility. Because many health professionals consider the assumption of personal responsibility as being crucial for the long-term success of obesity treatment, any action that releases people from assuming personal responsibility is counterproductive.

Opponents to granting disease status to obesity predict that the financial ramifications would be devastating for taxpayers and the health insurance industry. Health-care costs, which increase every year, would skyrocket. Antiobesity programs would drive insurance premiums even higher and place unreasonable burdens on the already overburdened Medicare and Medicaid programs. Employers, especially small businesses, might be forced by high health-care costs to drop employee coverage altogether.

**FIGURE 2.1**

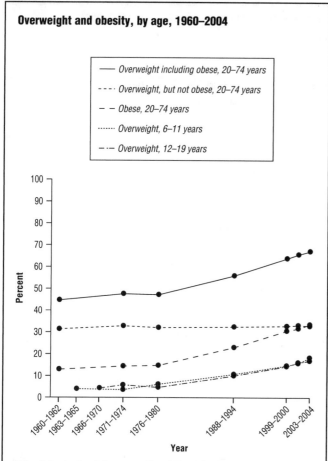

**Overweight and obesity, by age, 1960–2004**

— Overweight including obese, 20–74 years
---- Overweight, but not obese, 20–74 years
– – Obese, 20–74 years
······ Overweight, 6–11 years
–·– Overweight, 12–19 years

Notes: Estimates for adults are age adjusted. For adults: overweight including obese is defined as a body mass index (BMI) greater than or equal to 25, overweight but not obese as a BMI greater than or equal to 25 but less than 30, and obese as a BMI greater than or equal to 30. For children: overweight is defined as a BMI at or above the sex- and age-specific 95th percentile BMI cut points from the 2000 CDC Growth Charts: United States. Obese is not defined for children.

SOURCE: "Figure 13. Overweight and Obesity by Age: United States, 1960–2004," in *Health, United States, 2006, with Chartbook on Trends in the Health of Americans*, Centers for Disease Control and Prevention, National Center for Health Statistics, 2006, http://www.cdc.gov/nchs/data/hus/hus06.pdf#summary (accessed October 8, 2007)

expenses related to obesity, just as they can for expenditures related to heart disease, cancer, diabetes, and other illnesses.

In July 2004 the federal Medicare program discarded its long-standing position that obesity is not a disease, which effectively removed a major roadblock for people seeking coverage for treatment of obesity. After years of review, the Centers for Medicare and Medicaid Services, which administers the health program for older adults and people who are disabled, announced in *CMS Manual System: Pub. 100-03 Medicare National Coverage Determinations* (October 1, 2004, http://www.cms.hhs.gov/transmittals/downloads/R23NCD.pdf) that it eliminated language— that "obesity itself cannot be considered an illness"—from its policy that had been used to deny coverage for weight-loss treatment. Even though the decision stopped short of declaring obesity a disease and does not automatically imply coverage for any specific treatment, it enables individuals, physicians, and companies to apply to Medicare for reimbursement for a variety of weight-loss therapies. Because private insurance companies often use Medicare as a model for their coverage and benefits, the Medicare decision has pressured them to expand coverage for weight-loss treatments. Ironically, the Medicare decision was announced at the same time that many private insurers intended to eliminate or sharply curtail coverage of weight-loss surgery.

## THE GENETICS OF BODY WEIGHT AND OBESITY

Genetics, the study of single genes and their effects, explains how and why traits such as hair color and blood types run in families. In the early twenty-first century the scientific community agreed that body shape and body weight are also regulated traits, that genes govern much of this regulation, and that altering genetically predetermined set points for body weight is often difficult. Genomics, a discipline that emerged during the 1980s, is the study of more than single genes; it considers the functions and interactions of all the genes in the genome. In terms of understanding genetics as a risk factor for obesity, genomics has broader applicability than does genetics because it is likely that humans carry dozens of genes that are directly related to body size and that most obesity is multifactorial—resulting from the complex interactions of multiple genes and environmental factors.

Because genomics is a relatively new discipline, many questions are still unanswered about how genes influence the ability to balance energy input and energy expenditure, and why individuals vary in their abilities to perform this critical body function. Table 2.5 summarizes what is known and what remains to be learned about variations in body weight, energy metabolism, and inherited obesity syndromes.

A related concern is the lack of universally accepted, effective treatment for obesity. If obesity is classified as a disease, which treatment or therapies should be covered? For example, if exercise is deemed beneficial, then health insurers might be required to pay for gym memberships. Furthermore, some opponents believe that it is not necessary to designate obesity as a disease to encourage Americans to seek treatment. They cite the more than $50 billion spent annually on weight-loss programs and services as evidence that Americans are not reluctant to seek treatment.

Even though the debate has not been fully resolved, obesity is rapidly acquiring recognition as a disease. In 2002 the Internal Revenue Service ruled that for tax purposes, obesity is a disease, allowing Americans for the first time to claim a deduction for some health-care

TABLE 2.3

## Overweight, obesity, and healthy weight, by sex, age, race, Hispanic origin, and poverty level, selected years, 1960–2004

[Data are based on measured height and weight of a sample of the civilian noninstitutionalized population]

| Sex, age, race and Hispanic origin[a], and poverty level | Overweight[b] | | | | |
| --- | --- | --- | --- | --- | --- |
| | 1960–1962 | 1971–1974 | 1976–1980[c] | 1988–1994 | 2001–2004 |
| **20–74 years, age adjusted[d]** | | | Percent of population | | |
| Both sexes[e] | 44.8 | 47.7 | 47.4 | 56.0 | 66.0 |
| Male | 49.5 | 54.7 | 52.9 | 61.0 | 70.7 |
| Female | 40.2 | 41.1 | 42.0 | 51.2 | 61.4 |
| **Not Hispanic or Latino:** | | | | | |
| White only, male | — | — | 53.8 | 61.6 | 71.1 |
| White only, female | — | — | 38.7 | 47.2 | 57.1 |
| Black or African American only, male | — | — | 51.3 | 58.2 | 66.8 |
| Black or African American only, female | — | — | 62.6 | 68.5 | 79.5 |
| Mexican male | — | — | 61.6 | 69.4 | 75.8 |
| Mexican female | — | — | 61.7 | 69.6 | 73.2 |
| **Percent of poverty level:[f]** | | | | | |
| Below 100% | — | 49.3 | 50.0 | 59.8 | 63.9 |
| 100%–less than 200% | — | 50.9 | 49.0 | 58.2 | 66.2 |
| 200% or more | — | 46.7 | 46.6 | 54.5 | 66.1 |
| **20 years and over, age adjusted[d]** | | | | | |
| Both sexes[e] | — | — | — | 56.0 | 66.0 |
| Male | — | — | — | 60.9 | 70.5 |
| Female | — | — | — | 51.4 | 61.6 |
| **Not Hispanic or Latino:** | | | | | |
| White only, male | — | — | — | 61.6 | 71.0 |
| White only, female | — | — | — | 47.5 | 57.6 |
| Black or African American only, male | — | — | — | 57.8 | 67.0 |
| Black or African American only, female | — | — | — | 68.2 | 79.6 |
| Mexican male | — | — | — | 68.9 | 74.6 |
| Mexican female | — | — | — | 68.9 | 73.0 |
| **Percent of poverty level:[f]** | | | | | |
| Below 100% | — | — | — | 59.6 | 63.4 |
| 100%–less than 200% | — | — | — | 58.0 | 66.2 |
| 200% or more | — | — | — | 54.8 | 66.1 |
| **20 years and over, crude** | | | | | |
| Both sexes[e] | — | — | — | 54.9 | 66.1 |
| Male | — | — | — | 59.4 | 70.4 |
| Female | — | — | — | 50.7 | 61.9 |
| **Not Hispanic or Latino:** | | | | | |
| White only, male | — | — | — | 60.6 | 71.6 |
| White only, female | — | — | — | 47.4 | 58.7 |
| Black or African American only, male | — | — | — | 56.7 | 66.3 |
| Black or African American only, female | — | — | — | 66.0 | 79.1 |
| Mexican male | — | — | — | 63.9 | 71.8 |
| Mexican female | — | — | — | 65.9 | 71.4 |
| **Percent of poverty level:[f]** | | | | | |
| Below 100% | — | — | — | 56.8 | 61.4 |
| 100%–less than 200% | — | — | — | 55.7 | 65.3 |
| 200% or more | — | — | — | 54.2 | 67.1 |
| **Male** | | | | | |
| 20–34 years | 42.7 | 42.8 | 41.2 | 47.5 | 59.0 |
| 35–44 years | 53.5 | 63.2 | 57.2 | 65.5 | 72.9 |
| 45–54 years | 53.9 | 59.7 | 60.2 | 66.1 | 78.5 |
| 55–64 years | 52.2 | 58.5 | 60.2 | 70.5 | 77.3 |
| 65–74 years | 47.8 | 54.6 | 54.2 | 68.5 | 76.1 |
| 75 years and over | — | — | — | 56.5 | 66.8 |
| **Female** | | | | | |
| 20–34 years | 21.2 | 25.8 | 27.9 | 37.0 | 51.6 |
| 35–44 years | 37.2 | 40.5 | 40.7 | 49.6 | 60.1 |
| 45–54 years | 49.3 | 49.0 | 48.7 | 60.3 | 67.4 |
| 55–64 years | 59.9 | 54.5 | 53.7 | 66.3 | 69.9 |
| 65–74 years | 60.9 | 55.9 | 59.5 | 60.3 | 71.5 |
| 75 years and over | — | — | — | 52.3 | 63.7 |

TABLE 2.3

**Overweight, obesity, and healthy weight, by sex, age, race, Hispanic origin, and poverty level, selected years, 1960–2004** [CONTINUED]

[Data are based on measured height and weight of a sample of the civilian noninstitutionalized population]

| Sex, age, race and Hispanic origin[a], and poverty level | Obesity[g] | | | | |
|---|---|---|---|---|---|
| | 1960–1962 | 1971–1974 | 1976–1980[c] | 1988–1994 | 2001–2004 |
| **20–74 years, age adjusted[d]** | | | Percent of population | | |
| Both sexes[e] | 13.3 | 14.6 | 15.1 | 23.3 | 32.1 |
| Male | 10.7 | 12.2 | 12.8 | 20.6 | 30.2 |
| Female | 15.7 | 16.8 | 17.1 | 26.0 | 34.0 |
| **Not Hispanic or Latino:** | | | | | |
| White only, male | — | — | 12.4 | 20.7 | 31.0 |
| White only, female | — | — | 15.4 | 23.3 | 31.5 |
| Black or African American only, male | — | — | 16.5 | 21.3 | 31.2 |
| Black or African American only, female | — | — | 31.0 | 39.1 | 51.6 |
| Mexican male | — | — | 15.7 | 24.4 | 30.5 |
| Mexican female | — | — | 26.6 | 36.1 | 40.3 |
| **Percent of poverty level:[f]** | | | | | |
| Below 100% | — | 20.7 | 21.9 | 29.2 | 34.9 |
| 100%–less than 200% | — | 18.4 | 18.7 | 26.6 | 34.6 |
| 200% or more | — | 12.4 | 12.9 | 21.4 | 30.6 |
| **20 years and over, age adjusted[d]** | | | | | |
| Both sexes[e] | — | — | — | 22.9 | 31.4 |
| Male | — | — | — | 20.2 | 29.5 |
| Female | — | — | — | 25.5 | 33.2 |
| **Not Hispanic or Latino:** | | | | | |
| White only, male | — | — | — | 20.3 | 30.2 |
| White only, female | — | — | — | 22.9 | 30.7 |
| Black or African American only, male | — | — | — | 20.9 | 30.8 |
| Black or African American only, female | — | — | — | 38.3 | 51.1 |
| Mexican male | — | — | — | 23.8 | 29.1 |
| Mexican female | — | — | — | 35.2 | 39.4 |
| **Percent of poverty level:[f]** | | | | | |
| Below 100% | — | — | — | 28.1 | 33.7 |
| 100%–less than 200% | — | — | — | 26.1 | 33.6 |
| 200% or more | — | — | — | 21.1 | 30.0 |
| **20 years and over, crude** | | | | | |
| Both sexes[e] | — | — | — | 22.3 | 31.5 |
| Male | — | — | — | 19.5 | 29.5 |
| Female | — | — | — | 25.0 | 33.3 |
| **Not Hispanic or Latino:** | | | | | |
| White only, male | — | — | — | 19.9 | 30.5 |
| White only, female | — | — | — | 22.7 | 31.2 |
| Black or African American only, male | — | — | — | 20.7 | 30.7 |
| Black or African American only, female | — | — | — | 36.7 | 51.1 |
| Mexican male | — | — | — | 20.6 | 27.8 |
| Mexican female | — | — | — | 33.3 | 38.5 |
| **Percent of poverty level:[f]** | | | | | |
| Below 100% | — | — | — | 25.9 | 33.0 |
| 100%–less than 200% | — | — | — | 24.3 | 32.6 |
| 200% or more | — | — | — | 20.9 | 30.7 |
| **Male** | | | | | |
| 20–34 years | 9.2 | 9.7 | 8.9 | 14.1 | 23.2 |
| 35–44 years | 12.1 | 13.5 | 13.5 | 21.5 | 33.8 |
| 45–54 years | 12.5 | 13.7 | 16.7 | 23.2 | 31.8 |
| 55–64 years | 9.2 | 14.1 | 14.1 | 27.2 | 36.0 |
| 65–74 years | 10.4 | 10.9 | 13.2 | 24.1 | 32.1 |
| 75 years and over | — | — | — | 13.2 | 19.9 |
| **Female** | | | | | |
| 20–34 years | 7.2 | 9.7 | 11.0 | 18.5 | 28.6 |
| 35–44 years | 14.7 | 17.7 | 17.8 | 25.5 | 33.3 |
| 45–54 years | 20.3 | 18.9 | 19.6 | 32.4 | 38.0 |
| 55–64 years | 24.4 | 24.1 | 22.9 | 33.7 | 39.0 |
| 65–74 years | 23.2 | 22.0 | 21.5 | 26.9 | 37.9 |
| 75 years and over | — | — | — | 19.2 | 23.2 |

**TABLE 2.3**

**Overweight, obesity, and healthy weight, by sex, age, race, Hispanic origin, and poverty level, selected years, 1960–2004** [CONTINUED]

[Data are based on measured height and weight of a sample of the civilian noninstitutionalized population]

| Sex, age, race and Hispanic origin[a], and poverty level | Healthy weight[h] | | | | |
|---|---|---|---|---|---|
| | 1960–1962 | 1971–1974 | 1976–1980[c] | 1988–1994 | 2001–2004 |
| **20–74 years, age adjusted[d]** | | | Percent of population | | |
| Both sexes[e] | 51.2 | 48.8 | 49.6 | 41.7 | 32.2 |
| Male | 48.3 | 43.0 | 45.4 | 37.9 | 28.1 |
| Female | 54.1 | 54.3 | 53.7 | 45.3 | 36.2 |
| **Not Hispanic or Latino:** | | | | | |
| White only, male | — | — | 45.3 | 37.4 | 27.8 |
| White only, female | — | — | 56.7 | 49.2 | 40.2 |
| Black or African American only, male | — | — | 46.6 | 40.0 | 31.3 |
| Black or African American only, female | — | — | 35.0 | 28.9 | 18.9 |
| Mexican male | — | — | 37.1 | 29.8 | 24.2 |
| Mexican female | — | — | 36.4 | 29.0 | 26.3 |
| **Percent of poverty level:[f]** | | | | | |
| Below 100% | — | 45.8 | 45.1 | 37.3 | 33.7 |
| 100%–less than 200% | — | 45.1 | 47.6 | 39.2 | 31.8 |
| 200% or more | — | 50.2 | 51.0 | 43.4 | 32.4 |
| **20 years and over, age adjusted[d]** | | | | | |
| Both sexes[e] | — | — | — | 41.6 | 32.3 |
| Male | — | — | — | 37.9 | 28.3 |
| Female | — | — | — | 45.0 | 36.1 |
| **Not Hispanic or Latino:** | | | | | |
| White only, male | — | — | — | 37.3 | 28.0 |
| White only, female | — | — | — | 48.7 | 39.8 |
| Black or African American only, male | — | — | — | 40.1 | 30.8 |
| Black or African American only, female | — | — | — | 29.2 | 18.9 |
| Mexican male | — | — | — | 30.2 | 25.3 |
| Mexican female | — | — | — | 29.7 | 26.5 |
| **Percent of poverty level:[f]** | | | | | |
| Below 100% | — | — | — | 37.5 | 34.3 |
| 100%–less than 200% | — | — | — | 39.3 | 31.9 |
| 200% or more | — | — | — | 43.1 | 32.4 |
| **20 years and over, crude** | | | | | |
| Both sexes[e] | — | — | — | 42.6 | 32.2 |
| Male | — | — | — | 39.4 | 28.4 |
| Female | — | — | — | 45.7 | 35.8 |
| **Not Hispanic or Latino:** | | | | | |
| White only, male | — | — | — | 38.2 | 27.4 |
| White only, female | — | — | — | 48.8 | 38.8 |
| Black or African American only, male | — | — | — | 41.5 | 31.5 |
| Black or African American only, female | — | — | — | 31.2 | 19.3 |
| Mexican male | — | — | — | 35.2 | 28.1 |
| Mexican female | — | — | — | 32.4 | 28.0 |
| **Percent of poverty level:[f]** | | | | | |
| Below 100% | | | | 39.8 | 36.2 |
| 100%–less than 200% | — | — | — | 41.5 | 32.6 |
| 200% or more | — | — | — | 43.6 | 31.6 |
| **Male** | | | | | |
| 20–34 years | 55.3 | 54.7 | 57.1 | 51.1 | 38.3 |
| 35–44 years | 45.2 | 35.2 | 41.3 | 33.4 | 26.5 |
| 45–54 years | 44.8 | 38.5 | 38.7 | 33.6 | 21.2 |
| 55–64 years | 44.9 | 38.3 | 38.7 | 28.6 | 22.2 |
| 65–74 years | 46.2 | 42.1 | 42.3 | 30.1 | 23.1 |
| 75 years and over | — | — | — | 40.9 | 32.1 |
| **Female** | | | | | |
| 20–34 years | 67.6 | 65.8 | 65.0 | 57.9 | 44.2 |
| 35–44 years | 58.4 | 56.7 | 55.6 | 47.1 | 38.3 |
| 45–54 years | 47.6 | 49.3 | 48.7 | 37.2 | 31.0 |
| 55–64 years | 38.1 | 41.1 | 43.5 | 31.5 | 29.2 |
| 65–74 years | 36.4 | 40.6 | 37.8 | 37.0 | 27.0 |
| 75 years and over | — | — | — | 43.0 | 34.6 |

**TABLE 2.3**

**Overweight, obesity, and healthy weight, by sex, age, race, Hispanic origin, and poverty level, selected years, 1960–2004** [CONTINUED]

[Data are based on measured height and weight of a sample of the civilian noninstitutionalized population]

—Data not available.

[a]Persons of Mexican origin may be of any race. Starting with 1999 data, race-specific estimates are tabulated according to the 1997 Revisions to the Standards for the Classification of Federal Data on Race and Ethnicity and are not strictly comparable with estimates for earlier years. The two non-Hispanic race categories shown in the table conform to the 1997 standards. Starting with 1999 data, race-specific estimates are for persons who reported only one racial group. Prior to data year 1999, estimates were tabulated according to the 1977 standards. Estimates for single race categories prior to 1999 included persons who reported one race or, if they reported more than one race, identified one race as best representing their race.

[b]Body mass index (BMI) greater than or equal to 25.

[c]Data for Mexicans are for 1982–1984.

[d]Age adjusted to the 2000 standard population using five age groups: 20–34 years, 35–44 years, 45–54 years, 55–64 years, and 65 years and over (65–74 years for estimates for 20–74 years). Age-adjusted estimates in this table may differ from other age-adjusted estimates based on the same data and presented elsewhere if different age groups are used in the adjustment procedure.

[e]Includes persons of all races and Hispanic origins, not just those shown separately.

[f]Poverty level is based on family income and family size. Persons with unknown poverty level are excluded.

[g]Body mass index (BMI) greater than or equal to 30.

[h]BMI of 18.5 to less than 25 kilograms/meter².

Notes: Percents do not sum to 100 because the percent of persons with BMI less than 18.5 is not shown and the percent of persons with obesity is a subset of the percent with overweight. Height was measured without shoes; two pounds were deducted from data for 1960–1962 to allow for weight of clothing. Excludes pregnant women.

SOURCE: "Table 73. Overweight, Obesity, and Healthy Weight among Person 20 Years of Age and Over, by Sex, Age, Race, Hispanic Origin and Poverty Level: United States, 1960–1962 through 2001–2004," in *Health, United States, 2006, with Chartbook on Trends in the Health of Americans*, Centers for Disease Control and Prevention, National Center for Health Statistics, 2006, http://www.cdc.gov/nchs/data/hus/hus06.pdf#summary (accessed October 8, 2007)

## Single Mutant Genes Cause Obesity

Even though most obesity in humans is not due to mutations (alterations or changes) in single genes, there are obesity syndromes caused by variations in single genes, and these account for approximately 5% of all obesity, according to the Centers for Disease Control and Prevention (CDC), in "Obesity and Genetics: A Public Perspective" (November 27, 2007, http://www.cdc.gov/genomics/training/perspectives/files/obesedit.htm). In rare cases of severe obesity that begin during childhood, a single gene has a major effect in determining the occurrence of obesity, with environmental factors playing a lesser role. The mutations occur in genes that encode proteins related to the regulation of food intake. One example is mutations of the leptin gene (on chromosome 7) and its receptor. The circulating hormone leptin (leptos means thin) sends the brain a satiety signal to decrease appetite. Obese mice of the ob/ob strain produce no leptin and tend to overeat; when given leptin, the mice stop eating and lose weight. However, experiments have failed to replicate these findings in humans. Blood concentrations of leptin are usually elevated in obese people, suggesting that they may be insensitive or resistant to leptin, rather than leptin deficient. Most obese individuals appear to have normal genetic sequences for leptin and its receptor, although people with a demonstrable genetic leptin deficiency suffer from extreme obesity.

Melanocortin 4 receptor (MC4R) deficiency is the most commonly occurring monogenic (single gene) form of obesity. Inheriting one copy of certain variants of the gene causes obesity in some families. In "Clinical Spectrum of Obesity and Mutations in the Melanocortin 4 Receptor Gene" (*New England Journal of Medicine*, vol. 348, no. 12, March 20, 2003), I. Sadaf Farooqi et al. report that mutations in MC4R produce a distinct obesity syndrome that is inherited. They conclude that these mutant receptors play a pivotal role in the control of eating behavior—that the regulation of body weight in humans is sensitive to variations in the amount of functional MC4R.

Timothy M. Frayling et al. identify in "A Common Variant in the FTO Gene Is Associated with Body Mass Index and Predisposes to Childhood and Adult Obesity" (*Science*, vol. 316, no. 5826, May 2007) a variant of a gene called FTO, located on chromosome 16, that is linked to obesity. The researchers find that people with one or two copies of the gene's variant were more likely to be overweight than those who had no copies at all.

## Multiple Gene Variants Involved in Body Weight and Obesity

Heritability studies, which seek to determine the proportion of variance of a particular trait that is attributable to genetic factors and the proportion that is attributable to environmental factors, indicate that genetic factors may account for as much as 75% of the variability in human body weight and approximately 33% of the variation in the overall body mass index (BMI; body weight in kilograms divided by the height in meters squared). Genetic factors affect the variations in resting metabolic rate, body fat distribution, and weight gain related to overfeeding, which explains in part why some individuals are more susceptible than others to weight gain or weight loss. To ensure survival in times of scarce food supplies, the human body has evolved to resist any loss of body fat. This biological drive to maintain weight is coordinated through central nervous system pathways, with the involvement of many neuropeptides. (Peptides are released by neurons as intercellular messengers.

**TABLE 2.4**

**Health consequences of overweight and obesity**

**Premature death**

- An estimated 300,000 deaths per year may be attributable to obesity.
- The risk of death rises with increasing weight.
- Even moderate weight excess (10 to 20 pounds for a person of average height) increases the risk of death, particularly among adults aged 30 to 64 years.
- Individuals who are obese (body mass index (BMI) > 30) have a 50 to 100% increased risk of premature death from all causes, compared to individuals with a healthy weight.

**Heart disease**

- The incidence of heart disease (heart attack, congestive heart failure, sudden cardiac death, angina or chest pain, and abnormal heart rhythm) is increased in persons who are overweight or obese (BMI > 25).
- High blood pressure is twice as common in adults who are obese than in those who are at a healthy weight.
- Obesity is associated with elevated triglycerides (blood fat) and decreased high density lipoprotein (HDL) cholesterol ("good cholesterol").

**Diabetes**

- A weight gain of 11 to 18 pounds increases a person's risk of developing type 2 diabetes to twice that of individuals who have not gained weight.
- Over 80% of people with diabetes are overweight or obese.

**Cancer**

- Overweight and obesity are associated with an increased risk for some types of cancer including endometrial (cancer of the lining of the uterus), colon, gall bladder, prostate, kidney, and postmenopausal breast cancer.
- Women gaining more than 20 pounds from age 18 to midlife double their risk of postmenopausal breast cancer, compared to women whose weight remains stable.

**Breathing problems**

- Sleep apnea (interrupted breathing while sleeping) is more common in obese persons.
- Obesity is associated with a higher prevalence of asthma.

**Arthritis**

- For every 2-pound increase in weight, the risk of developing arthritis is increased by 9 to 13%.
- Symptoms of arthritis can improve with weight loss.

**Reproductive complications**

Complications of pregnancy
- Obesity during pregnancy is associated with increased risk of death in both the baby and the mother and increases the risk of maternal high blood pressure by 10 times.
- In addition to many other complications, women who are obese during pregnancy are more likely to have gestational diabetes and problems with labor and delivery.
- Infants born to women who are obese during pregnancy are more likely to be high birthweight and, therefore, may face a higher rate of Cesarean section delivery and low blood sugar (which can be associated with brain damage and seizures).
- Obesity during pregnancy is associated with an increased risk of birth defects, particularly neural tube defects, such as spina bifida.
- Obesity in premenopausal women is associated with irregular menstrual cycles and infertility.

**Additional health consequences**

- Overweight and obesity are associated with increased risks of gall bladder disease, incontinence, increased surgical risk, and depression.
- Obesity can affect the quality of life through limited mobility and decreased physical endurance as well as through social, academic, and job discrimination.

**Children and adolescents**

- Risk factors for heart disease, such as high cholesterol and high blood pressure, occur with increased frequency in overweight children and adolescents compared to those with a healthy weight.
- Type 2 diabetes, previously considered an adult disease, has increased dramatically in children and adolescents. Overweight and obesity are closely linked to type 2 diabetes.
- Overweight adolescents have a 70% chance of becoming overweight or obese adults. This increases to 80% if one or more parent is overweight or obese.
- The most immediate consequence of overweight, as perceived by children themselves, is social discrimination.

SOURCE: "Overweight and Obesity: Health Consequences," in *The Surgeon General's Call to Action to Prevent and Decrease Overweight and Obesity*, U.S. Department of Health and Human Services, Office of the Surgeon General, 2001, http://www.surgeongeneral.gov/topics/obesity/calltoaction/fact_consequences.htm (accessed October 8, 2007)

Many neuropeptides are also hormones outside of the nervous system.) Evidence from twin, adoption, and family studies reveals that biological relatives exhibit similarities in the maintenance of body weight. First-degree relatives of moderately obese people are at three to four times the risk of obesity relative to the general population. First-degree relatives of severely obese people are at five times greater risk. Genetic predisposition to obesity does not mean that developing the condition is inevitable; however, research indicates that inherited genetic variation is an important risk factor for obesity.

Genetic factors have been implicated in the development of eating disorders such as anorexia and bulimia and appear to be involved in the extent to which diet and exercise are effective strategies for weight reduction.

Furthermore, genetic variations among individuals may promote different food preferences and eating patterns that interact with environmental conditions to maintain healthy body weight or promote obesity.

These genetic risk factors tend to be familial but are not inherited in a simple manner; they may reflect many genetic variations, and each variation may contribute a small amount of risk and may interact with environmental elements to produce obesity. The twelfth update of the human obesity gene map by the Pennington Biomedical Research Center (http://obesitygene.pbrc.edu/), which was completed in October 2005, contains more than six hundred genes, markers, and chromosomal regions that are associated or linked to human obesity. Besides offering direction for future efforts to prevent and treat obe-

TABLE 2.5

## Obesity and genetics

| What we know: | What we don't know: |
|---|---|
| Biological relatives tend to resemble each other in many ways, including body weight. Individuals with a family history of obesity may be predisposed to gain weight and interventions that prevent obesity are especially important. | Why are biological relatives more similar in body weight? What genes are associated with this observation? Are the same genetic associations seen in every family? How do these genes affect energy metabolism and regulation? |
| In an environment made constant for food intake and physical activity, individuals respond differently. Some people store more energy as fat in an environment of excess; others lose less fat in an environment of scarcity. The different responses are largely due to genetic variation between individuals. | Why are interventions based on diet and exercise more effective for some people than others? What are the biological differences between these high and low responders? How do we use these insights to tailor interventions to specific needs? |
| Fat stores are regulated over long periods of time by complex systems that involve input and feedback from fatty tissues, the brain, and endocrine glands like the pancreas and the thyroid. Overweight and obesity can result from only a very small positive energy input imbalance over a long period of time. | What elements of energy regulation feedback systems are different in individuals? How do these differences affect energy metabolism and regulation? |
| Rarely, people have mutations in single genes that result in severe obesity that starts in infancy. Studying these individuals is providing insight into the complex biological pathways that regulate the balance between energy input and energy expenditure. | Do additional obesity syndromes exist that are caused by mutations in single genes? If so, what are they? What are the natural history, management strategy, and outcome for affected individuals? |
| Obese individuals have genetic similarities that may shed light on the biological differences that predispose to gain weight. This knowledge may be useful in preventing or treating obesity in predisposed people. | How do genetic variations that are shared by obese people affect gene expression and function? How do genetic variation and environmental factors interact to produce obesity? What are the biological features associated with the tendency to gain weight? What environmental factors are helpful in countering these tendencies? |
| Pharmaceutical companies are using genetic approaches (pharmacogenomics) to develop new drug strategies to treat obesity | Will pharmacologic approaches benefit most people affected with obesity? Will these drugs be accessible to most people? |
| The tendency to store energy in the form of fat is believed to result from thousands of years of evolution in an environment characterized by tenuous food supplies. In other words, those who could store energy in times of plenty, were more likely to survive periods of famine and to pass this tendency to their offspring. | How can thousands of years of evolutionary pressure be countered? Can specific factors in the modern environment (other than the obvious) be identified and controlled to more effectively counter these tendencies? |

SOURCE: "Obesity and Genetics: What We Know, What We Don't Know and What It Means," in *Public Health Perspectives*, Centers for Disease Control and Prevention, National Office of Public Health Genomics, 2007, http://www.cdc.gov/genomics/training/file/print/perspectives/obeseknow.pdf (accessed October 8, 2007)

sity, mounting genetic evidence offers a compelling argument that obesity is not a personal failing, and that in most cases obesity involves multiple genetic and environmental components that affect endocrine, metabolic, and regulatory mechanisms.

### Genetic Susceptibility and Environmental Influences

Even though genetics may largely predetermine adult body weight absent specific environmental triggers or influences, genetic destiny in terms of body weight may not necessarily be realized. For example, an individual with a strong genetic predisposition for obesity will not become obese in the absence of sufficient food (caloric) intake. Similarly, when people genetically predisposed to normal body weight consume a largely high-fat diet, they may become overweight or obese because they may be more inclined to overeat. This is in part because the brain has difficulty conveying the satiety signal—the message to stop eating—when fatty foods are being consumed.

Besides caloric intake and physical activity, both of which are able to modify body weight, environmental influences before birth also significantly influence adult health and body weight. Research demonstrates that the pregnant mother's nutritional status affects the metabolism of her unborn child. Women who are severely malnourished during pregnancy stimulate the fetus to modify its metabolism to conserve and store energy, a survival practice that can promote overweight when the food supply is ample.

Societal and cultural norms can also cause environmental influences such as lifestyle and behavior to override genetic programming. For example, in the United States many young women with genetic predisposition to normal body weight or even overweight sharply limit their caloric intake and exercise vigorously to achieve "model thin" bodies. Similarly, in cultures where overweight is perceived as an indication of prosperity and is admired and coveted, people may override genetic tendencies to be normal weight by increasing caloric intake in an effort to achieve the culturally established ideal.

### HEALTH RISKS AND CONSEQUENCES OF OVERWEIGHT AND OBESITY

The *Surgeon General's Call to Action to Prevent and Decrease Overweight and Obesity, 2001* (2001, http://www.surgeongeneral.gov/topics/obesity/calltoaction/CalltoAction.pdf) predicted that "overweight and obesity may soon cause as much preventable disease and death as cigarette smoking" and that failure to address these conditions "could wipe out the gains we have made in areas such as heart disease, diabetes, several forms of cancer, and other chronic health problems." By 2007 the scientific community acknowledged that the health consequences of overweight and obesity threatened to erode Americans' life span.

People who are overweight or obese are at higher risk of developing one or more serious medical conditions, and obesity is associated with increases in deaths from all causes. Overweight and obesity significantly increase the risk for Type 2 diabetes; hypercholesterolemia (high cholesterol), hypertension, heart disease, and stroke; gall-bladder disease; osteoarthritis, chronic joint pain, and back injury; sleep apnea (interrupted breathing while sleeping) and other respiratory problems; and several cancers. According James A. Greenberg, in "Correcting Biases in Estimates of Mortality Attributable to Obesity" (*Obesity*, vol. 14, no. 11, 2006), obesity is a contributing cause in at least four hundred thousand deaths per year.

## Hypercholesterolemia, Hypertension, Heart Disease, and Stroke

Overweight, obesity, and excess abdominal fat are directly related to cardiovascular risk factors, including high levels of total serum cholesterol, LDL-cholesterol (a fatlike substance often called bad cholesterol because high levels increase the risk for heart disease), triglycerides, blood pressure, fibrinogen, and insulin, and low levels of HDL-cholesterol (often called good cholesterol because high levels appear to protect against heart disease). The association between total serum cholesterol and coronary heart disease is largely due to low-density lipoprotein (LDL). A high-risk LDL-cholesterol is greater than or equal to 160 milligrams per deciliter (mg/dL) with a 10 mg/dL rise in LDL-cholesterol corresponding to approximately a 10% increase in risk. The percent of the population suffering from high serum cholesterol levels fell from 19.7% in 1988–94 to 16.5% in 2001–04. (See Table 2.6.) The overall decline in high serum cholesterol occurred in response to the increasing use of effective cholesterol-lowering statin drugs.

The percent of the population suffering from hypertension increased between 1988–94 and 2001–04, from 21.7% to 26.7% of the population. (See Table 2.7.) The highest rates for those aged twenty to seventy-four during the 2001–04 period were reported among African-American females (40.3%). Both men and women were increasingly likely to have hypertension as they aged. Hypertension is approximately three times more common in obese than in normal-weight people, and the relationship between weight and blood pressure is clearly one of cause and effect because when weight increases, so does blood pressure, and when weight decreases, blood pressure falls.

The physiological processes that produce the hypertension associated with obesity include sodium retention and increases in vascular resistance, blood volume, and cardiac output (the volume of blood pumped, measured in liters per minute). Even though it is not known precisely how weight loss results in a decrease in blood pressure, it is known that weight loss is associated with a reduction in vascular resistance and total blood volume and cardiac

output. Weight loss also results in improvement in insulin resistance, a reduction in sympathetic nervous system activity, and suppression of the renin-angiotensin-aldosterone system, a group of hormones that are responsible for the opening and narrowing of blood vessels and the retention of fluids.

Obesity increases the risk for coronary artery disease, which in turn increases the risk for future heart failure. Congestive heart failure is not a disease but a condition that occurs when the heart is unable to pump enough blood to meet the needs of the body's tissues. When the heart fails, it is unable to pump out all the blood that enters its chambers. Congestive heart failure is a frequent complication of severe obesity and a major cause of death. The duration of obesity is a strong predictor of congestive heart failure because over time elevated total blood volume and high cardiac output cause the left ventricle of the heart to increase in size (known as left ventricular hypertrophy) beyond that expected from normal growth. Even though left ventricular hypertrophy is frequently identified in cardiac patients with obesity and in part results from hypertension, abnormalities in left ventricular mass and function also occur in the absence of hypertension and may be related to the severity of obesity.

Inflammation in blood vessels and throughout the body is thought to increase the risk for heart disease and stroke. People with more body fat have higher blood levels of substances such as plasminogen activator inhibitor-1—an enzyme produced in the kidneys that inhibits the conversion of plasminogen to plasmin and initiates fibrinolysis. Fibrinolysis leads to the breakdown of fibrin, which is responsible for the semisolid character of a blood clot that can occlude (block) blood vessels. This is the mechanism believed to account for the finding that obesity is associated with an increased risk of blood clot formation. Occluded arteries may produce myocardial infarction (heart attack) or stroke (sudden injury to the brain due to a compromised blood and oxygen supply). Overweight increases the risk for ischemic stroke—resulting from a clot or blockage—but does not appear to increase the risk for hemorrhagic stroke (bleeding inside the brain), which, in general, is associated with more fatality. According to the National Heart, Lung, and Blood Institute, in *Guidelines on Overweight and Obesity: Electronic Textbook* (1998, http://www.nhlbi.nih.gov/guidelines/obe sity/e_txtbk/index.htm), the risk of stroke increases as BMI rises. For example, the risk of ischemic stroke is 75% higher in women with a BMI greater than 27, and 137% higher in women with a BMI greater than 32, compared to women having a BMI less than 21.

In "Association of Overweight with Increased Risk of Coronary Heart Disease Partly Independent of Blood Pressure and Cholesterol Levels: A Meta-analysis of 21 Cohort Studies Including More Than 300,000 Persons" (*Archives of Internal Medicine*, vol. 167, no. 16, September

**TABLE 2.6**

## Serum cholesterol levels among persons 20 years of age and over, by demographic characteristics, selected years, 1960–2004

[Data are based on physical examinations of a sample of the civilian noninstitutionalized population]

| Sex, age, race and Hispanic origin[a], and poverty level | 1960–1962 | 1971–1974 | 1976–1980[b] | 1988–1994 | 2001–2004 |
|---|---|---|---|---|---|
| **20–74 years, age adjusted[c]** | Percent of population with high serum total cholesterol | | | | |
| Both sexes[d] | 33.3 | 28.6 | 27.8 | 19.7 | 16.5 |
| Male | 30.6 | 27.9 | 26.4 | 18.8 | 16.6 |
| Female | 35.6 | 29.1 | 28.8 | 20.5 | 16.2 |
| **Not Hispanic or Latino:** | | | | | |
| White only, male | — | — | 26.4 | 18.7 | 16.5 |
| White only, female | — | — | 29.6 | 20.7 | 16.7 |
| Black or African American only, male | — | — | 25.5 | 16.4 | 14.4 |
| Black or African American only, female | — | — | 26.3 | 19.9 | 14.3 |
| Mexican male | — | — | 20.3 | 18.7 | 17.0 |
| Mexican female | — | — | 20.5 | 17.7 | 12.8 |
| **Percent of poverty level:[e]** | | | | | |
| Below 100% | — | 24.4 | 23.5 | 19.3 | 18.9 |
| 100%–less than 200% | — | 28.9 | 26.5 | 19.4 | 17.5 |
| 200% or more | — | 28.9 | 29.0 | 19.6 | 16.0 |
| **20 years and over, age adjusted[c]** | | | | | |
| Both sexes[d] | — | — | — | 20.8 | 16.7 |
| Male | — | — | — | 19.0 | 16.1 |
| Female | — | — | — | 22.0 | 16.8 |
| **Not Hispanic or Latino:** | | | | | |
| White only, male | — | — | — | 18.8 | 16.0 |
| White only, female | — | — | — | 22.2 | 17.4 |
| Black or African American only, male | — | — | — | 16.9 | 14.2 |
| Black or African American only, female | — | — | — | 21.4 | 14.8 |
| Mexican male | — | — | — | 18.5 | 16.9 |
| Mexican female | — | — | — | 18.7 | 14.0 |
| **Percent of poverty level:[e]** | | | | | |
| Below 100% | — | — | — | 20.6 | 19.3 |
| 100%–less than 200% | — | — | — | 20.6 | 17.8 |
| 200% or more | — | — | — | 20.4 | 15.9 |
| **20 years and over, crude** | | | | | |
| Both sexes[d] | — | — | — | 19.6 | 16.7 |
| Male | — | — | — | 17.7 | 16.4 |
| Female | — | — | — | 21.3 | 17.0 |
| **Not Hispanic or Latino:** | | | | | |
| White only, male | — | — | — | 18.0 | 16.5 |
| White only, female | — | — | — | 22.5 | 18.1 |
| Black or African American only, male | — | — | — | 14.7 | 13.8 |
| Black or African American only, female | — | — | — | 18.2 | 13.5 |
| Mexican male | — | — | — | 15.4 | 15.1 |
| Mexican female | — | — | — | 14.3 | 10.8 |
| **Percent of poverty level:[e]** | | | | | |
| Below 100% | — | — | — | 17.6 | 17.3 |
| 100%–less than 200% | — | — | — | 19.8 | 16.4 |
| 200% or more | — | — | — | 19.5 | 16.6 |
| **Male** | | | | | |
| 20–34 years | 15.1 | 12.4 | 11.9 | 8.2 | 9.0 |
| 35–44 years | 33.9 | 31.8 | 27.9 | 19.4 | 21.2 |
| 45–54 years | 39.2 | 37.5 | 36.9 | 26.6 | 23.1 |
| 55–64 years | 41.6 | 36.2 | 36.8 | 28.0 | 19.9 |
| 65–74 years | 38.0 | 34.7 | 31.7 | 21.9 | 11.0 |
| 75 years and over | — | — | — | 20.4 | 9.9 |
| **Female** | | | | | |
| 20–34 years | 12.4 | 10.9 | 9.8 | 7.3 | 9.3 |
| 35–44 years | 23.1 | 19.3 | 20.7 | 12.3 | 11.4 |
| 45–54 years | 46.9 | 38.7 | 40.5 | 26.7 | 20.0 |
| 55–64 years | 70.1 | 53.1 | 52.9 | 40.9 | 27.6 |
| 65–74 years | 68.5 | 57.7 | 51.6 | 41.3 | 26.3 |
| 75 years and over | — | — | — | 38.2 | 23.8 |

**TABLE 2.6**

**Serum cholesterol levels among persons 20 years of age and over, by demographic characteristics, selected years, 1960–2004 [CONTINUED]**

[Data are based on physical examinations of a sample of the civilian noninstitutionalized population]

| Sex, age, race and Hispanic origin[a], and poverty level | 1960–1962 | 1971–1974 | 1976–1980[b] | 1988–1994 | 2001–2004 |
|---|---|---|---|---|---|
| **20–74 years, age adjusted[c]** | | | Mean serum cholesterol level, mg/dL | | |
| Both sexes[d] | 222 | 216 | 215 | 205 | 202 |
| Male | 220 | 216 | 213 | 204 | 201 |
| Female | 224 | 217 | 216 | 205 | 201 |
| **Not Hispanic or Latino:** | | | | | |
| White only, male | — | — | 213 | 204 | 201 |
| White only, female | — | — | 216 | 206 | 202 |
| Black or African American only, male | — | — | 211 | 201 | 198 |
| Black or African American only, female | — | — | 216 | 204 | 198 |
| Mexican male | — | — | 209 | 206 | 202 |
| Mexican female | — | — | 209 | 204 | 199 |
| **Percent of poverty level:[e]** | | | | | |
| Below 100% | — | 211 | 211 | 203 | 202 |
| 100%–less than 200% | — | 217 | 213 | 203 | 202 |
| 200% or more | — | 217 | 216 | 206 | 202 |
| **20 years and over, age adjusted[c]** | | | | | |
| Both sexes[d] | — | — | — | 206 | 202 |
| Male | — | — | — | 204 | 201 |
| Female | — | — | — | 207 | 202 |
| **Not Hispanic or Latino:** | | | | | |
| White only, male | — | — | — | 205 | 201 |
| White only, female | — | — | — | 208 | 203 |
| Black or African American only, male | — | — | — | 202 | 198 |
| Black or African American only, female | — | — | — | 207 | 199 |
| Mexican male | — | — | — | 206 | 201 |
| Mexican female | — | — | — | 206 | 200 |
| **Percent of poverty level:[e]** | | | | | |
| Below 100% | — | — | — | 205 | 203 |
| 100%–less than 200% | — | — | — | 205 | 202 |
| 200% or more | — | — | — | 207 | 202 |
| **20 years and over, crude** | | | | | |
| Both sexes[d] | — | — | — | 204 | 202 |
| Male | — | — | — | 202 | 201 |
| Female | — | — | — | 206 | 203 |
| **Not Hispanic or Latino:** | | | | | |
| White only, male | — | — | — | 203 | 201 |
| White only, female | — | — | — | 208 | 205 |
| Black or African American only, male | — | — | — | 198 | 197 |
| Black or African American only, female | — | — | — | 201 | 196 |
| Mexican male | — | — | — | 199 | 198 |
| Mexican female | — | — | — | 198 | 194 |
| **Percent of poverty level:[e]** | | | | | |
| Below 100% | — | — | — | 200 | 199 |
| 100%–less than 200% | — | — | — | 202 | 200 |
| 200% or more | — | — | — | 205 | 203 |
| **Male** | | | | | |
| 20–34 years | 198 | 194 | 192 | 186 | 186 |
| 35–44 years | 227 | 221 | 217 | 206 | 210 |
| 45–54 years | 231 | 229 | 227 | 216 | 213 |
| 55–64 years | 233 | 229 | 229 | 216 | 208 |
| 65–74 years | 230 | 226 | 221 | 212 | 194 |
| 75 years and over | — | — | — | 205 | 194 |
| **Female** | | | | | |
| 20–34 years | 194 | 191 | 189 | 184 | 186 |
| 35–44 years | 214 | 207 | 207 | 195 | 198 |
| 45–54 years | 237 | 232 | 232 | 217 | 209 |
| 55–64 years | 262 | 245 | 249 | 235 | 219 |
| 65–74 years | 266 | 250 | 246 | 233 | 219 |
| 75 years and over | — | — | — | 229 | 213 |

2007), Rik P. Bogers et al. reveal that even moderately overweight people have an increased risk of heart disease, independent of the impact of their weight on blood pressure and cholesterol levels. The results of this study are important because they show that even when overweight people are treated for high blood pressure and

**TABLE 2.6**

**Serum cholesterol levels among persons 20 years of age and over, by demographic characteristics, selected years, 1960–2004** [CONTINUED]

[Data are based on physical examinations of a sample of the civilian noninstitutionalized population]

*Estimates are considered unreliable.
—Data not available.
ªPersons of Mexican origin may be of any race. Starting with 1999 data, race-specific estimates are tabulated according to the 1997 Revisions to the Standards for the Classification of Federal Data on Race and Ethnicity and are not strictly comparable with estimates for earlier years. The two non-Hispanic race categories shown in the table conform to the 1997 standards. Starting with 1999 data, race-specific estimates are for persons who reported only one racial group. Prior to data year 1999, estimates were tabulated according to the 1977 standards. Estimates for single race categories prior to 1999 included persons who reported one race or, if they reported more than one race, identified one race as best representing their race.
ᵇData for Mexicans are for 1982–1984.
ᶜAge adjusted to the 2000 standard population using five age groups: 20–34 years, 35–44 years, 45–54 years, 55–64 years, and 65 years and over (65–74 years for estimates for 20–74 years). Age-adjusted estimates may differ from other age-adjusted estimates based on the same data and presented elsewhere if different age groups are used in the adjustment procedure.
ᵈIncludes persons of all races and Hispanic origins, not just those shown separately.
ᵉPoverty level is based on family income and family size. Persons with unknown poverty level are excluded.
Notes: High serum cholesterol is defined as greater than or equal to 240 mg/dL (6.20 mmol/L). Borderline high serum cholesterol is defined as greater than or equal to 200 mg/dL and less than 240 mg/dL.

SOURCE: "Table 70. Serum Total Cholesterol Levels among Persons 20 Years of Age and Over, by Sex, Age, Race and Hispanic Origin, and Poverty Status: United States, 1960–1962 through 2001–2004," in *Health, United States, 2006, with Chartbook on Trends in the Health of Americans*, Centers for Disease Control and Prevention, National Center for Health Statistics, 2006, http://www.cdc.gov/nchs/data/hus/hus06.pdf#summary (accessed October 8, 2007)

high cholesterol, they are still at increased risk for heart disease.

Figure 2.2 shows the process, known as a treatment algorithm, that is used to assess and treat overweight individuals, based on their body weight, abdominal fat, and the risk factors for cardiovascular morbidity and mortality.

**Type 2 Diabetes**

Diabetes is a disease that affects the body's use of food, causing blood glucose (sugar levels in the blood) to become too high. Normally, the body converts sugars, starches, and proteins into a form of sugar called glucose. The blood then carries glucose to all the cells throughout the body. In the cells, with the help of the hormone insulin, the glucose is either converted into energy for use immediately or stored for the future. Beta cells of the pancreas, a small organ located behind the stomach, manufacture the insulin. The process of turning food into energy via glucose (blood sugar) is important because the body depends on glucose for every function.

With diabetes, the body can convert food to glucose, but there is a problem with insulin. In one type of diabetes (insulin-dependent diabetes or Type 1), the pancreas does not manufacture enough insulin, and in another type (noninsulin dependent or Type 2), the body has insulin but cannot use the insulin effectively (this latter condition is called insulin resistance). When insulin is either absent or ineffective, glucose cannot get into the cells to be used for energy. Instead, the unused glucose builds up in the bloodstream and circulates through the kidneys. If a person's blood-glucose level rises high enough, the excess glucose "spills" over into the urine, causing frequent urination. This, in turn, leads to an increased feeling of thirst as the body tries to compensate for the fluid lost through urination.

Noninsulin-dependent diabetes (also known as Type 2) is most often seen in adults and is the most common type of diabetes in the United States. In this type, the pancreas produces insulin, but it is not used effectively, and the body resists responding to it. Heredity may be a predisposing factor in the genesis of Type 2 diabetes, but because the pancreas continues to produce insulin, the disease is considered more of a problem of insulin resistance, in which the body is not using the hormone efficiently.

Because diabetes deprives body cells of the glucose needed to function properly, several complications can develop to threaten the lives of diabetics further. The healing process of the body is slowed or impaired, and the risk of infection increases. Complications of diabetes include higher risk and rates of heart disease; circulatory problems, especially in the legs, are often severe enough to require surgery or even amputation; diabetic retinopathy, a condition that can cause blindness; kidney disease that may require dialysis; dental problems; and problems with pregnancy.

The U.S. Department of Health and Human Services notes in the press release "Diet and Exercise Dramatically Delay Type 2 Diabetes" (August 6, 2001, http://www.nih.gov/news/pr/aug2001/niddk-08.htm) that more than 80% of people with Type 2 diabetes are overweight, and in people prone to Type 2 diabetes becoming overweight can trigger the onset of the disease. It is not known precisely how overweight contributes to the causation of this disease. One hypothesis is that being overweight causes cells to change, making them less effective at using sugar from the blood. This then stresses the cells that produce insulin, causing them to gradually fail. Maintaining a healthy weight and keeping physically fit can usually prevent or delay the onset of Type 2 diabetes.

**TABLE 2.7**

## Hypertension and elevated blood pressure among persons 20 years of age and over, by demographic characteristics, 1988–94 and 2001–04

[Data are based on interviews and physical examinations of a sample of the civilian noninstitutionalized population]

| Sex, age, race and Hispanic origin[a], and poverty level | Hypertension[b, c] | | Elevated blood pressure[b] | |
|---|---|---|---|---|
| | 1988–1994 | 2001–2004 | 1988–1994 | 2001–2004 |
| **20–74 years, age-adjusted[d]** | | Percent of population | | |
| Both sexes[e] | 21.7 | 26.7 | 15.4 | 15.9 |
| Male | 23.4 | 26.9 | 18.2 | 16.1 |
| Female | 20.0 | 26.2 | 12.6 | 15.5 |
| **Not Hispanic or Latino:** | | | | |
| White only, male | 22.6 | 26.0 | 17.3 | 14.9 |
| White only, female | 18.4 | 24.1 | 11.2 | 14.1 |
| Black or African American only, male | 34.3 | 37.8 | 27.9 | 25.1 |
| Black or African American only, female | 35.0 | 40.3 | 23.5 | 24.0 |
| Mexican male | 23.4 | 22.1 | 19.1 | 14.9 |
| Mexican female | 21.0 | 25.1 | 16.5 | 16.8 |
| **Percent of poverty level:[f]** | | | | |
| Below 100% | 27.5 | 30.8 | 19.0 | 19.5 |
| 100%–less than 200% | 22.6 | 30.1 | 15.8 | 18.7 |
| 200% or more | 20.4 | 25.4 | 14.6 | 14.8 |
| **20 years and over, age-adjusted[d]** | | | | |
| Both sexes[e] | 25.5 | 30.9 | 18.5 | 19.0 |
| Male | 26.4 | 30.3 | 20.6 | 18.3 |
| Female | 24.4 | 31.0 | 16.4 | 19.2 |
| **Not Hispanic or Latino:** | | | | |
| White only, male | 25.6 | 29.3 | 19.7 | 17.1 |
| White only, female | 23.0 | 29.0 | 15.1 | 17.9 |
| Black or African American only, male | 37.5 | 41.5 | 30.3 | 27.8 |
| Black or African American only, female | 38.3 | 44.3 | 26.4 | 26.9 |
| Mexican male | 26.9 | 26.1 | 22.2 | 17.8 |
| Mexican female | 25.0 | 29.7 | 20.4 | 20.5 |
| **Percent of poverty level:[f]** | | | | |
| Below 100% | 31.7 | 35.5 | 22.5 | 23.1 |
| 100%–less than 200% | 26.6 | 34.0 | 19.3 | 21.5 |
| 200% or more | 23.9 | 29.4 | 17.5 | 17.7 |
| **20 years and over, crude** | | | | |
| Both sexes[e] | 24.1 | 30.8 | 17.6 | 18.7 |
| Male | 23.8 | 29.0 | 18.7 | 17.6 |
| Female | 24.4 | 32.5 | 16.5 | 19.8 |
| **Not Hispanic or Latino:** | | | | |
| White only, male | 24.3 | 29.9 | 18.7 | 17.4 |
| White only, female | 24.6 | 32.9 | 16.4 | 20.2 |
| Black or African American only, male | 31.1 | 36.7 | 25.5 | 24.7 |
| Black or African American only, female | 32.5 | 41.6 | 22.2 | 24.4 |
| Mexican male | 16.4 | 15.8 | 13.9 | 11.4 |
| Mexican female | 15.9 | 19.0 | 12.7 | 12.8 |
| **Percent of poverty level:[f]** | | | | |
| Below 100% | 25.7 | 28.3 | 18.7 | 18.3 |
| 100%–less than 200% | 26.7 | 34.6 | 19.8 | 21.6 |
| 200% or more | 22.2 | 29.9 | 16.2 | 17.8 |
| **Male** | | | | |
| 20–34 years | 7.1 | 7.0 | 6.6 | 6.1 |
| 35–44 years | 17.1 | 19.2 | 15.2 | 11.9 |
| 44–54 years | 29.2 | 35.9 | 21.9 | 23.0 |
| 55–64 years | 40.6 | 47.5 | 28.4 | 25.7 |
| 65–74 years | 54.4 | 61.7 | 39.9 | 30.2 |
| 75 years and over | 60.4 | 67.1 | 49.7 | 45.0 |
| **Female** | | | | |
| 20–34 years | 2.9 | *2.7 | *2.4 | * |
| 35–44 years | 11.2 | 14.0 | 6.4 | 6.8 |
| 45–54 years | 23.9 | 35.2 | 13.7 | 22.3 |
| 55–64 years | 42.6 | 54.4 | 27.0 | 29.6 |
| 65–74 years | 56.2 | 72.9 | 38.2 | 48.3 |
| 75 years and over | 73.6 | 82.0 | 59.9 | 58.5 |

# TABLE 2.7

**Hypertension and elevated blood pressure among persons 20 years of age and over, by demographic characteristics, 1988–94 and 2001–04** [CONTINUED]

[Data are based on interviews and physical examinations of a sample of the civilian noninstitutionalized population]

*Estimates are considered unreliable.

aPersons of Mexican origin may be of any race. Starting with 1999 data, race-specific estimates are tabulated according to the 1997 Revisions to the Standards for the Classification of Federal Data on Race and Ethnicity and are not strictly comparable with estimates for earlier years. The two non-Hispanic race categories shown in the table conform to the 1997 standards. Starting with 1999 data, race-specific estimates are for persons who reported only one racial group. Prior to data year 1999, estimates were tabulated according to the 1977 standards. Estimates for single-race categories prior to 1999 included persons who reported one race or, if they reported more than one race, identified one race as best representing their race.

bHypertension is defined as having elevated blood pressure and/or taking antihypertensive medication. Elevated blood pressure is defined as having systolic pressure of at least 140 mmHg or diastolic pressure of at least 90 mmHg. Those with elevated blood pressure may be taking prescribed medicine for high blood pressure.

cRespondents were asked, "Are you now taking prescribed medicine for your high blood pressure?"

dAge-adjusted to the 2000 standard population using five age groups: 20–34 years, 35–44 years, 45–54 years, 55–64 years, and 65 years and over (65–74 years for estimates for 20–74 years). Age-adjusted estimates may differ from other age-adjusted estimates based on the same data and presented elsewhere if different age groups are used in the adjustment procedure.

eIncludes persons of all races and Hispanic origins, not just those shown separately.

fPoverty level is based on family income and family size. Persons with unknown poverty level are excluded.

Notes: Percents are based on the average of blood pressure measurements taken. In 2001–2004, 77% of participants had three blood pressure readings. Excludes pregnant women.

SOURCE: "Table 69. Hypertension and Elevated Blood Pressure among Persons 20 Years of Age and Over, by Sex, Age, Race, and Hispanic Origin, and Poverty Level: United States, 1988–1994 and 2001–2004," in *Health, United States, 2006, with Chartbook on Trends in the Health of Americans*, Centers for Disease Control and Prevention, National Center for Health Statistics, August 2007, http://www.cdc.gov/nchs/data/hus/hus06.pdf#summary (accessed October 8, 2007)

From 1980 through 2005 the number of Americans with diabetes nearly tripled, from 5.6 million to 15.8 million. (See Figure 2.3.) Worse still, these numbers may significantly underestimate the true prevalence of diabetes in the United States in view of National Health and Nutrition Survey findings that show sizeable numbers of adults have undiagnosed diabetes.

**"DIABESITY" AND "DOUBLE DIABETES."** The recognition of obesity-dependent diabetes prompted scientists and physicians to coin a new term to describe this condition: diabesity. The term was first used in the 1990s and has gained widespread acceptance. Even though diabesity is attributed to the same causes as Type 2 diabetes—insulin resistance and pancreatic cell dysfunction—researchers are beginning to link the inflammation associated with obesity to the development of diabetes and cardiovascular disease.

Francine R. Kaufman contends in *Diabesity: The Obesity-Diabetes Epidemic That Threatens America—And What We Must Do to Stop It* (2005) that the diabesity epidemic "imperils human existence as we now know it" and observes that more than one-third of American children born in 2000 will develop diabetes in their lifetime. Kaufman warns that unless drastic measures are taken, by 2020 there will be a 72% increase in the number of diabetics in the United States.

Another recent phenomenon is patients diagnosed with both Type 1 and Type 2 diabetes simultaneously. Dubbed "double diabetes," it has been reported in both children and adults. Among children, it often results when children with Type 1 diabetes who rely on insulin injections to control their diabetes gain weight and develop the insulin resistance that characterizes Type 2 diabetes. Adults who have been diagnosed with Type

2 diabetes, but fail to respond to treatment have been found to also suffer from the Type 1, insulin-dependent form of the disease.

## Osteoarthritis and Joint Injury

*Being only ten pounds overweight increases the force on the knee by 30–60 pounds with each step.... Even small amounts of weight loss reduce the risk of developing knee Osteoarthritis (OA). Preliminary studies suggest weight loss decreases pain substantially in those with knee OA.*

—Susan Bartlett, "Osteoarthritis Weight Management" (2007)

The word *arthritis* literally means joint inflammation. The name applies to more than one hundred related diseases known as rheumatic diseases. A joint is any point where two bones meet. When a joint becomes inflamed, swelling, redness, pain, and loss of motion occur. In the most serious forms of the disease, the loss of motion can be physically disabling. Arthritis is the leading cause of disability and the leading cause of limitation of activity among working-age adults in the United States. (See Figure 2.4.)

People who are overweight or obese are at increased risk for osteoarthritis, which is not an inflammatory arthritis. Osteoarthritis, sometimes called degenerative arthritis, causes the breakdown of bones and cartilage (connective tissue attached to bones), and usually causes pain and stiffness in the fingers, knees, feet, hips, and back. Extra weight places extra pressure on joints and cartilage, causing them to erode. Furthermore, people with more body fat may have higher blood levels of substances that cause inflammation. Inflammation at the joints may increase the risk for osteoarthritis.

According to the CDC, in "Arthritis: Data and Statistics" (July 30, 2007, http://www.cdc.gov/arthritis/

**FIGURE 2.2**

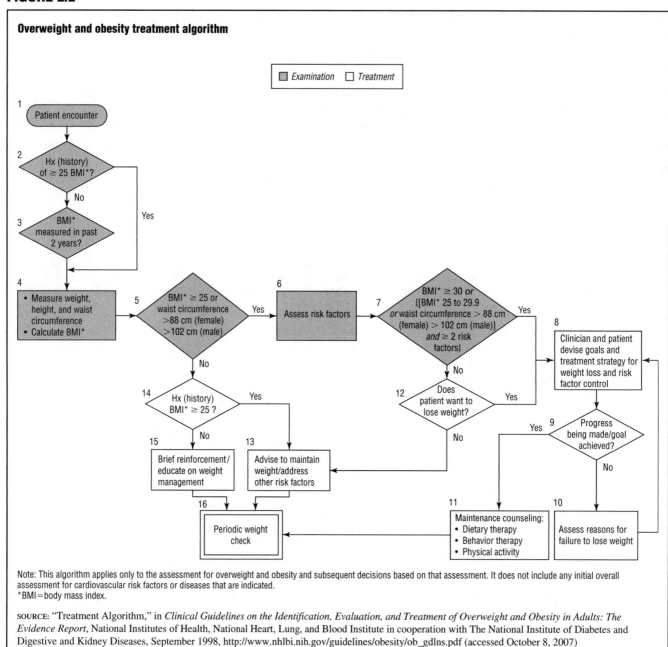

**Overweight and obesity treatment algorithm**

☐ Examination (grey) ☐ Treatment (white)

1. Patient encounter

2. Hx (history) of ≥ 25 BMI*?
No →

3. BMI* measured in past 2 years?
Yes (2 & 3 → to box 5)

4. • Measure weight, height, and waist circumference • Calculate BMI*

5. BMI* ≥ 25 or waist circumference >88 cm (female) >102 cm (male)
Yes → 6. Assess risk factors
No →

6. Assess risk factors →

7. BMI* ≥ 30 or {[BMI* 25 to 29.9 or waist circumference > 88 cm (female) > 102 cm (male)] and ≥ 2 risk factors}
Yes →
No →

8. Clinician and patient devise goals and treatment strategy for weight loss and risk factor control

9. Progress being made/goal achieved?
Yes → 11
No → 10

10. Assess reasons for failure to lose weight

11. Maintenance counseling: • Dietary therapy • Behavior therapy • Physical activity

12. Does patient want to lose weight?
Yes → (to 8)
No → 13

13. Advise to maintain weight/address other risk factors

14. Hx (history) BMI* ≥ 25 ?
Yes → 13
No → 15

15. Brief reinforcement/ educate on weight management

16. Periodic weight check

Note: This algorithm applies only to the assessment for overweight and obesity and subsequent decisions based on that assessment. It does not include any initial overall assessment for cardiovascular risk factors or diseases that are indicated.
*BMI=body mass index.

SOURCE: "Treatment Algorithm," in *Clinical Guidelines on the Identification, Evaluation, and Treatment of Overweight and Obesity in Adults: The Evidence Report*, National Institutes of Health, National Heart, Lung, and Blood Institute in cooperation with The National Institute of Diabetes and Digestive and Kidney Diseases, September 1998, http://www.nhlbi.nih.gov/guidelines/obesity/ob_gdlns.pdf (accessed October 8, 2007)

data_statistics/epi_briefs/osteoarthritis.htm), osteoarthritis affects approximately 20.7 million Americans, usually after age forty-five. Data from the National Health and Nutrition Examination Survey, 2005–2006 (2007, http://www.cdc.gov/nchs/about/major/nhanes/nhanes2005-2006/nhanes05_06.htm) reveal that obese women have about four times the risk of knee osteoarthritis, compared to women of healthy weight, and for obese men, the risk is nearly five times greater. People with clinically severe obesity—those in the highest fifth of body weight—have a tenfold risk of developing knee osteoarthritis, compared to those in the lowest fifth.

Weight loss may decrease the likelihood of developing osteoarthritis in the knees, hips, and lower back and

has been shown to relieve the symptoms of osteoarthritis. Marlene Fransen reports in "Dietary Weight Loss and Exercise for Obese Adults with Knee Osteoarthritis: Modest Weight Loss Targets, Mild Exercise, Modest Effects" (*Arthritis and Rheumatism*, vol. 50, no. 5, May 2004) that a decrease in BMI of two units or greater during a ten-year period decreased the risk of developing knee osteoarthritis by more than 50%. In another study, Stephen Messier et al. find in "Exercise and Dietary Weight Loss in Overweight and Obese Older Adults with Knee Osteoarthritis: The Arthritis, Diet, and Activity Promotion Trial" (*Arthritis and Rheumatism*, vol. 50, no. 5, May 2004) that people with osteoarthritis who lost weight had improved range of motion and less joint pain.

FIGURE 2.3

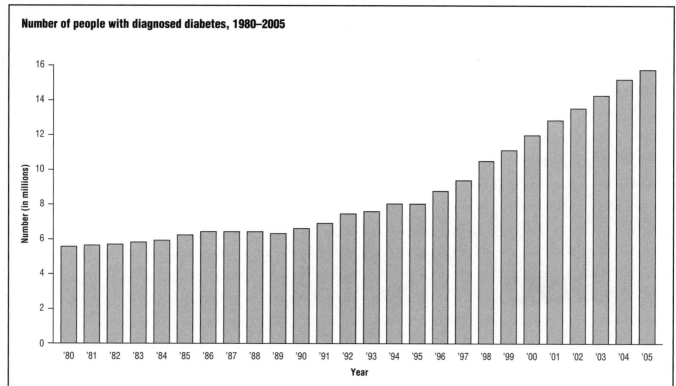

**Number of people with diagnosed diabetes, 1980–2005**

SOURCE: "Number (in Millions) of Persons with Diagnosed Diabetes, United States, 1980–2005," in *Diabetes Data and Trends*, Centers for Disease Control and Prevention, National Center for Chronic Disease Prevention and Health Promotion, Division of Diabetes Translation, March 26, 2007, http://www.cdc .gov/diabetes/statistics/prev/national/figpersons.htm (accessed October 9, 2007)

Gregory M. Ford et al. state in "Associations of Body Mass Index with Meniscal Tears" (*American Journal of Preventive Medicine*, vol. 28, no. 4, 2005) that overweight is also linked to cartilage tears in the knee. The study's subjects were men and women aged fifty to seventy-nine who had surgery to repair the meniscus, the shock-absorbing cartilage in the knee. Ford et al. find that people with a BMI even slightly over the healthy range were three times more likely to have a cartilage tear. The heaviest men were fifteen times more likely to have torn knee cartilage, and the heaviest women were twenty-five times more likely to have torn cartilage than those in the healthy weight ranges. One possible explanation for this finding may be that obese people have circulation problems that reduce the blood supply to the cartilage. Ford et al. conclude that overweight probably accounts for more than half of the nation's 850,000 annual operations to repair cartilage tears in the knee.

**Gallbladder Disease**

Gallstones are small, hard pellets that can form when bile in the gallbladder—a muscular saclike organ that lies under the liver in the right side of the abdomen—precipitates (becomes solid out of the bile solution). Bile contains water, cholesterol, fats, bile salts, proteins, and bilirubin. The gallbladder stores and concentrates the bile

produced in the liver that is not immediately needed for digestion. Bile is released from the gallbladder into the small intestine in response to food. The pancreatic duct joins the common bile duct at the small intestine, adding enzymes to aid in digestion. (See Figure 2.5.) If bile contains too much cholesterol, bile salts, or bilirubin, under certain conditions it can harden into stones. Most gallstones are formed primarily from cholesterol.

People who are overweight are at a higher risk for developing gallstones because the liver overproduces cholesterol and delivers it into the bile, which then becomes supersaturated. The National Heart, Lung, and Blood Institute indicates in "Gallstones" (1998, http:// www.nhlbi.nih.gov/guidelines/obesity/e_txtbk/ratnl/2217 .htm) that the risk of gallstones is as high as twenty per one thousand women per year when BMI is above 40, compared to three per one thousand among women with BMI less than 24.

Analysis of data from the third National Health and Nutrition Examination Survey (NHANES III; http:// www.cdc.gov/nchs/about/major/nhanes/nh3data.htm), which was conducted between 1988 and 1994, reveals that the prevalence of gallstone disease among women increased from 9.4% in the first quartile of BMI to 25.5% in the fourth quartile of BMI. Among men, the prevalence of

FIGURE 2.4

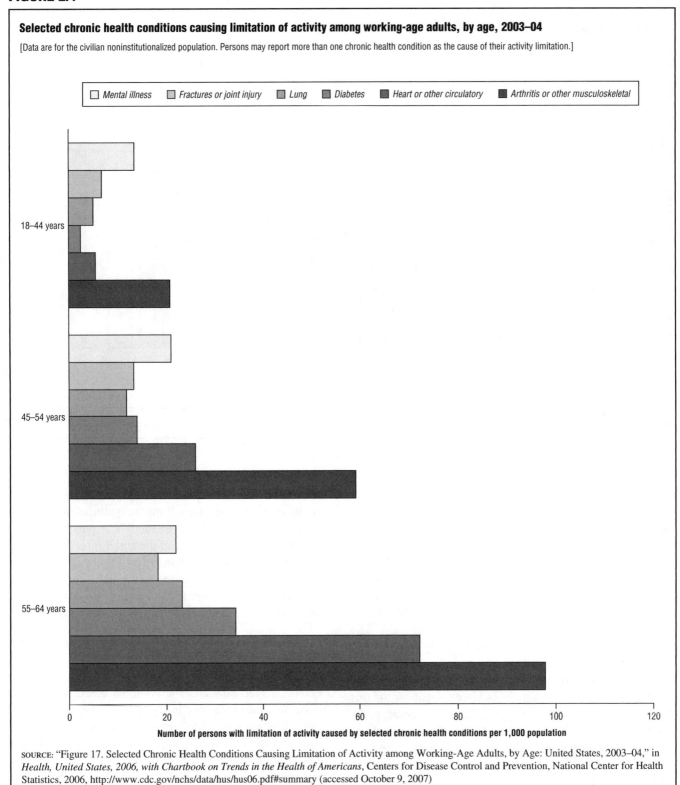

**Selected chronic health conditions causing limitation of activity among working-age adults, by age, 2003–04**

[Data are for the civilian noninstitutionalized population. Persons may report more than one chronic health condition as the cause of their activity limitation.]

□ Mental illness  ▨ Fractures or joint injury  ▨ Lung  ▨ Diabetes  ▨ Heart or other circulatory  ■ Arthritis or other musculoskeletal

Number of persons with limitation of activity caused by selected chronic health conditions per 1,000 population

SOURCE: "Figure 17. Selected Chronic Health Conditions Causing Limitation of Activity among Working-Age Adults, by Age: United States, 2003–04," in *Health, United States, 2006, with Chartbook on Trends in the Health of Americans*, Centers for Disease Control and Prevention, National Center for Health Statistics, 2006, http://www.cdc.gov/nchs/data/hus/hus06.pdf#summary (accessed October 9, 2007)

gallstone disease increased from 4.6% in the first quartile of BMI to 10.8% in the fourth quartile of BMI. Rapid weight loss or weight cycling (the repeated loss and regain of body weight) further increases cholesterol production in the liver, with resulting supersaturation and risk for gallstone formation.

**Fatty Liver Disease**

Fatty liver is defined as an excess accumulation of fat in the liver, usually exceeding 5% of the total liver weight. More than 50% of the excess fat deposit in the liver is triglyceride. The enlargement of the liver is caused by the reduction of fatty acid oxidation in the

FIGURE 2.5

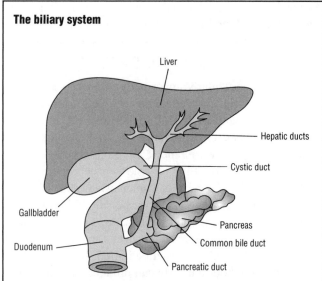

**The biliary system**

Liver

Hepatic ducts

Cystic duct

Gallbladder

Pancreas

Duodenum

Common bile duct

Pancreatic duct

The gallbladder and the ducts that carry bile and other digestive enzymes from the liver, gallbladder, and pancreas to the small intestine are called the biliary system.

SOURCE: "The Gallbladder and the Ducts that Carry Bile and Other Digestive Enzymes from the Liver, Gallbladder, and Pancreas to the Small Intestine are Called the Biliary System," in *Gallstones*, National Institutes of Health, National Institute of Diabetes and Digestive and Kidney Diseases, National Digestive Diseases Information Clearinghouse, July 2007, http://digestive.niddk.nih.gov/ddiseases/pubs/gallstones/index.htm (accessed October 9, 2007)

liver, resulting in excess accumulation of fat. It causes injury and inflammation in the liver and may lead to severe liver damage, cirrhosis (buildup of scar tissue that blocks proper blood flow in the liver), or liver failure. Heiner Wedemeyer and Michael P. Manns report in "Fatty Liver Disease—It's More Than Alcohol and Obesity" (*Medscape Gastroenterology*, vol. 5, no. 2, 2003) that an estimated nine million individuals in the United States suffer from nonalcoholic fatty liver disease.

People with diabetes or with higher than normal blood sugar levels (but not yet in the diabetic range) are more likely to have fatty liver disease than those with normal blood sugar levels. It is not known why some people who are overweight or diabetic get fatty liver and others do not. Losing weight reduces the buildup of fat in the liver and prevents further injury; however, weight loss should not exceed 2.2 pounds per week because more rapid weight loss may exacerbate the disease.

Interestingly, Hong Ji and Mark I. Friedman of the Monell Chemical Senses Center in Philadelphia, Pennsylvania, assert in "Reduced Capacity for Fatty Acid Oxidation in Rats with Inherited Susceptibility to Diet-Induced Obesity" (*Metabolism: Clinical and Experimental*, vol. 56, no. 8, August 2007) that a genetic defect that impairs fatty acid oxidation—fat metabolism in the liver—may be a factor in obesity. Animal studies suggest that it may be harder for people prone to obesity to use

energy from the fats they eat and to burn their own body fat for energy than their healthy-weight counterparts. This in turn prompts them to overeat in an effort to get enough energy.

## Cancer

Cancer encompasses a large group of diseases characterized by the uncontrolled growth and spread of abnormal cells. These cells may grow into masses of tissue called malignant tumors. The dangerous aspect of cancer is that cancer cells invade and destroy normal tissue.

The spread of cancer cells occurs either by local growth of the tumor or by some of the cells becoming detached and traveling through the blood and lymph systems to start additional tumors in other parts of the body. Metastasis (the spread of cancer cells) may be confined to a region of the body, but if left untreated (and often despite treatment), the cancer cells can spread throughout the entire body, eventually causing death. It is perhaps the rapid, invasive, and destructive nature of cancer that makes it, arguably, the most feared of all diseases, even though it is second to heart disease as the leading cause of death in the United States. (See Table 2.2.)

Overweight increases the risk of developing several types of cancer, including cancers of the colon, esophagus, gallbladder, kidney, pancreas, liver, and prostate, as well as uterine (specifically cancer of the lining of the uterus) and postmenopausal breast cancer. Excessive weight gain during adult life increases the risk for several of these cancers. For example, according to the National Heart, Lung, and Blood Institute, in "Cancer" (1998, http://www.nhlbi.nih.gov/guidelines/obesity/e_txtbk/ratnl/22110.htm), a gain of more than twenty pounds from age eighteen to midlife doubles a woman's risk of breast cancer, and even more modest weight gains are associated with increased risk. The International Agency for Research on Cancer estimates in *Handbooks of Cancer Prevention: Weight Control and Physical Activity* (2002) that 40% of uterine cancers, up to 25% of kidney cancers, and about 10% of breast and colon cancers would not develop if people avoided gaining excess weight. The agency also asserts that obesity increases the risk of breast cancer in postmenopausal women by as much as 40%.

In "Obesity, Recreational Physical Activity, and Risk of Pancreatic Cancer in a Large U.S. Cohort" (*Cancer Epidemiology Biomarkers and Prevention*, vol. 14, February 2005), Alpa V. Patel et al. find a relationship between obesity and pancreatic cancer. The researchers obtained information on current weight and weight at age eighteen, location of weight gain, and recreational physical activity to obtain a baseline using a self-administered questionnaire for 145,627 men and women who were cancer-free at the start of the study. During the seven years of follow-up, 242 cases of pancreatic cancer were diagnosed in the research subjects. Analysis of these data revealed an increased risk of pancreatic cancer among obese men and women, compared to men and women of normal weight. The risk of pancreatic cancer was also higher among men and women who reported a tendency for central (abdominal) weight gain, compared to subjects who reported a tendency for peripheral weight gain.

Julie A. Ross et al. suggest in "Body Mass Index and Risk of Leukemia in Older Women" (*Cancer Epidemiology Biomarkers and Prevention*, vol. 13, November 2004) that about 30% of cases of acute myelogenous leukemia (AML) in older women might be linked to overweight or obesity. An analysis of data from the Iowa Women's Health Study, which gathered data from more than forty thousand Iowa women aged fifty-five to sixty-nine from 1986 to 2001 found that the risk for AML was 90% higher among women fifty-five and older who were overweight (BMI of 25 to 29). The risk was 140% higher among obese (BMI of 30 or greater) women in this age group.

Overweight may also increase the risk of dying from some cancers. In "Weight Loss in Breast Cancer Patient Management" (*Journal of Clinical Oncology*, vol. 20, no. 4, February 2002), Rowan T. Chlebowski, Erin Aiello, and Anne McTiernan conclude that "women with breast cancer who are overweight or gain weight after diagnosis are found to be at greater risk for breast cancer recurrence and death compared with lighter women. Obesity is also associated with hormonal profiles likely to stimulate breast cancer growth."

It is not known exactly how being overweight increases cancer risk, recurrence, or mortality. It may be that fat cells make or influence hormones that affect cell growth—such as a high-fat, high-caloric diet—or physical inactivity that promote overweight contribute to cancer risk.

## Sleep Apnea and Sleep Disorders

Sleep apnea is a condition in which breathing becomes shallow or stops completely for short periods during sleep. Each pause lasts about ten to twenty seconds or longer and pauses can occur twenty times or more an hour. Sleep apnea can increase the risk of developing high blood pressure, heart attack, or stroke. Untreated sleep apnea can increase the risk of diabetes and daytime sleepiness and difficulty concentrating can increase the risk for work-related accidents and automobile accidents.

The most common type of sleep apnea, and the type that is linked to overweight and obesity, is obstructive sleep apnea (OSA). During sleep there is insufficient airflow into the lungs through the mouth and nose, and the amount of oxygen in the blood may drop because the airway is transiently occluded (blocked). The National Heart, Lung, and Blood Institute notes in "Who Is At Risk for Obstructive Sleep Apnea?" (May 14, 2007, http://www.nhlbi.nih.gov/health/dci/Diseases/SleepApnea/SleepApnea_WhoIsAtRisk.html) that over twelve million Americans have OSA and that one in twenty-five men and one in fifty women over age forty have debilitating sleep apnea that causes them to be sleepy during the day.

Obesity, particularly upper-body obesity, is a risk factor for sleep apnea and is related to its severity. Most people with sleep apnea have a BMI greater than 30. In general, men whose neck circumference is seventeen inches or greater and women with neck circumference of sixteen inches or greater are at higher risk for sleep apnea. Large neck girth in both men and women who snore is highly predictive of sleep apnea because people with large neck girth store more fat around their necks, which may compromise their airway. A smaller airway can make breathing difficult or stop it altogether. In addition, fat stored in the neck and throughout the body can produce substances that cause inflammation, and inflammation in the neck may be a risk factor for sleep apnea. Weight loss usually resolves or significantly improves sleep apnea by decreasing neck size and reducing inflammation.

Too little sleep is also linked to obesity. Robert D. Vorona et al. of the Eastern Virginia Medical School find in "Overweight and Obese Patients in a Primary Care Population Report Less Sleep than Patients with a Normal Body Mass Index" (*Archives of Internal Medicine*, vol. 165, no. 1, January 10, 2005) that people who were overweight or obese reported that they slept less per week than their normal-weight counterparts. Total sleep time decreased as BMI increased except in the extremely obese group. The difference averaged sixteen minutes per day between normal-weight and overweight subjects, totaling nearly two hours per week. Vorona et al. speculate that lost sleep might have metabolic and hormonal consequences. For example, sleep restriction may reduce levels of leptin, a hormone involved in appetite regulation, which could account for the relationship between diminished sleep and obesity. They observe that their findings do not establish a cause-and-effect relationship between sleep and obesity; however, they "suggest that an extra twenty minutes of sleep per night seems to be associated with a lower BMI."

## Women's Reproductive Health

Besides increased risk of breast and endometrial cancers (the endometrium is the lining of the uterus), women who are overweight or obese may suffer from infertility (difficulty or inability to conceive a child) and other gynecological or pregnancy-related medical problems. Obesity is associated with menstrual irregularities such as abnormally heavy menstrual periods and amenorrhea (cessation of menstruation), and has been found to affect ovulation, response to fertility treatment, pregnancy rates, and pregnancy outcomes.

In "The Importance of Diagnosing the Polycystic Ovary Syndrome" (*Annals of Internal Medicine*, vol. 132, no. 12, June 20, 2000), Rogerio A. Lobo and Enrico Carmina note that abdominal obesity in women is linked to polycystic ovarian syndrome (PCOS), an endocrine condition that afflicts approximately 6% to 10% of premenopausal women. PCOS is characterized by the accumulation of cysts (fluid-filled sacs) on the ovaries, chronic anovulation (absent ovulation), and other metabolic disturbances. Symptoms include excess facial and body hair, acne, obesity, irregular menstrual cycles, insulin resistance, and infertility. A key characteristic of PCOS is hyperandrogenism—excessive production of male hormones (androgens), particularly testosterone, by the ovaries—which is responsible for the acne, male-pattern hair growth, and baldness seen in women with PCOS. Hyperandrogenism has been linked with insulin resistance and hyperinsulinemia (high blood insulin levels), both of which are common in PCOS. Women with PCOS have an increased risk of early-onset heart disease, hypertension, diabetes, and reproductive cancers and a higher incidence of miscarriage and infertility. In overweight women, modest weight loss (as little as 5%) through diet and exercise may correct hyperandrogenism and restore ovulation and fertility.

Obesity during pregnancy is associated with increased morbidity for both the expectant mother and the unborn child. Obese pregnant women are significantly more likely to suffer from hypertension and gestational diabetes (glucose intolerance of variable severity that starts or is first recognized during pregnancy) than normal-weight expectant mothers. Obesity is also associated with difficulties in managing labor and delivery, leading to higher rates of cesarean section (delivery of a fetus by surgical incision through the abdominal wall and uterus). Risks associated with anesthesia are higher in obese women, as there is greater tendency toward hypoxemia (abnormal lack of oxygen in the blood) and greater difficulty administering local or general anesthesia.

The children of women who are obese during pregnancy are at increased risk of birth defects—congenital malformations, particularly of neural tube defects. Neural tube defects are abnormalities of the brain and spinal cord resulting from the failure of the neural tube to develop properly during early pregnancy. The neural tube is the embryonic nerve tissue that eventually develops into the brain and the spinal cord.

Research indicates that boys born to higher-weight mothers may be more likely to develop testicular cancer. In "Is There an Association between Maternal Weight and the Risk of Testicular Cancer? An Epidemiologic Study of Norwegian Data with an Emphasis on World War II" (*International Journal of Cancer*, vol. 116, no. 2, 2005), Elin L. Aschim et al. indicate that exposure of unborn males

to high levels of estrogen is believed to be involved in the subsequent development of testicular cancer. Boys born to higher-weight mothers are more likely to have been exposed to high estrogen levels than boys born to lower-weight mothers because higher weight results in higher insulin levels and lower levels of the protein that normally binds estrogen. As a result, higher levels of estrogen are able to cross the placenta and affect the male fetus.

Furthermore, women who are obese before pregnancy appear to have a higher risk of stillbirth and of having an infant die soon after birth. In "Pre-pregnancy Weight and the Risk of Stillbirth and Neonatal Death" (*British Journal of Gynecology*, vol. 112, no. 4, 2005), Janni Kristensen et al. find that compared to normal-weight women, those who were obese before pregnancy had twice the risk of stillbirth and newborn deaths. Even though earlier researchers reported comparable risks attributed in part to the higher rates of diabetes and high blood pressure observed in overweight pregnant women, Kristensen et al.'s study finds that diabetes and high blood pressure were not responsible for their findings. When women with diabetes or high blood pressure were excluded from their analysis, the risks of stillbirth and newborn death linked to obesity were still significantly higher than the risks for normal-weight or overweight women. Even though it is not yet known how obesity increases the risk of stillbirth and early infant death, Kristensen et al. posit that obesity influences the hormonal system and the metabolism of blood fats that in turn may compromise blood flow to the placenta (an organ that forms during pregnancy and functions as a filter between the mother and fetus).

**WEIGHT GAIN DURING PREGNANCY.** Weight gain during pregnancy is expected and beneficial. The fetus, expanded blood volume, the enlarged uterus, breast tissue growth, and other products of conception generate approximately thirteen to seventeen pounds of extra weight. Weight gain beyond this anticipated amount is largely maternal adipose tissue that is often retained after pregnancy. The challenge health professionals face when developing recommendations about weight gain during pregnancy is achieving a balance between gains intended to produce high-birth-weight infants, who may then require delivery by cesarean section, and low-birth-weight infants with a higher infant mortality rate. Analysis of data from the CDC's Pregnancy Nutrition Surveillance System shows that extremely overweight women benefit from reduced weight gain during pregnancy to help decrease the risk for high-birth-weight infants. Table 2.8 shows the recommended amount of weight gain during pregnancy based on pre-pregnancy BMI.

In "Gestational Weight Gain and Pregnancy Outcomes in Obese Women" (*Obstetrics and Gynecology*, vol. 110, no. 4, October 2007), a study that analyzed the pregnancies of more than 120,000 obese women to see

**TABLE 2.8**

**Recommended weight gain during pregnancy**

| BMI | Kilograms | Pounds |
|---|---|---|
| <19.8 | 12.5 to 18 | 28 to 40 |
| >19.8 to 26 | 11.5 to 16 | 25 to 35 |
| >26 to 29 | 7 to 11.5 | 15 to 25 |
| >29 | ≤6 | ≤13 |

SOURCE: "Weight Gain during Pregnancy," in *Guidelines on Overweight and Obesity: Electronic Textbook*, National Institutes of Health, National Heart, Lung, and Blood Institute in cooperation with The National Institute of Diabetes and Digestive and Kidney Diseases, 1998, http://www.nhlbi.nih.gov/guidelines/obesity/e_txtbk/ratnl/22111.htm (accessed October 9, 2007)

how weight gain affected pregnancy-related high blood pressure, cesarean delivery, and birth weight, Deborah W. Kiel et al. of St. Louis University conclude that obese women who gained little or no weight during pregnancy as well as those who lost weight, fared best. By not gaining weight during pregnancy, they reduced their risk for high blood pressure, had fewer cesarean deliveries, and were more likely to have babies of normal weight.

**Metabolic Syndrome**

Charles M. Alexander et al. estimate in "NCEP-Defined Metabolic Syndrome, Diabetes, and Prevalence of Coronary Heart Disease among NHANES III Participants Age 50 Years and Older" (*Diabetes*, vol. 52, no. 5, 2003) that 44% of Americans over age fifty exhibit a cluster of medical conditions characterized by insulin resistance and the presence of obesity, abdominal fat, high blood sugar and triglycerides, high blood cholesterol, and high blood pressure. This constellation of symptoms, called metabolic syndrome, was first defined in the *Third Report of the National Cholesterol Education Program (NCEP) Expert Panel on Detection, Evaluation, and Treatment of High Blood Cholesterol in Adults (Adult Treatment Panel III)* (September 2002, http://www.nhlbi.nih.gov/guidelines/cholesterol/atp3full.pdf). The report concludes that for most affected people, metabolic syndrome results from poor diet and insufficient physical activity.

The diagnosis of metabolic syndrome, which is also known as syndrome X, requires that people meet at least three of the following criteria:

- Waistline measurement (waist circumference) of forty inches or more for men and thirty-five inches or more for women

- Blood pressure of 130/85 millimeters of mercury (mmHg) or higher

- Fasting blood glucose level greater than 100 mg/dL

- Serum triglyceride level above 150 mg/dL

- HDL level less than 40 mg/dL for men or under 50 mg/dL for women

According to the American Heart Association, three groups of people are most likely to be diagnosed with metabolic syndrome: diabetics, people with hypertension and hyperinsulinemia, and people who have suffered heart attacks and have hyperinsulinemia without glucose intolerance.

Even though research shows that the signs of metabolic syndrome are common among family members, until recently a definitive genetic link had not been identified. Ruth J. F. Loos et al. demonstrate in "Genome-Wide Linkage Scan for the Metabolic Syndrome in the HERITAGE Family Study" (*Journal of Clinical Endocrinology and Metabolism*, vol. 88, no. 12, 2003) the existence of genetic regions that may signal a predisposition to metabolic syndrome. Loos et al. report that they found evidence of genetic linkages to metabolic syndrome in both African-American and white patients.

The exact origins and mechanism of metabolic syndrome are not fully known; regardless, affected individuals experience a series of biochemical changes that, in time, lead to the development of potentially harmful medical conditions. The biochemical changes begin when insulin loses its ability to cause cells to absorb glucose from the blood (insulin resistance). As a result, glucose levels remain high after food is consumed, and the pancreas, sensing a high glucose level in the blood, continues to secrete insulin. Loss of insulin sensitivity may be genetic or may be in response to high fat levels with fatty deposits in the pancreas.

Moderate weight loss, in the range of 5% to 10% of body weight, can help restore the body's sensitivity to insulin and greatly reduce the chance that the syndrome will progress into a more serious illness. Increased activity alone has also been shown to improve insulin sensitivity.

John K. Ninomiya et al. investigated how each of the factors associated with metabolic syndrome influences cardiovascular risk and reported their findings in "Association of the Metabolic Syndrome with History of Myocardial Infarction and Stroke in the Third National Health and Nutrition Examination Survey" (*Circulation*, vol. 109, January 2004). Using data from 10,357 NHANES III participants, they found that having the metabolic syndrome doubled the risk. High triglycerides independently increased this risk by 66%, high blood pressure raised it by 44%, low HDL cholesterol raised it by 35%, and insulin resistance raised it by 30%. Obesity alone did not increase the risk, but Ninomiya et al. observe that obesity places people at increased risk for the other four conditions.

Ninomiya et al. conclude that metabolic syndrome was significantly associated with myocardial infarction and stroke in both men and women, and assert that "these findings re-affirm the clinical importance of the metabolic syndrome as a significant risk factor for cardiovascular disease and the need to develop strategies for controlling this syndrome and its component conditions."

Metabolic syndrome is also a risk factor for cognitive decline—loss of memory, thinking, and reasoning skills. In "The Metabolic Syndrome, Inflammation, and Risk of Cognitive Decline" (*Journal of the American Medical Association*, vol. 292, no. 18, November 10, 2004), Kristine Yaffe et al. analyze data from 2,632 nondemented (cognitively normal) participants in the Health, Aging, and Body Composition study. The subjects were aged seventy to seventy-nine and were followed from 1997 through 2002. Subjects with metabolic syndrome were more likely than those without to exhibit cognitive decline. Yaffe et al. hypothesize that metabolic syndrome hastened atherosclerosis (a hardening of the walls of the arteries caused by the buildup of fatty deposits on the inner walls of the arteries that interferes with blood flow) or inflammation, which led to cognitive decline.

**REDEFINING THE METABOLIC SYNDROME.** In an effort to standardize diagnosis, prevention, screening, and treatment, the International Diabetes Federation presented in *The IDF Consensus Worldwide Definition of the Metabolic Syndrome* (2006, http://www.idf.org/webdata/docs/IDF_Meta_def_final.pdf) a new worldwide definition of metabolic syndrome. The diagnostic criteria are central obesity, defined as waist equal to or more than 94 centimeters (37 inches) for males and 80 centimeters (31.5 inches) for females of European descent, and ethnic-specific levels for Chinese, Japanese, and South Asians; along with two of the following: triglycerides of at least 150 mg/dL; low HDL-cholesterol, defined as less than 40 mg/dL in males and less than 50 mg/dL in females; blood pressure of at least 130/85 mmHg; fasting hyperglycemia, defined as glucose equal to or greater than 100 mg/dL; previous diagnosis of diabetes; or impaired glucose tolerance. The new definition of the metabolic syndrome, which includes diabetes or prediabetes, abdominal obesity, unfavorable lipid profile, and hypertension, triples the risk of myocardial infarction (heart attack) and stroke and doubles mortality from these conditions. It also increases the risk of developing Type 2 diabetes, if not already present, fivefold. Using this definition, Earl S. Ford calculates in "Prevalence of the Metabolic Syndrome Defined by the International Diabetes Federation among Adults in the U.S." (*Diabetes Care*, vol. 28, no. 11, 2005) that about one-quarter of the U.S. adult population may be diagnosed as having the metabolic syndrome.

**SOME QUESTION THE DIAGNOSIS OF METABOLIC SYNDROME.** Two leading diabetes organizations, the American Diabetes Association and the European Association for the Study of Diabetes, question the utility of the diagnosis of metabolic syndrome. Representatives of these organizations say they feel the syndrome is neither a distinct disease nor well established by scientific research. In "The Metabolic Syndrome: Time for a Critical Appraisal" (*Diabetes Care*, vol. 28, no. 9, 2005), Richard Kahn et al. state, "While there is no question that certain CVD [cardiovascular disease] risk factors are prone to cluster, we found that the metabolic syndrome has been imprecisely defined, there is a lack of certainty regarding its pathogenesis, and there is considerable doubt regarding its value as a CVD risk marker. Our analysis indicates that too much critically important information is missing to warrant its designation as a 'syndrome.'"

Kahn et al. advise physicians against classifying metabolic syndrome as a disease. Instead, they encourage them to screen and treat high triglyceride levels, high blood pressure, low levels of HDL cholesterol, and high blood glucose as separate conditions to reduce the risk of heart disease. Their specific recommendations are that:

1. Adults with any major cardiovascular disease risk factor should be evaluated for the presence of other cardiovascular disease risk factors.

2. Patients with cardiovascular disease risk variables above the cut point for normal should receive counseling for lifestyle modification, and at cut points indicative of frank disease, treatment should correspond to established guidelines.

3. Providers should avoid labeling patients with the term *metabolic syndrome*, as this might create the impression that the metabolic syndrome denotes a greater risk than its components, that it is more serious than other cardiovascular disease risk factors, or that the underlying pathophysiology is clear.

4. All cardiovascular disease risk factors should be individually and aggressively treated.

5. Until randomized controlled trials have been completed, there is no appropriate pharmacological treatment for the metabolic syndrome, nor should it be assumed that pharmacological therapy to reduce insulin resistance will be beneficial to patients with the metabolic syndrome.

# CHAPTER 3

# THE INFLUENCES OF MENTAL HEALTH AND CULTURE ON WEIGHT AND EATING DISORDERS

That diet and appetite are closely linked to psychological health and emotional well-being is widely recognized. Psychological factors often influence eating habits. Many people overeat when they are bored, stressed, angry, depressed, or anxious. Psychological distress can aggravate weight problems by triggering impulses to overeat. Emotional discomfort drives many people to overeat as a way to relieve anxiety and improve mood. Some people revert to the "comfort foods of their youth"—the meals or treats offered to them when they were sick or foods that evoke memories of the carefree days of childhood. Others rely on chocolate and other sweets, which actually contain chemicals known to have a soothing effect on mood. Over time, the associations between emotions, food, and eating can become firmly fixed.

Emotional arousal may also sabotage healthy self-care efforts such as resolutions to diet and exercise. Anxiety and depression can produce feelings of helplessness and hopelessness about efforts to lose weight that undermine the best intentions, prompt detrimental food choices and inactivity, and over time cause many people to give up trying entirely. Because overweight and obesity often contribute to emotional stress and psychological disorders, a cycle develops that couples increasing weight gain with progressively more severe emotional difficulties.

Emotional disturbance alone is rarely the causative factor of overweight or obesity. However, for people with a genetic susceptibility or predisposition to obesity and exposure to environmental factors that promote obesity, emotional and psychological stress can trigger or exacerbate the problem. Even efforts to lose weight can backfire—serving to increase rather than alleviate emotional stress. For example, people who fail to lose weight or those who succeed in losing weight only to regain it may suffer from frustration and diminished feelings of competence and self-worth. Similarly, being overweight or obese and feeling self-conscious about it or suffering from weight-based discrimination or prejudice can be ongoing sources of stress and frustration. Feelings of helplessness, frustration, and continuous emotional stress can cause or worsen mental health problems such as anxiety and depression.

Many mental health and medical professionals view overweight as both a cause and consequence of disturbances in physical and mental health. Even though it may be important to determine whether a metabolic disturbance caused an individual to become overweight or resulted from excessive weight gain, or whether depression triggered behaviors leading to obesity or resulted from problems associated with obesity, it is often impossible to distinguish whether overweight is a symptom of another disorder or the causative factor.

## THE ORIGINS OF EATING DISORDERS

Despite the challenges of compromised self-esteem and societal prejudice, the National Institute of Diabetes and Digestive and Kidney Diseases indicates that most overweight people have about the same number of psychological problems as people of average weight. However, the Weight-control Information Network (WIN), in "Binge Eating Disorder" (September 2004, http://win.niddk.nih.gov/publications/binge.htm), states that an estimated 10% to 15% of people who are mildly obese and try to lose weight repetitively suffer from eating disorders such as binge eating, and those with the most severe eating disorders are more likely to have symptoms of depression and low self-esteem. Binge eaters have lost control of their eating behaviors and consume abnormal quantities of food in short periods of time. Binge-eating disorders are thought to be even more common in people who are severely obese.

Even though depression and stress may contribute to a substantial percent of cases of obesity, they are considered the leading causes of eating disorders. Most mental health professionals concur that the origins of eating disorders can be traced to behavioral or psychological difficulties. Anger and impulsive behavior have been associated with binge-eating disorders, but even mild mental health or social problems such as shyness or lack of self-confidence can lead to social withdrawal, isolation, and a sedentary lifestyle that promotes weight gain and ultimately obesity. James I. Hudson et al. explain in "The Prevalence and Correlates of Eating Disorders in the National Comorbidity Survey Replication" (*Biological Psychiatry*, vol. 61, 2007) that eating disorders frequently coexist with other mental disorders, including depression, substance abuse, and anxiety disorders.

At first glance, eating disorders appear to center on preoccupations with food and weight; however, mental health professionals believe these disorders are often about more than simply food. Besides psychological factors that may predispose people to eating disorders, including diminished self-esteem, depression, anxiety, loneliness, or feelings of lack of control, a variety of interpersonal and social factors have been implicated as causal factors for these disorders. Interpersonal issues that may increase the risk for developing eating disorders include troubled family and personal relationships; difficulty expressing emotions; a history of physical or sexual abuse; or the experience of being teased, taunted, or ridiculed about body size, shape, or weight. Social factors that may contribute to eating disorders include sharply restricted, rigid definitions of beauty that exclude people who do not conform to a particular body weight and shape; cultures that glorify thinness and overemphasize the importance of obtaining a "perfect body"; and cultures that judge and value people based on external physical appearance rather than on internal qualities such as character, intellect, generosity, and kindness. Appearance-driven concerns, rather than health needs, continue to motivate many obese individuals to lose weight. Societal pressures reinforce these appearance-driven concerns by portraying obese individuals in a negative manner.

A related consideration that further complicates pinpointing the origins of eating disorders is the extent to which temperament interacts with interpersonal and social factors to promote eating disorders. Researchers and mental health professionals observe that temperamental tendencies such as perfectionism, compulsivity, impulsivity, and other behavioral, cognitive, and emotional leanings seem to predispose to eating disorders.

**Binge-Eating Disorders**

Binge eating is a common problem among people who are overweight and obese. Besides consuming unusually large amounts of food in a single sitting, binge eaters generally suffer from low mood and low alertness, and experience uncontrollable compulsions to eat. They experience food cravings before binge episodes and feelings of discontent, dissatisfaction, and restlessness following binges.

In "Binge Eating Disorder," the WIN describes binge-eating disorder as the most common eating disorder, affecting approximately 2% of American adults, or about four million people in the United States. Melissa Spearing of the National Institute of Mental Health (NIMH) estimates in *Eating Disorders: Facts about Eating Disorders and the Search for Solutions* (2001) that between 2% and 5% of Americans experience binge-eating disorder in a six-month period. In the *BodyWise Handbook*, October 2005, http://www.womenshealth.gov/bodyimage/kids/bodywise/), the National Women's Health Information Center cites research suggesting that binge-eating disorder is the most common eating disorder, affecting about one-third of obese participants in weight-loss programs. Steven Reinberg reports in "Binge Eating Tops Other Eating Disorders: Survey" (*HealthDay*, February 1, 2007) that the first national survey on eating disorders named binge eating the most common eating disorder affecting Americans, afflicting 3.5% of women and 2% of men.

Even though the disorder is more common in people who are severely obese, normal-weight people also develop the disorder. People who suffer from binge eating often:

- Feel that eating is out of their ability to control

- Eat amounts of food most people would think are unusually large

- Eat much more quickly than usual during binge episodes

- Eat until the point of physical discomfort

- Consume large amounts of food, even when they are not hungry

- Eat alone because they feel embarrassed about the amount of food they eat

- Feel disgusted, depressed, or guilty after overeating

In an effort to identify the risk factors for binge-eating disorder, Ruth H. Striegel-Moore et al. of Wesleyan University compared women diagnosed with binge-eating disorder to those with no history of an eating disorder and reported their findings in "Toward an Understanding of Risk Factors for Binge-Eating Disorder in Black and White Women: A Community-Based Case-Control Study" (*Psychological Medicine*, vol. 35, no. 6, 2005). They conclude that childhood obesity and the presence of eating problems among other family members are reliable, specific risk factors for binge-eating disorder. Subjects with binge-eating disorder also reported more family discord and felt they had more parental demands placed on them than the subjects with no history of an eating disorder.

## TABLE 3.1

### Health consequences of eating disorders

- Eating disorders are serious, potentially life-threatening conditions that affect a emotional and physical health.
- Eating disorders are not just a "fad" or a "phase." People do not just "catch" an eating disorder for a period of time. They are real, complex, and devastating conditions that can have serious consequences for health, productivity, and relationships.
- People struggling with an eating disorder need to seek professional help. The earlier a person with an eating disorder seeks treatment, the greater the likelihood of physical and emotional recovery.

**Health consequences of anorexia nervosa:** In anorexia nervosa's cycle of self-starvation, the body is denied the essential nutrients it needs to function normally. Thus, the body is forced to slow down all of its processes to conserve energy, resulting in serious medical consequences:
  – Abnormally slow heart rate and low blood pressure, which mean that the heart muscle is changing. The risk for heart failure rises as the heart rate and blood pressure level sink lower and lower.
  – Reduction of bone density (osteoporosis), which results in dry, brittle bones.
  – Muscle loss and weakness.
  – Severe dehydration, which can result in kidney failure.
  – Fainting, fatigue, and overall weakness.
  – Dry hair and skin; hair loss is common.
  – Growth of a downy layer of hair called lanugo all over the body, including the face, in an effort to keep the body warm.

**Health consequences of bulimia nervosa:** The recurrent binge-and-purge cycles of bulimia can affect the entire digestive system and can lead to electrolyte and chemical imbalances in the body that affect the heart and other major organ functions. Some of the health consequences of bulimia nervosa include:
  – Electrolyte imbalances that can lead to irregular heartbeats and possibly heart failure and death. Electrolyte imbalance is caused by dehydration and loss of potassium and sodium from the body as a result of purging behaviors.
  – Potential for gastric rupture during periods of bingeing.
  – Inflammation and possible rupture of the esophagus from frequent vomiting.
  – Tooth decay and staining from stomach acids released during frequent vomiting.
  – Chronic irregular bowel movements and constipation as a result of laxative abuse.
  – Peptic ulcers and pancreatitis.

**Health consequences of binge eating disorder:** Binge eating disorder often results in many of the same health risks associated with clinical obesity. Some of the potential health consequences of binge eating disorder include:
  – High blood pressure.
  – High cholesterol levels.
  – Heart disease as a result of elevated triglyceride levels.
  – Secondary diabetes.
  – Gallbladder disease.

SOURCE: "Health Consequences of Eating Disorders," National Eating Disorders Association, 2002, http://www.nationaleatingdisorders.org/nedaDir/files/documents/handouts/HlthCnsq. pdf (accessed October 9, 2007)

## FIGURE 3.1

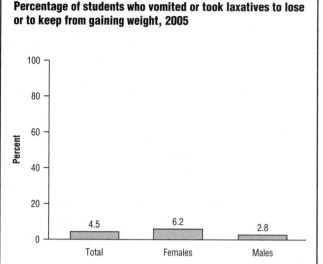

### Percentage of students who vomited or took laxatives to lose or to keep from gaining weight, 2005

SOURCE: "Percentage of Students Who Vomited or Took Laxatives to Lose Weight or to Keep from Gaining Weight during the Past 30 Days," in *YRBSS Youth Online: Comprehensive Results*, Centers for Disease Control and Prevention, National Center for Chronic Disease Prevention and Health Promotion, Division of Adolescent and School Health, March 19, 2007, http://apps.nccd.cdc.gov/yrbss/displaygraphV.asp?path=byHT&colval=2005&rowval1=Sex&rowval2=None&compval=&Graphval=yes&cat=5&loc=XX&year=2005&quest=Q70&Byvar=Q2&ByResvar=CI (accessed October 11, 2007)

According to the NIMH, dieting plays a role in the onset of two serious eating disorders: anorexia nervosa and bulimia. Preteens, teens, and college-age women are at special risk. In fact, the National Women's Health Resource Center states in "Eating Disorders" (March 8, 2007, http://www.healthywomen.org/healthtopics/eating disorders) that more than 90% of those who develop an eating disorder are young women between the ages of twelve and twenty-five, although researchers are beginning to report rising rates of anorexia and bulimia among men. Studies suggest that for every ten women with an eating disorder, one male is afflicted. According to the NIMH, in *Eating Disorders* (2007, http://www.nimh.nih.gov/health/publications/eating-disorders/nimheatingdisorders.pdf), 5% to 15% of those with anorexia or bulimia and 35% of people with binge-eating disorders are male.

Hudson et al. report that of the 2,980 adults they surveyed about eating disorders, 0.9% of women and 0.3% of men suffered from anorexia and 1.5% of women and 0.5% of men had bulimia. A 2005 survey of high school students found that 6.2% of teenaged girls and 2.8% of teenaged boys reported behaviors that may be symptoms of eating disorders such as vomiting or taking laxatives to lose weight or keep from gaining weight. (See Figure 3.1.)

### Anorexia Nervosa

Anorexia nervosa involves severe weight loss—a minimum of 15% below normal body weight. Anorexic

## Some Dieters Are Consumed by Eating Disorders

In the twenty-first century, American society is preoccupied with body image. Americans are constantly bombarded with images of thin, beautiful young women and lean, muscular men in magazines, on billboards, on the Internet, on television, and in movies. Advertising implies that to be thin and beautiful is to be happy. Many prominent weight-loss programs reinforce this suggestion. Well-balanced, low-fat food plans, or other diets that restrict carbohydrates or calories combined with exercise can help many overweight people achieve a healthier weight and lifestyle. Dieting to achieve a healthy weight is quite different from dieting obsessively to become "model" thin, which can have consequences ranging from mildly harmful to life-threatening. Table 3.1 enumerates the health consequences of eating disorders.

people literally starve themselves, even though they may be very hungry. For reasons that researchers do not yet fully understand, anorexics become terrified of gaining weight. Both food and weight become obsessions. They often develop strange eating habits, refuse to eat with other people, and exercise strenuously to burn calories and prevent weight gain. Anorexic individuals continue to believe they are overweight even when they are dangerously thin.

This condition often begins when a young woman who is slightly overweight or normal weight starts to diet to lose weight. After achieving the desired weight loss, she redoubles her efforts to lose weight, and dieting becomes an obsession that may eclipse other interests. Affected individuals take pleasure in how well they can avoid food consumption and measure their self-worth by their ability to lose weight. Eating and weight gain are perceived as weaknesses and personal failures.

The medical complications of anorexia are similar to starvation. When the body attempts to protect its most vital organs, the heart and the brain, it goes into "slow gear." Menstrual periods stop, and breathing, pulse, blood pressure, and thyroid function slow down. The nails and hair become brittle, the skin dries, and the lack of body fat produces an inability to withstand cold temperatures. Depression, weakness, and a constant obsession with food are also symptoms of the disease. In addition, personality changes may occur. The person suffering from anorexia may have outbursts of anger and hostility or may withdraw socially. In the most serious cases, death can result.

Scientists often describe eating disorders as addictions, and Alexandra Jean et al. support this notion in "Anorexia Induced by Activation of Serotonin 5-HT4 Receptors Is Mediated by Increases in CART in the Nucleus Accumbens" (*Proceedings of the National Academy of Sciences*, vol. 104, no, 40, October 3, 2007). They find that anorexia and MDMA, a psychoactive drug known as ecstasy, share a common signaling pathway in the brain. Jean et al. also note that by stimulating serotonin 5-HT4 receptors, which are involved in addictive behavior, they could decrease the amount of food mice consumed and diminish the urge to eat in food-deprived mice.

### Bulimia

People who suffer from bulimia eat compulsively and then purge (get rid of the food) through self-induced vomiting, use of laxatives, diuretics, strict diets, fasts, exercise, or a combination of several of these compensatory behaviors. Bulimia often begins when a person is disgusted with the excessive amount of "bad" food consumed and vomits to rid the body of the calories.

Many bulimics are at a normal body weight or above because of their frequent binge-purge behavior, which can occur from once or twice a week to several times per day. Those bulimics who maintain normal weight may manage to keep their eating disorder a secret for years. As with anorexia, bulimia usually begins during adolescence, but many bulimics do not seek help until they are in their thirties or forties.

Binge eating and purging is dangerous. In rare cases, bingeing can cause esophageal ruptures, and purging can result in life-threatening cardiac (heart) conditions because the body loses vital minerals. The acid in vomit wears down tooth enamel and the stomach lining and can cause scarring on the hands when fingers are pushed down the throat to induce vomiting. The esophagus may become inflamed, and glands in the neck may become swollen.

Bulimics often talk of being "hooked" on certain foods and needing to feed their "habits." This addictive behavior carries over into other areas of their lives, including the likelihood of alcohol and drug abuse. Many bulimic people suffer from coexisting medical or mental health problems, such as severe depression, which increases their risk of committing suicide.

## CAUSES OF EATING DISORDERS

Evidence suggests a genetic component to susceptibility to eating disorders. For example, in the general population the chance of developing anorexia is about one out of two hundred, but when a family member has the disorder, the risk increases to one out of thirty. Twin studies demonstrate that when one twin is affected, there is a 50% chance the other will develop an eating disorder. Tom Vink et al. examined the deoxyribonucleic acid of 145 anorexia patients and reported their findings in "Association between an Agouti-Related Protein Gene Polymorphism and Anorexia Nervosa" (*Molecular Psychiatry*, vol. 6, May 2001). They find that 11% carried the same genetic mutation. The mutation was of a gene that manufactures agouti-related protein, which stimulates the desire to eat. Vink et al. hypothesize that a deficiency of the protein may be involved in anorexia.

Kelly L. Klump et al. of Michigan State University indicate in "Puberty Moderates Genetic Influences on Disordered Eating" (*Psychological Medicine*, vol. 37, no. 5, May 2007), a study of 510 fourteen-year-old twins, that during puberty there is an increased risk of developing eating disorders. The researchers find that before the onset of puberty, environmental factors made the strongest contribution to the risk of developing eating disorders; however, during and after puberty, genetic factors predominated, accounting for more than half of the risk. These findings suggest that puberty influences the expression of genes for disordered eating.

Besides a genetic predisposition, bulimics and anorexics seem to have different temperaments. Bulimics are likely to be impulsive (acting without thought of the consequences) and are more likely to abuse alcohol and drugs. Anorexics tend to be perfectionists, good students, and competitive athletes. They usually keep their feelings to themselves and rarely disobey their parents. Bulimics and anorexics share certain traits: they lack self-esteem, have feelings of helplessness, and fear gaining weight. In both disorders, the eating problems appear to develop as a way of handling stress and anxiety.

Bulimics consume huge amounts of food (often junk food) in a search for comfort and stress relief. The binge-ing, however, brings only guilt and depression. By contrast, anorexics restrict food to gain a sense of control and mastery over some aspect of their life. Controlling their weight seems to offer two advantages: they can take control of their body, and they can gain approval from others.

Psychological theories that explain the origins of bulimia include conflicted relationships between mothers and daughters, attempts to control one's own body in the face of seemingly uncontrollable family or other inter-personal relationships, or ambivalence about sexual development and attention. The latter theory has also been used to explain overweight and obesity in teenaged girls and young women—as protection from or defense against attention from males that may make them fearful or uncomfortable.

## OCCURRENCE OF EATING DISORDERS

Maria Makino, Koji Tsuboi, and Lorraine Denner-stein note in "Prevalence of Eating Disorders: A Com-parison of Western and Non-Western Countries" (*Med-scape General Medicine*, vol. 6, no. 3, 2004) that "the prevalence of eating disorders in non-Western countries is lower than that of the Western countries but appears to be increasing." Thinness is not necessarily admired among all people throughout the world, especially in countries where hunger is not a matter of choice.

The NIMH states in "The Numbers Count: Mental Disorders in America" (January 9, 2008, http://www.nimh .nih.gov/health/publications/the-numbers-count-mental-disorders-in-america.shtml#Eating) that in their lifetime, an estimated 0.5% to 3.7% of women suffer from ano-rexia, and 1.1% to 4.2% suffer from bulimia. The National Women's Health Information Center observes that eating disorders often coexist with other high-risk health behaviors such as tobacco, alcohol and drug use, delinquency, unprotected sexual activity, and suicide attempts.

In "Pathological Dieting and Alcohol Use in College Women—A Continuum of Behaviors" (*Eating Behav-iors*, vol. 6, no. 1, January 2005), Dean D. Krahn et al. examined the relationship between dieting, binge eating disorder, and alcohol use in female college students.

They find a relationship between dieting and bingeing severity and the frequency, intensity, and negative con-sequences of alcohol use in the students. Dieting and bingeing were more closely associated with alcohol use than were factors such as depression, age at which drink-ing began, or parents' drinking history. Furthermore, the severity of the eating disorder behavior was linked to the occurrence of negative consequences of alcohol use, including blackouts and unintended sexual activity. Krahn et al. conclude that destructive eating behaviors are often associated with harmful alcohol use.

According to the National Eating Disorders Associa-tion (NEDA), in *Statistics: Eating Disorders and Their Precursors* (2006, http://www.nationaleatingdisorders .org/p.asp?WebPage_ID=286&Profile_ID=41138), con-servative estimates of the prevalence of eating disorders in the United States project that as many as ten million women and one million men are affected. An estimated 35% of normal dieters progress to the pathological, extreme dieting that is a precursor of eating disorders. The Eating Disorders Coalition for Research, Policy, and Action reports in "Statistics and Study Findings" (2008, http://www.eatingdisorderscoalition.org/reports/statistics .html) that the incidence of eating disorders is increasing in younger age groups and is becoming more common in diverse ethnic and sociocultural groups.

## TREATMENT OF EATING DISORDERS

Generally, a physician treats the medical complica-tions of the disorder, whereas a nutritionist advises the affected individual about specific diet and eating plans. To help the person with an eating disorder face his or her underlying problems and emotional issues, psychother-apy is usually necessary. Sometimes the challenge is to convince people with eating disorders to seek and obtain treatment; other times it is difficult to gain their adher-ence to treatment. Many anorexics deny their illness, and getting and keeping anorexic patients in treatment can be difficult. Treating bulimia is also difficult. Many bulim-ics are easily frustrated and want to leave treatment if their symptoms are not quickly relieved.

Several approaches are used to treat eating disorders. Cognitive-behavioral therapy (CBT) teaches people how to monitor their eating and change unhealthy eating hab-its. It also teaches them how to change the way they respond in stressful situations. CBT is based on the prem-ise that thinking influences emotions and behavior—that feelings and actions originate with thoughts. CBT posits that it is possible to change the way people feel and act even if their circumstances do not change. It teaches the advantages of feeling calm when faced with undesirable situations. CBT clients learn that they will confront unde-sirable events and circumstances whether they become troubled about them or not. When they are troubled about events or circumstances, they have two problems: the troubling event or circumstance, and the troubling feelings about the event or circumstance. Clients learn that when they do not become troubled about trying events and circumstances, they can reduce the number of problems they face by half.

Interpersonal psychotherapy (IPT) helps people look at their relationships with friends and family and make changes to resolve problems. IPT is short-term therapy that has demonstrated effectiveness for the treatment of depression. According to the International Society for Interpersonal Psychotherapy, IPT emphasizes that mental health and emotional problems occur within an interpersonal context. For this reason, the therapy aims to intervene specifically in social functioning to relieve symptoms.

Group therapy has been found helpful for bulimics, who are relieved to find that they are not alone or unique in their eating behaviors. A combination of behavioral therapy and family systems therapy is often the most effective with anorexics. Family systems therapy considers the family as the unit of treatment and focuses on relationships and communication patterns within the family rather than on the personality traits or symptoms displayed by individual family members. Problems are addressed by modifying the system rather than by trying to change an individual family member. People with eating disorders who also suffer from depression may benefit from antidepressant and antianxiety medications to help relieve coexisting mental health problems.

Recovery from eating disorders is uneven. The Eating Disorders Coalition for Research, Policy, and Action characterizes recovery as a process that frequently entails multiple rehospitalizations, limited ability to work or attend school, and limited capacity for interpersonal relationships. In "Natural Course of Bulimia Nervosa and of Eating Disorder Not Otherwise Specified: 5-Year Prospective Study of Remissions, Relapses, and the Effects of Personality Disorder Psychopathology" (*Journal of Clinical Psychiatry*, vol. 68, no. 5, May 2007), Carlos M. Grilo et al. report that about one-third of sufferers recover after an initial episode and treatment, another third fluctuate between recovery and relapse, and the remaining one-third suffer chronic decline and deterioration.

In part, eating disorders are difficult to treat effectively because many sufferers resist entering treatment and/or fail to complete treatment programs. Cecilia Bergh et al. recognize in "Randomized Controlled Trial of a Treatment for Anorexia and Bulimia Nervosa" (*PNAS*, vol. 99, no. 14, July 9, 2002) that even though hospital treatment of people with anorexia is often successful, 30% to 50% of patients suffer relapses within the first year. In "Eating Behavior among Women with Anorexia Nervosa" (*American Journal of Clinical Nutrition*, vol. 82, no. 2, August 2005), Robin Sysko et al. of Rutgers University sought to find out whether current treatment for anorexia successfully addresses severe caloric restriction and other characteristic features of anorexia nervosa. To do this, they scrutinized eating behavior among people with anorexia nervosa before and immediately after treatment that restored their weight and compared these behaviors to those of control subjects.

Sysko et al. observed twelve anorexic patients and twelve individuals without eating disorders who were asked to consume a strawberry yogurt shake, which they were told would be their lunch for the day. They were also told to consume as much as they wanted. The yogurt shake was in an opaque container and was drunk with a straw so that the subjects could not see the shake. They were also not told the contents of the shake or how many calories it contained. The anorexic patients were tested when they were admitted for treatment, and retested after they had reached 90% of their ideal body weight.

Before treatment, anorexic patients consumed an average of 3.8 ounces of the shake, which increased to an average of 6.3 ounces after treatment. However, in both instances control subjects consumed significantly more than anorexic patients, at an average of 17.3 ounces. The researchers observed that subjects with anorexia found the experiment difficult and anxiety provoking because they were unable to see the shake and control their calorie intake. This was despite the fact that subjects treated for anorexia displayed significant decreases in psychological and eating-disordered symptoms after they had regained weight. Sysko et al. believe their findings underscore the need to intervene with anorexics who have left an intensive treatment program. They hope to devise strategies to help normalize patients' eating behavior outside the hospital, for example by helping reduce their anxiety and fear about eating unknown quantities of food.

## New Directions in Research and Treatment

The NIMH, in *Eating Disorders: Facts about Eating Disorders and the Search for Solutions*, reports that the results of its research are aiding both the understanding of eating disorders and their treatment. Research on intervening in the binge-eating cycle demonstrates that initiating structured patterns of eating enables people with eating disorders to experience less hunger, less deprivation, and fewer negative feelings about food and eating. When the two key predictors of bingeing—hunger and negative feelings—are reduced, the frequency of binges declines.

Continued study of the human genome promises the identification of susceptibility genes (genes that indicate an individual's increased risk for developing eating disorders) that will help develop more effective treatments for these disorders. Other research is investigating the relationship between brain functions and emotional and social behavior related to eating disorders and the role of the brain in feeding behavior. Scientists have learned that both appetite and energy expenditure are regulated by a highly complex network of nerve cells and intercellular messengers called neuropeptides. The role of sex hormones, known as gonadal steroids, in the development of eating disorders is suggested by gender and the onset of puberty as a risk for these disorders. These discoveries provide insight into the biochemical mechanisms of eating disorders and offer potential direction for the development of new drugs and treatments for these disorders.

In "A Review and Primer of Molecular Genetic Studies of Anorexia Nervosa" (*International Journal of Eating Disorders*, vol. 37, supplement, July 2005), Kelly L. Klump and Kyle L. Gobrogge summarize recent findings about the genetic underpinnings of eating disorders. They report that research reveals some role for the brain system that involves the chemical serotonin in the development of anorexia nervosa; serotonin is a neurotransmitter involved in the regulation of mood and certain mental disorders, such as depression and anxiety. Genomic regions on chromosomes 1 and 10 are likely to harbor susceptibility genes for anorexia as well as other eating disorders. The findings from these genetic studies support those of neurobiologic studies indicating that alterations in serotonin functioning may contribute to the development of eating disorders.

## PREVENTING EATING DISORDERS

Conventional public health definitions describe primary prevention as the prevention of new cases and secondary prevention as the prevention of recurrence of a disease or prevention of its progression. Primary prevention measures fall into two categories: actions to protect against disease and disability and actions to promote health such as good nutrition and hygiene, adequate exercise and rest, and avoidance of environmental and health risks. Health promotion also includes education about other interdependent dimensions of health known as wellness. Examples of health promotion programs aimed at preventing eating disorders include programs to enhance self-esteem, nutrition education classes, and programs to support children and teens to resist unhealthy pressures to conform to unrealistic body weight.

Secondary prevention programs are intended to identify and detect disease in its earliest stages, when it is most likely to be successfully treated. With early detection and diagnosis, it may be possible to cure the disease, slow its progression, prevent or minimize complications, and limit disability. Secondary prevention of eating disorders includes efforts to identify affected individuals to intervene early and prevent the development of serious and potentially life-threatening consequences.

Tertiary prevention programs aim to improve the quality of life for people with various diseases by limiting

complications and disabilities, reducing the severity and progression of the disease, and providing rehabilitation (therapy to restore function and self-sufficiency). Unlike primary and secondary prevention, tertiary prevention involves actual treatment for the disease, and in the case of eating disorders it is conducted primarily by medical and mental-health practitioners rather than by public health or social service agencies. An example of tertiary prevention is a program that monitors people with eating disorders to ensure that they maintain appropriate body weight and adhere to healthy diets and other prescribed medication or treatment. Because the treatment of eating disorders is not always effective or lasting, many health professionals contend that initiatives directed at controlling or eliminating the disorders by treating each affected individual or by training enough professionals as interventionists are ill advised. Instead, they advocate redirecting time, energy, and resources to primary and secondary prevention efforts.

Table 3.2 lists the basic principles for the prevention of eating disorders prepared by the NEDA. These principles underscore the complexity of addressing the problem and the need for comprehensive, community-wide prevention programs that address the social and cultural issues promoting the rise of these disorders. The NEDA also urges parents to spearhead efforts to prevent eating disorders by practicing positive, healthy attitudes and behaviors and encouraging children to resist media stereotypes about body shape and weight. Furthermore, it outlines the philosophies and actions parents can adopt and the behaviors they can model to help their children cultivate healthy attitudes about food, eating, exercise, and body weight. Table 3.3 outlines steps parents can take to help prevent eating disorders.

### Changing Social and Cultural Norms

Cultural idealization of thinness as a standard of female beauty and worth and the societal acceptance of dieting as a female ritual have been widely cited as sociocultural causes of eating disorders. The widespread misperception that the body is readily reshaped and that one can, and should, strive to change its size and form to correspond with aesthetic preferences also contributes to distorted perceptions and unrealistic expectations.

Media images that create, reflect, communicate, and reinforce cultural definitions of attractiveness, especially female beauty, are often acknowledged as factors that contribute to the rise of eating disorders. They exert powerful influences on values, attitudes, and practices for body image, diet, and activity. The role of the media, in conjunction with the fashion and entertainment industries, especially those targeting women and girls, in promoting unrealistic standards of female beauty and unhealthy eating habits has been named as a causative

## TABLE 3.2

**Eating disorders prevention**

**What is eating disorders prevention?**

Prevention is any systematic attempt to change the circumstances that promote, initiate, sustain, or intensify problems like eating disorders.

- **Primary prevention** refers to programs or efforts that are designed to prevent the occurrence of eating disorders before they begin. Primary prevention is intended to help promote healthy development.
- **Secondary prevention** (sometimes called "targeted prevention") refers to programs or efforts that are designed to promote the early identification of an eating disorder—to recognize and treat an eating disorder before it spirals out of control. The earlier an eating disorder is discovered and addressed, the better the chance for recovery.

**Basic principles for the prevention of eating disorders**

1. Eating disorders are serious and complex problems. We need to be careful to avoid thinking of them in simplistic terms, like "anorexia is just a plea for attention," or "bulimia is just an addiction to food." Eating disorders arise from a variety of physical, emotional, social, and familial issues, all of which need to be addressed for effective prevention and treatment.
2. Eating disorders are not just a "woman's problem" or "something for the girls." Males who are preoccupied with shape and weight can also develop eating disorders as well as dangerous shape control practices like steroid use. In addition, males play an important role in prevention. The objectification and other forms of mistreatment of women by others contribute directly to two underlying features of an eating disorder: obsession with appearance and shame about one's body.
3. Prevention efforts will fail, or worse, inadvertently encourage disordered eating, if they concentrate solely on warning the public about the signs, symptoms, and dangers of eating disorders. Effective prevention programs must also address:
    - Our cultural obsession with slenderness as a physical, psychological, and moral issue.
    - The roles of men and women in our society.
    - The development of people's self-esteem and self-respect in a variety of areas (school, work, community service, hobbies) that transcend physical appearance.
4. Whenever possible, prevention programs for schools, community organizations, etc., should be coordinated with opportunities for participants to speak confidentially with a trained professional with expertise in the field of eating disorders, and, when appropriate, receive referrals to sources of competent, specialized care.

SOURCE: Michael Levine and Margo Maine, "Eating Disorders Can Be Prevented!" National Eating Disorders Association, 2006, http://www.nationaleatingdisorders.org/p.asp?WebPage_ID=286&Profile_ID=41169 (accessed October 9, 2007)

## TABLE 3.3

**Ten things parents can do to prevent eating disorders**

1. Consider your thoughts, attitudes, and behaviors toward your own body and the way that these beliefs have been shaped by the forces of weightism and sexism. Then educate your children about
   (a) the genetic basis for the natural diversity of human body shapes and sizes, and
   (b) the nature and ugliness of prejudice.
   - Make an effort to maintain positive, healthy attitudes & behaviors. Children learn from the things you say and do!
2. Examine closely your dreams and goals for your children and other loved ones. Are you overemphasizing beauty and body shape, particularly for girls?
   - Avoid conveying an attitude which says in effect, "I will like you more if you lose weight, don't eat so much, look more like the slender models in ads, fit into smaller clothes, etc."
   - Decide what you can do and what you can stop doing to reduce the teasing, criticism, blaming, staring, etc. that reinforce the idea that larger or fatter is "bad" and smaller or thinner is "good."
3. Learn about and discuss with your sons and daughters (a) the dangers of trying to alter one's body shape through dieting, (b) the value of moderate exercise for health, and (c) the importance of eating a variety of foods in well-balanced meals consumed at least three times a day.
   - Avoid categorizing foods into "good/safe /no-fat or low-fat" vs."bad/dangerous/fattening."
   - Be a good role model in regard to sensible eating, exercise, and self-acceptance.
4. Make a commitment not to avoid activities (such as swimming, sunbathing, dancing, etc.) simply because they call attention to your weight and shape. Refuse to wear clothes that are uncomfortable or that you don't like but wear simply because they divert attention from your weight or shape.
5. Make a commitment to exercise for the joy of feeling your body move and grow stronger, not to purge fat from your body or to compensate for calories eaten.
6. Practice taking people seriously for what they say, feel, and do, not for how slender or "well put together" they appear.
7. Help children appreciate and resist the ways in which television, magazines, and other media distort the true diversity of human body types and imply that a slender body means power, excitement, popularity, or perfection.
8. Educate boys and girls about various forms of prejudice, including weightism, and help them understand their responsibilities for preventing them.
9. Encourage your children to be active and to enjoy what their bodies can do and feel like. Do not limit their caloric intake unless a physician requests that you do this because of a medical problem.
10. Do whatever you can to promote the self-esteem and self-respect of all of your children in intellectual, athletic, and social endeavors. Give boys and girls the same opportunities and encouragement. Be careful not to suggest that females are less important than males, e.g., by exempting males from housework or child care. A well-rounded sense of self and solid self-esteem are perhaps the best antidotes to dieting and disordered eating.

SOURCE: Michael Levine and Linda Smolak, "Ten Things Parents Can Do to Prevent Eating Disorders," National Eating Disorders Association, 2006, http://www.nationaleatingdisorders.org/p.asp?WebPage_ID=286&Profile_ID=41171 (accessed October 9, 2007)

factor for body dissatisfaction, unhealthy dieting behavior, and the rise of eating disorders.

The NEDA reports in "The Media, Body Image, and Eating Disorders" (2006, http://www.nationaleatingdisorders.org/p.asp?WebPage_ID=286&Profile_ID=41166) that women's magazines contained 10.5 times more advertisements and articles promoting diet and weight loss than were found in men's magazines. It also reports that a study of 4,294 network television commercials revealed that 1 out of every 3.8 commercials conveyed some sort of attractiveness message—advising viewers about qualities that were attractive or unattractive.

Even though media messages portraying thinness as a desirable attribute do not directly cause eating disorders, they help create the context in which people learn to place a value on the size and shape of their body. To the extent that media advertising defines cultural values about that which is beautiful and desirable, the media have potent power over the development of self-esteem

and body image. Even if the media were to present more diverse and realistic images of people, this change would be unlikely to immediately reduce or eliminate eating disorders. However, many observers do believe it would reduce the pressures to conform to one ideal, lessen feelings of body dissatisfaction, and ultimately decrease the potential for eating disorders.

According to many health professionals and media observers, besides promoting unrealistic and unattainable body weight, media coverage of health, nutrition, diet, overweight, and inactivity does not fulfill its potential to educate people about how to make healthful changes in their life. The *Surgeon General's Call to Action to Prevent and Decrease Overweight and Obesity, 2001* (2001, http://www.surgeongeneral.gov/topics/obesity/calltoaction/CalltoAction.pdf), a report that outlines strategies to address the increasing prevalence of overweight and obesity in the United States, also identifies the media as having a key role in prevention efforts. The report recommends a range of proactive interventions intended to educate the public and change Americans' eating behavior and exercise patterns. It takes direct aim at preventing eating disorders by calling for media actions to "promote the recognition of inappropriate weight change" and enumerates the efforts necessary to reorient the media including:

- Communicating to media professionals that the primary concern of overweight and obesity is one of health rather than appearance.

- Informing media professionals about the prevalence of overweight and obesity in low-income and racial and ethnic minority populations and the need for culturally sensitive health messages.

- Communicating the importance of prevention of overweight by paying attention to the caloric intake and being physically active at all ages.

- Building awareness of the social and environmental influences on making healthy decisions regarding diet and physical activity.

- Providing professional education for media professionals on policy areas related to diet and physical activity.

- Emphasizing to media professionals the need to develop uniform health messages about physical activity and nutrition that are consistent with the *Dietary Guidelines for Americans, 2005* (January 2005, http://www.health.gov/dietaryguidelines/dga2005/document/pdf/DGA2005.pdf).

The surgeon general report also describes specific actions the media can take to help Americans change their attitudes and behaviors, including:

- Launching a national campaign to increase public awareness of the health benefits of regular physical activity, healthful dietary choices, and maintaining a healthy weight, based on the *Dietary Guidelines for Americans*.

- Educating consumers about realistic and reasonable goals for weight-loss programs and weight-management products.

- Incorporating messages about proper nutrition, including eating at least five servings of fruits and vegetables per day, and regular physical activity in youth-oriented television programming.

- Training nutrition and exercise scientists and specialists in media advocacy skills that will enable them to disseminate their knowledge to a broad audience.

- Encouraging a balance between advertising campaigns that endorse the consumption of excess calories and inactivity with messages promoting the benefits of healthy diets and exercise.

- Advocating that media celebrities use their influence as role models to demonstrate eating and physical activity lifestyles for health rather than for appearance.

- Encouraging the media to employ actors of diverse sizes.

**ADVERTISING CAMPAIGN EMPHASIZING REALISTIC BODIES DRAWS PRAISE AND CRITICISM.** In June 2005 Dove, a skin and hair care division of the Unilever company, launched the "Campaign for Real Beauty," which featured a purportedly unretouched photo of six smiling women of various sizes and ethnicities posing in plain white underwear to promote a skin-firming cream. The women, who were not models, ranged from a slim size 6 to a curvy size 14, and graced print advertisements and billboards. The campaign generated considerable discussion and debate in the media.

Dove claimed that it developed the campaign in response to the results of its "Real Truth about Beauty" survey, which were published in *"The Real Truth about Beauty: A Global Report": Findings of the Global Study on Women, Beauty, and Well-Being* (September 2004, http://www.campaignforrealbeauty.com/uploadedfiles/dove_white_paper_final.pdf). Conducted by researchers from Harvard University and the London School of Economics, the study interviewed thirty-two hundred women aged eighteen to sixty-four in ten countries. A scant 2% of the women surveyed considered themselves "beautiful" and only 13% were "very satisfied" with their body weight and shape.

According to the article "Dove Ads with 'Real' Women Get Attention" (Associated Press, July 29, 2005), Philippe Harousseau, the Dove marketing director, described the campaign as responsive to "our belief that beauty comes in different shapes, sizes and ages. Our mission is to make more women feel beautiful every day by broadening the definition of beauty." Industry observers wondered whether the company was in fact, broadening the definition of beauty and improving women's body image and self-esteem or simply launching a provocative advertising campaign. Even though the company has not disclosed just how much the advertisements have helped promote its products, it concedes that the campaign has been beneficial for all Dove products, not just the firming creams.

The Associated Press notes that the ads were not, however, universally well received. For example, the *Chicago Sun-Times* columnist Richard Roeper characterized the women as "chunky," which earned him angry letters from about a thousand readers. Some skeptics asserted that even though they endorsed the notion of featuring real women who feel good about their bodies in the ads, they believed the ads sent contradictory messages—promoting a product to reduce the curves the models are flaunting. The most impassioned detractors accused the company of appearing hypocritical because the ads aim to profit from "improving" the same curves the campaign exhorts women to celebrate.

The Dove advertising campaign was still under way in early 2008 and had already prompted some attitudinal change. In 2006 the company launched the Dove Self-Esteem Fund, a program that aims to challenge traditional media definitions of beauty—ultrathin bodies and perfectly symmetrical features—to help girls and women feel better about their looks.

# CHAPTER 4
# DIET, NUTRITION, AND WEIGHT ISSUES
# AMONG CHILDREN AND ADOLESCENTS

One of the most disturbing observations about overweight and obesity in the United States is the epidemic of supersized kids. Inge Lissau et al. find in "Body Mass Index and Overweight in Adolescents in Thirteen European Countries, Israel, and the United States" (*Archives of Pediatrics and Adolescent Medicine*, vol. 158, no. 1, January 2004) that the United States, followed by Greece and Portugal, had the highest percentage of overweight teens. In 2007 many more American children and adolescents were seriously overweight than were overweight just thirty years ago. The most accurate data about the prevalence of overweight among children and adolescents come from the Centers for Disease Control and Prevention's (CDC) 2003–04 National Health and Nutrition Examination Survey (NHANES), which finds that the percentage of overweight children aged six to eleven more than quadrupled (from 4% to 18.8%) and the percentage of teens aged twelve to nineteen nearly tripled (from 6.1% to 17.4%) between 1971–74 and 2003–04. (See Figure 4.1.)

With children and teens as well as adults, body mass index (BMI) is used to determine underweight, overweight, and at risk for overweight. Children's body fatness changes over the years as they grow, and girls and boys differ in their body fatness as they mature. In light of these differences, the BMI for children (also referred to as BMI-for-age) is gender and age specific. For example, Figure 4.2 shows BMI percentiles for boys aged two to twenty and demonstrates how different BMI numbers are interpreted for a ten-year-old boy. Figure 4.3 shows that children of different ages (and genders) may have the same BMI number, but that number will fall in a different percentile for each child, classifying the ten-year-old boy as overweight and the fifteen-year-old as at a healthy weight.

Overweight is defined as at or above the age- and gender-specific ninety-fifth percentile on the BMI. Still, even children at the eighty-fifth percentile are considered at risk for overweight- and obesity-induced illness and overweight throughout their adult lives. (See Table 4.1.)

Overweight children are much more likely to become overweight adults—William H Dietz indicates in "'Adiposity Rebound': Reality or Epiphenomenon?" (*The Lancet*, vol. 356, 2000) that an estimated 30% of adult obesity begins in childhood—unless they adopt and maintain healthier patterns of eating and exercise. The prevalence of overweight among adolescents is of particular concern because overweight adolescents are at even greater risk than overweight children of becoming overweight adults.

Like adults, children and adolescents are eating more than ever and exercising less. Even though the link between obesity and disease in adolescence is weaker than it is for obese adults, teens who are overweight are at a high risk of health problems later in life. Furthermore, in "Even Children Have Heart Disease—Especially Those Who Are Overweight" (*Medscape Cardiology*, vol. 8, no. 1, 2004), Victoria Porter explains that 50% to 80% of obese teens become obese adults. Type 2 diabetes, high blood lipid levels, and hypertension (high blood pressure) occur with increased frequency among overweight youth. Overweight children and teens are also at risk for psychosocial problems ranging from teasing and ostracism to social isolation and discrimination.

## PREVALENCE OF OVERWEIGHT TEENS BY RACE AND ETHNICITY

The 2003–04 NHANES finds that nearly one-fifth of white, Mexican-American, and non-Hispanic African-American male teens, aged twelve to nineteen, were overweight. (See Figure 4.4.) The largest increase in the prevalence of overweight was among non-Hispanic white (from 11.6% in 1988–94 to 19.1% in 2003–04) and African-American (from 10.7% to 18.5%) teens, compared to a

FIGURE 4.1

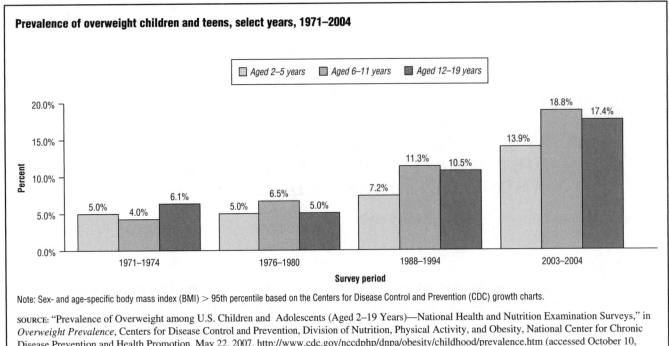

**Prevalence of overweight children and teens, select years, 1971–2004**

☐ Aged 2–5 years    ☐ Aged 6–11 years    ■ Aged 12–19 years

Note: Sex- and age-specific body mass index (BMI) > 95th percentile based on the Centers for Disease Control and Prevention (CDC) growth charts.

SOURCE: "Prevalence of Overweight among U.S. Children and Adolescents (Aged 2–19 Years)—National Health and Nutrition Examination Surveys," in *Overweight Prevalence*, Centers for Disease Control and Prevention, Division of Nutrition, Physical Activity, and Obesity, National Center for Chronic Disease Prevention and Health Promotion, May 22, 2007, http://www.cdc.gov/nccdphp/dnpa/obesity/childhood/prevalence.htm (accessed October 10, 2007)

smaller increase among Mexican-American (from 14.1% to 18.3%) teens.

From 1988–94 to 2003–04 non-Hispanic African-American (from 13.2% to 25.4%) teenaged girls, aged twelve to nineteen, experienced the largest increase in the prevalence of overweight, compared to non-Hispanic white (from 7.4% to 15.4%) and Mexican-American (from 9.2% to 14.1) teens. (See Figure 4.5.) The 2003–04 NHANES finding of increasing percentages of overweight teens suggests the likelihood of yet another generation of overweight adults who may be at risk for subsequent overweight and obesity-related health problems.

## WHY ARE SO MANY CHILDREN AND TEENS OVERWEIGHT?

Most children are overweight for the same reason as their adult counterparts: they consume more calories than they expend. Infants and toddlers appear to be effective regulators of caloric consumption, taking in only the calories needed for growth and development. By the time children are school age, this self-regulatory mechanism has weakened and when offered larger portions, they will eat them.

Heredity and environment play key roles in determining a child's risk of becoming overweight or obese. The American Academy of Child and Adolescent Psychiatry notes in "Obesity in Children and Teens" (January 2001, http://www.aacap.org/cs/root/facts_for_families/obesity _in_children_and_teens) that if one parent is obese,

then there is a 50% chance that a child will be obese, and when both parents are obese, a child has an 80% chance of being obese. Even though there is mounting evidence of genetic predisposition and susceptibility to overweight and obesity, childhood obesity is still considered largely an environmental problem—the result of behaviors, attitudes, and preferences learned early in life. Children's relationships with food develop in response to family and cultural values and practices as well as the influences of school, peers, and the media.

The question remains: Which environmental factors have given rise to the increasing prevalence of overweight children and teens during the past three decades? Many observers point to reliance on fat-laden convenience and fast foods, along with time spent watching television, playing videogames, and surfing the Internet, instead of being outdoors and getting physical activity. The CDC reports in *Physical Activity and Good Nutrition: Essential Elements to Prevent Chronic Diseases and Obesity, 2007* (April 2007, http://www.cdc.gov/nccdphp/ publications/aag/pdf/dnpa.pdf) that about two-thirds of high school students do not get the recommended levels of physical activity, with daily participation in physical education classes falling from 42% in 1991 to 33% in 2005. Television viewing, media advertising, dwindling school physical education programs, neighborhoods where it is unsafe for children to play outdoors, and even working mothers have been implicated.

**FIGURE 4.2**

**Body mass index (BMI) percentiles for boys, ages 2–20**

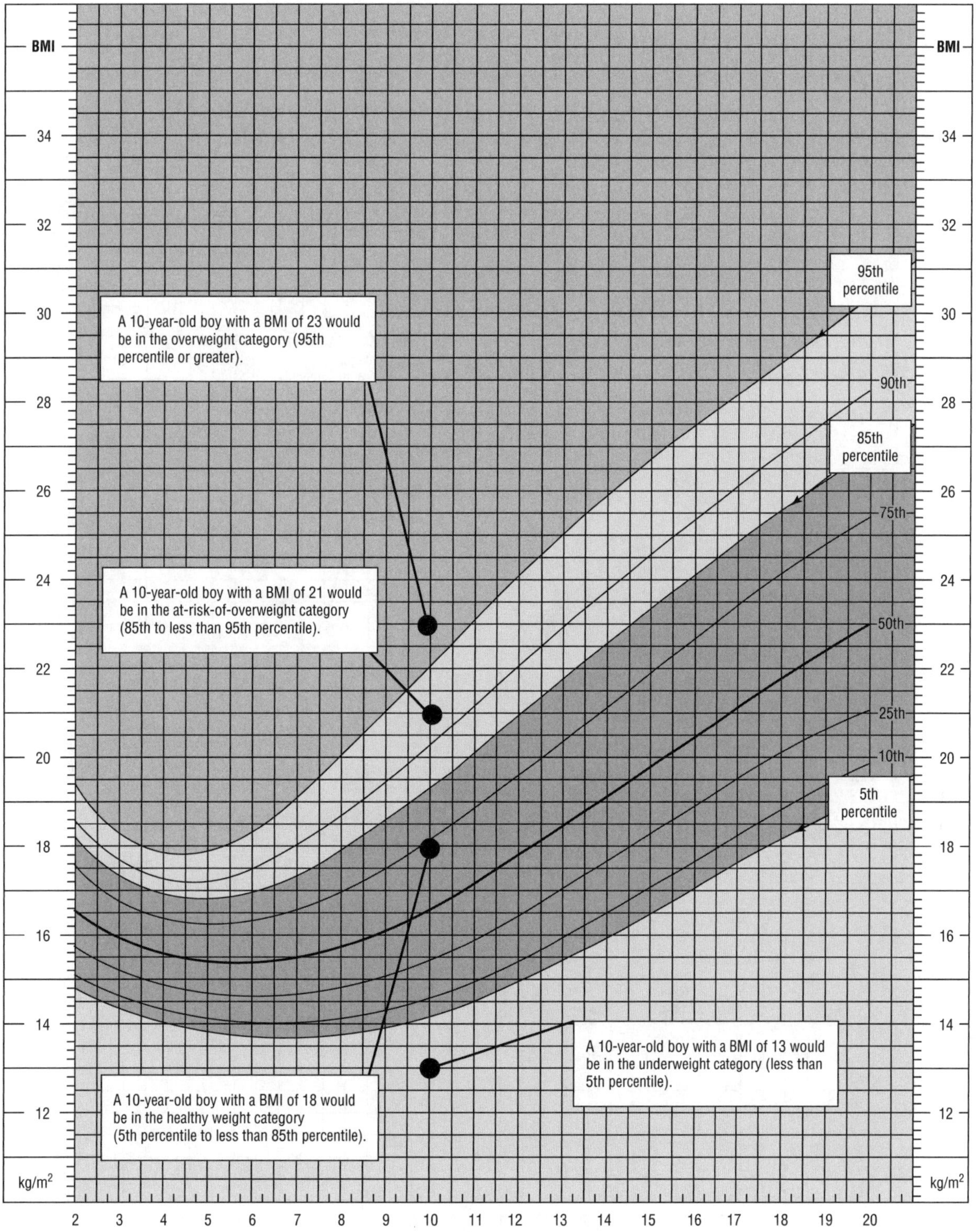
BMI
34
32
30
28
26
24
22
20
18
16
14
12
kg/m2
95th percentile
90th
85th percentile
75th
50th
25th
10th
5th percentile
A 10-year-old boy with a BMI of 23 would be in the overweight category (95th percentile or greater).
A 10-year-old boy with a BMI of 21 would be in the at-risk-of-overweight category (85th to less than 95th percentile).
A 10-year-old boy with a BMI of 18 would be in the healthy weight category (5th percentile to less than 85th percentile).
A 10-year-old boy with a BMI of 13 would be in the underweight category (less than 5th percentile).
2 3 4 5 6 7 8 9 10 11 12 13 14 15 16 17 18 19 20
Age (years)

SOURCE: "Body Mass Index-for-Age Percentiles: Boys, 2 to 20 Years," in *About BMI for Children and Teens*, Centers for Disease Control and Prevention, Division of Nutrition, Physical Activity, and Obesity, National Center for Chronic Disease Prevention and Health Promotion, May 22, 2007, http://cdc.gov/nccdphp/dnpa/bmi/childrens_BMI/about_childrens_BMI.htm (accessed October 10, 2007)

**FIGURE 4.3**

**The interpretation of body mass index (BMI) varies by age**

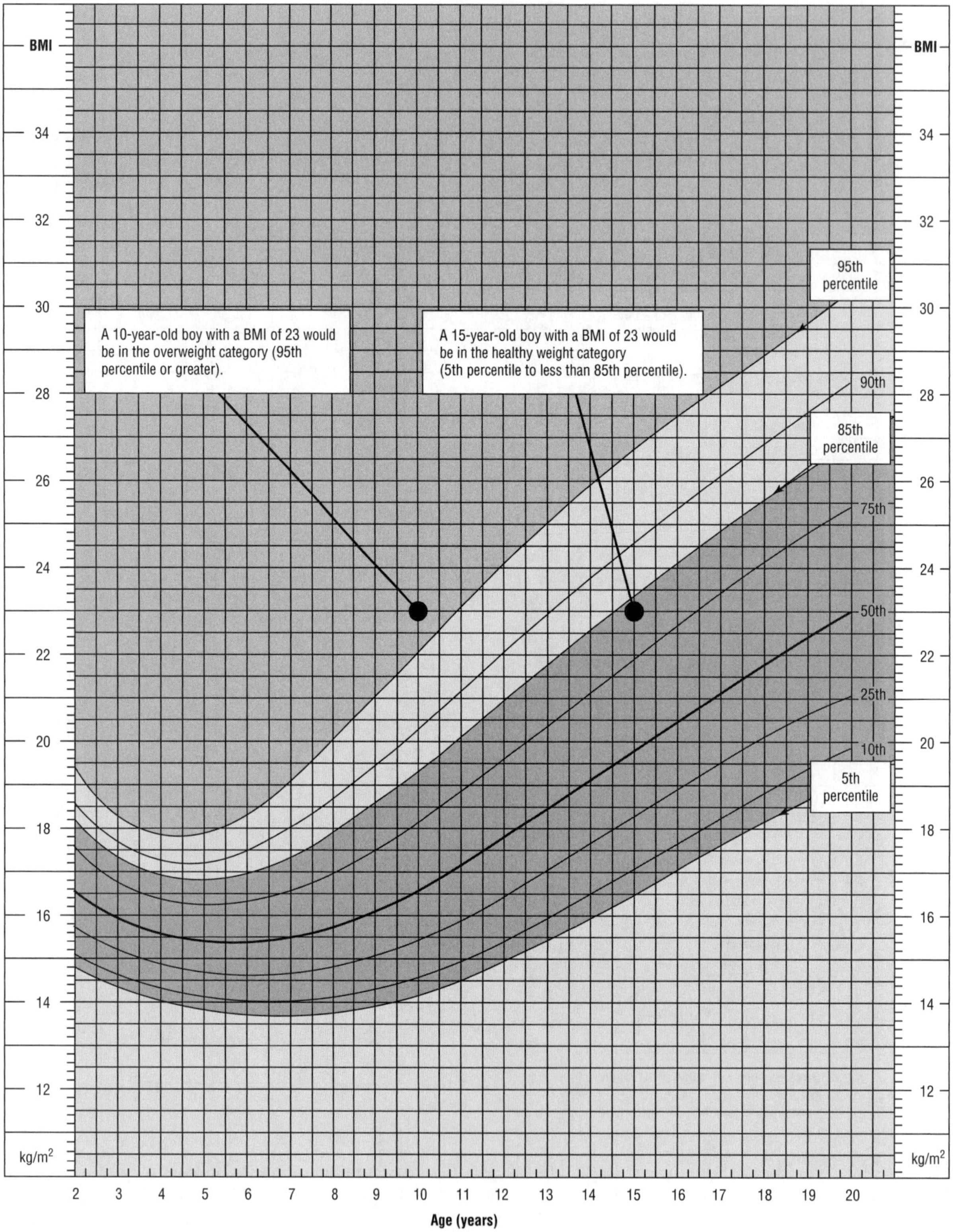

BMI
34
32
30
28
26
24
22
20
18
16
14
12
kg/m2
A 10-year-old boy with a BMI of 23 would be in the overweight category (95th percentile or greater).
A 15-year-old boy with a BMI of 23 would be in the healthy weight category (5th percentile to less than 85th percentile).
95th percentile
90th
85th percentile
75th
50th
25th
10th
5th percentile
2
3
4
5
6
7
8
9
10
11
12
13
14
15
16
17
18
19
20
Age (years)


SOURCE: "Body Mass Index-for-Age Percentiles: Boys, 2 to 20 years," in *About BMI for Children and Teens*, Centers for Disease Control and Prevention, Division of Nutrition, Physical Activity, and Obesity, National Center for Chronic Disease Prevention and Health Promotion, May 22, 2007, http://cdc.gov/nccdphp/dnpa/bmi/childrens_BMI/about_childrens_BMI.htm (accessed October 10, 2007)

## TABLE 4.1

**Weight status categories by BMI-for-age percentiles**

| Weight status category | Percentile range |
|---|---|
| Underweight | Less than the 5th percentile |
| Healthy weight | 5th percentile to less than the 85th percentile |
| At risk of overweight | 85th to less than the 95th percentile |
| Overweight | Equal to or greater than the 95th percentile |

SOURCE: "Weight Status Category," in *About BMI for Children and Teens*, Centers for Disease Control and Prevention, Division of Nutrition, Physical Activity, and Obesity, National Center for Chronic Disease Prevention and Health Promotion, May 22, 2007, http://www.cdc.gov/nccdphp/dnpa/bmi/childrens_BMI/about_childrens_BMI.htm (accessed October 10, 2007)

## FIGURE 4.4

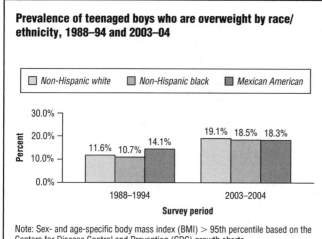

**Prevalence of teenaged boys who are overweight by race/ethnicity, 1988–94 and 2003–04**

Note: Sex- and age-specific body mass index (BMI) > 95th percentile based on the Centers for Disease Control and Prevention (CDC) growth charts.

SOURCE: "Adolescent Boys: Prevalence of Overweight by Race/Ethnicity (Aged 12–19 Years)—National Health and Nutrition Examination Surveys," in *Overweight Prevalence*, Centers for Disease Control and Prevention, Division of Nutrition, Physical Activity, and Obesity, National Center for Chronic Disease Prevention and Health Promotion, May 22, 2007, http://www.cdc.gov/nccdphp/dnpa/obesity/childhood/prevalence.htm (accessed October 10, 2007)

Working parents have been accused of a variety of nutritional and parenting infractions that have contributed to children's overindulgence in unhealthy foods. First, they leave children unsupervised and unable to satisfy their hunger with anything except cookies, chips, and soda. Some observers speculate that these children are starved emotionally—for time and attention—as well as nutritionally. They may also be hungry for information, because even though many adolescents are responsible for choosing and preparing their own food, they are often unprepared to make healthy choices.

Eating alone, in front of a television or computer, kids are more likely to overeat because they are lonely, bored, or susceptible to advertising cues. Overcome with guilt because they are not home to prepare meals, some working parents may intensify the problem by indulging their children with too many food treats. Stay-at-home

## FIGURE 4.5

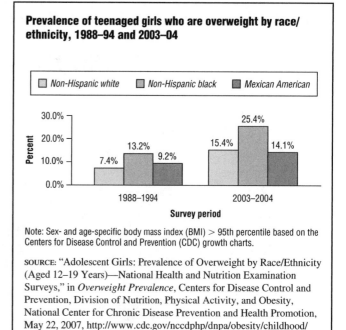

**Prevalence of teenaged girls who are overweight by race/ethnicity, 1988–94 and 2003–04**

Note: Sex- and age-specific body mass index (BMI) > 95th percentile based on the Centers for Disease Control and Prevention (CDC) growth charts.

SOURCE: "Adolescent Girls: Prevalence of Overweight by Race/Ethnicity (Aged 12–19 Years)—National Health and Nutrition Examination Surveys," in *Overweight Prevalence*, Centers for Disease Control and Prevention, Division of Nutrition, Physical Activity, and Obesity, National Center for Chronic Disease Prevention and Health Promotion, May 22, 2007, http://www.cdc.gov/nccdphp/dnpa/obesity/childhood/prevalence.htm (accessed October 10, 2007)

parents do not necessarily convey healthier attitudes about food, eating, and nutrition than parents who work outside the home. Both groups may use food, especially sweets, to reward good behavior or may pressure children to clean their plates. Though these suppositions remain unproven, it is known that parents with eating disorders, obsessive dieters, and those with unhealthy eating habits are powerful, negative role models for children.

## RESULTS FROM THE YOUTH RISK BEHAVIOR SURVEILLANCE

The "Youth Risk Behavior Surveillance System—United States, 2005" (*Morbidity and Mortality Weekly Report*, vol. 55, no. SS-5, June 9, 2006) is a national school-based survey conducted by the CDC. It examines health-risk behaviors among youth and young adults, including unhealthy dietary behaviors, physical inactivity, and overweight. This section summarizes key findings from the national survey of students in grades nine through twelve conducted in 2005.

Figure 4.6 shows that overweight is increasing among American youth. In 2003 just three states reported more that 15% to 19% of high school students were overweight (greater than or equal to the ninety-fifth percentile for BMI); by 2005 there were twice as many states with this percentage of overweight teens. Similarly, just four states reported that less than 10% of teens were overweight in 2005, compared to six in 2003.

Just one-fifth (20.1%) of students had eaten fruits and vegetables at least five times per day during the seven

**FIGURE 4.6**

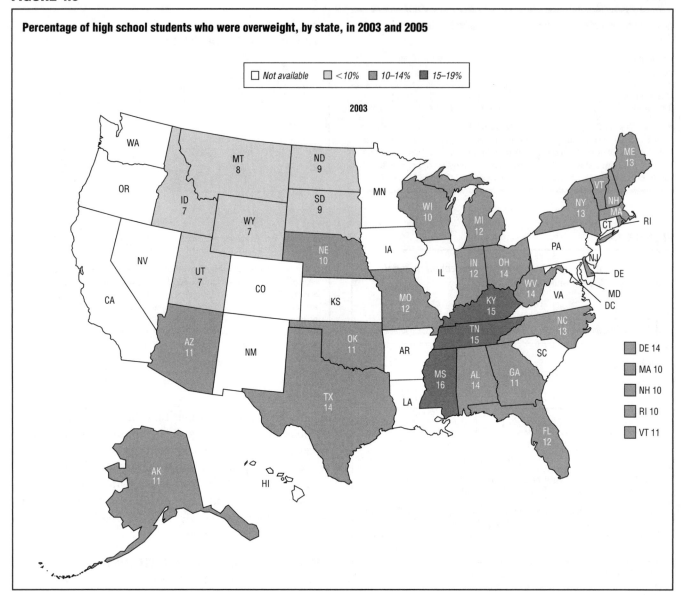

**Percentage of high school students who were overweight, by state, in 2003 and 2005**

Legend: ☐ Not available    ☐ <10%    ■ 10–14%    ■ 15–19%

2003

DE 14
MA 10
NH 10
RI 10
VT 11

days preceding the survey. (See Figure 4.7.) More male (21.4%) than female (18.7%) students reported having eaten fruits and vegetables five or more times per day.

Even fewer students (16.2%) had drunk at least three glasses of milk per day than had eaten the recommended servings of fruit and vegetables during the seven days preceding the survey. (See Figure 4.8.) The prevalence of having consumed at least three glasses of milk per day was higher among male (20.8%) than female (11.6%) students.

Figure 4.9 reveals that nearly one-third (31.5%) of high school students described themselves as "slightly" or "very" overweight. Many more teenaged girls (38.1%) than teenaged boys (25.1%) considered themselves overweight. Overall, almost half (45.6%) of the students said they were trying to lose weight, but twice as many female teens (61.7%) as male teens (29.9%) reported making an effort to lose weight. (See Figure 4.10.)

Strategies for losing weight varied. Figure 4.11 shows that more than two-thirds (67.4%) of teenaged girls and more than half (52.9%) of teen boys exercised to lose or maintain their weight. More than half (54.8%) of female teens and more than one-quarter (26.8%) of male teens said they had tried to lose or control their weight by eating less, counting calories, or choosing foods low in fat during the month preceding the survey. (See Figure 4.12.) However, 17% of teen girls and 7.6% of teen boys said they had gone without eating for twenty-four hours or more, and 8.1% of teen girls and 4.6% of teen boys had taken diet pills, powders, or liquids without a doctor's advice in an effort to lose weight. (See Figure 4.13 and Figure 4.14.)

FIGURE 4.6

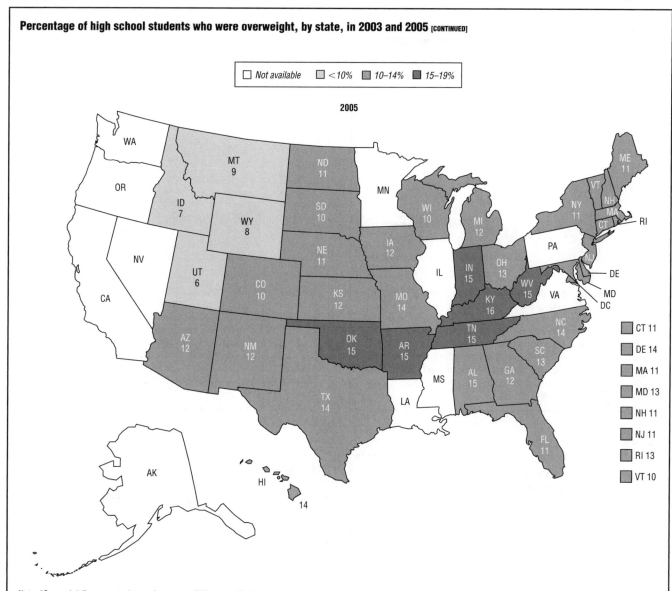

**Percentage of high school students who were overweight, by state, in 2003 and 2005** [CONTINUED]

Note: "Overweight" means students who were ≥ 95th percentile for body mass index, by age and sex, on the basis of reference data.

SOURCE: Adapted from "Percentage of High School Students Who Were Overweight*—Selected U.S. States, Youth Risk Behavior Survey, 2003," and "Percentage of High School Students Who Were Overweight*—Selected U.S. States, Youth Risk Behavior Survey, 2005," in *Childhood Overweight*, Centers for Disease Control and Prevention, National Center for Chronic Disease Prevention and Health Promotion, Division of Adolescent and School Health, July 5, 2006, http://www.cdc.gov/HealthyYouth/overweight/overweight-youth.htm (accessed October 10, 2007)

## Is Fast Food to Blame?

In "Effects of Fast-Food Consumption on Energy Intake and Diet Quality among Children in a National Household Survey" (*Pediatrics*, vol. 113, no. 1, January 2004), Shanthy A. Bowman et al. state that one-third of U.S. children eat fast food on any given day, consuming extra calories, sugar, and fat in the process. When the researchers looked at the diets of 6,212 children and teens, they found that children of all races, incomes, and U.S. regions commonly consumed fast-food meals. Bowman et al. indicate that on a typical day, more than 30% of U.S. children aged four to nineteen ate burgers, fries, and other fast-food fare. They report that children

who ate fast food consumed an average of 187 more calories than did those who did not eat fast food, and, on average, children ate 126 extra calories on the days they ate fast food, compared to fast food–free days. Bowman et al. calculate that the extra fast-food calories could result in an additional six pounds of weight gain in a year.

To determine how much soda and fast food California teenagers consume, Theresa A. Hastert et al. of the University of California, Los Angeles Center for Health Policy Research analyzed data from four thousand twelve- to seventeen-year-old participants in the 2003 California Health Interview Survey and reported their findings in *More California Teens Consume Soda and*

**FIGURE 4.7**

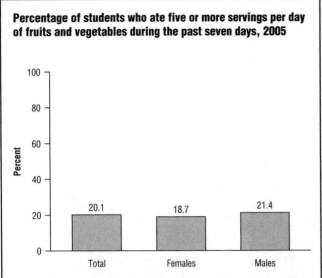

Percentage of students who ate five or more servings per day of fruits and vegetables during the past seven days, 2005

SOURCE: "Percentage of Students Who Ate Five or More Servings per Day of Fruits and Vegetables during the Past Seven Days," in *YRBSS Youth Online: Comprehensive Results*, Centers for Disease Control and Prevention, National Center for Chronic Disease Prevention and Health Promotion, Division of Adolescent and School Health, March 19, 2007, http://apps.nccd.cdc.gov/yrbss/displaygraphV.asp?path=byHT&colval=2005&rowval1=Sex&rowval2=None&compval=&Graphval=yes&cat=5&loc=XX&year=2005&quest=508&Byvar=Q2&ByResvar=CI (accessed October 10, 2007)

**FIGURE 4.8**

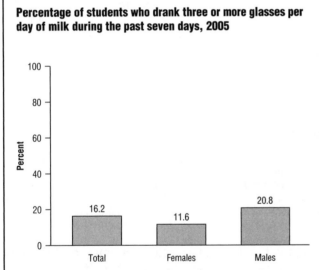

Percentage of students who drank three or more glasses per day of milk during the past seven days, 2005

SOURCE: "Percentage of Students Who Drank Three or More Glasses per Day of Milk during the Past Seven Days," in *YRBSS Youth Online: Comprehensive Results*, Centers for Disease Control and Prevention, National Center for Chronic Disease Prevention and Health Promotion, Division of Adolescent and School Health, March 30, 2007, http://apps.nccd.cdc.gov/yrbss/displaygraphV.asp?path=byHT&colval=2005&rowval1=Sex&rowval2=None&compval=&Graphval=yes&cat=5&loc=XX&year=2005&quest=Q77&Byvar=Q2&ByResvar=CI (accessed October 10, 2007)

**FIGURE 4.9**

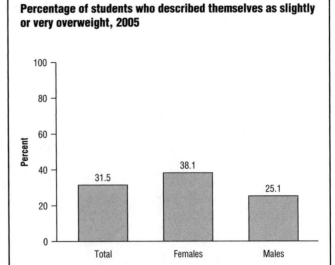

Percentage of students who described themselves as slightly or very overweight, 2005

SOURCE: "Percentage of Students Who Described Themselves as Slightly or Very Overweight," in *YRBSS Youth Online: Comprehensive Results*, Centers for Disease Control and Prevention, National Center for Chronic Disease Prevention and Health Promotion, Division of Adolescent and School Health, March 19, 2007, http://apps.nccd.cdc.gov/yrbss/displaygraphV.asp?path=byHT&colval=2005&rowval1=Sex&rowval2=None&compval=&Graphval=yes&cat=5&loc=XX&year=2005&quest=Q64&Byvar=Q2&ByResvar=CI (accessed October 10, 2007)

**FIGURE 4.10**

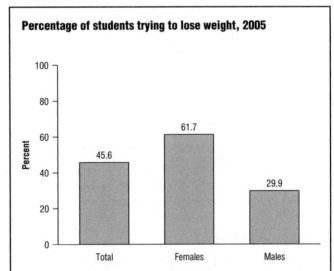

Percentage of students trying to lose weight, 2005

SOURCE: "Percentage of Students Who Were Trying to Lose Weight," in *YRBSS Youth Online: Comprehensive Results*, Centers for Disease Control and Prevention, National Center for Chronic Disease Prevention and Health Promotion, Division of Adolescent and School Health, March 19, 2007, http://apps.nccd.cdc.gov/yrbss/displaygraphV.asp?path=byHT&colval=2005&rowval1=Sex&rowval2=None&compval=&Graphval=yes&cat=5&loc=XX&year=2005&quest=Q65&Byvar=Q2&ByResvar=CI (accessed October 11, 2007)

*Fast Food Each Day than Five Servings of Fruits and Vegetables* (September 2005, http://www.healthpolicy.ucla.edu/pubs/files/teen_fastfood_PB.pdf). They find that more than 2 million California teens—66.3% of the total teen population in the state—drink soda every day and about 1.5 million (48%) eat fast food daily. The average

FIGURE 4.11

**Percentage of students who exercised to lose or keep from gaining weight, 2005**

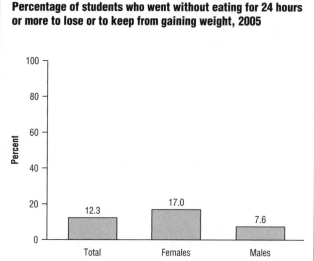

SOURCE: "Percentage of Students Who Exercised to Lose Weight or to Keep from Gaining Weight during the Past 30 Days," in *YRBSS Youth Online: Comprehensive Results*, Centers for Disease Control and Prevention, National Center for Chronic Disease Prevention and Health Promotion, Division of Adolescent and School Health, March 19, 2007, http://apps.nccd.cdc.gov/yrbss/displaygraphV.asp?path=byHT&colval=2005&rowval1=Sex&rowval2=None&compval=&Graphval=yes&cat=5&loc=XX&year=2005&quest=Q66&Byvar=Q2&ByResvar=CI (accessed October 11, 2007)

FIGURE 4.13

**Percentage of students who went without eating for 24 hours or more to lose or to keep from gaining weight, 2005**

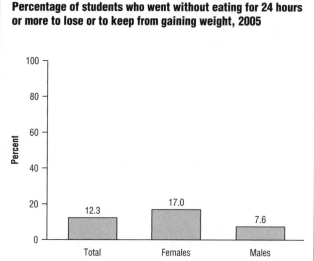

SOURCE: "Percentage of Students Who Went without Eating for 24 Hours or More to Lose Weight or to Keep from Gaining Weight during the Past 30 Days," in *YRBSS Youth Online: Comprehensive Results*, Centers for Disease Control and Prevention, National Center for Chronic Disease Prevention and Health Promotion, Division of Adolescent and School Health, March 19, 2007, http://apps.nccd.cdc.gov/yrbss/displaygraphV.asp?path=byHT&colval=2005&rowval1=Sex&rowval2=None&compval=&Graphval=yes&cat=5&loc=XX&year=2005&quest=Q68&Byvar=Q2&ByResvar=CI (accessed October 11, 2007)

FIGURE 4.12

**Percentage of students who ate less, counted calories, or chose foods low in fat to lose or keep from gaining weight, 2005**

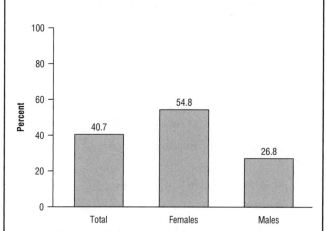

SOURCE: "Percentage of Students Who Ate Less Food, Fewer Calories, or Foods Low in Fat to Lose Weight or to Keep from Gaining Weight during the Past 30 Days," in *YRBSS Youth Online: Comprehensive Results*, Centers for Disease Control and Prevention, National Center for Chronic Disease Prevention and Health Promotion, Division of Adolescent and School Health, March 19, 2007, http://apps.nccd.cdc.gov/yrbss/displaygraphV.asp?path=byHT&colval=2005&rowval1=Sex&rowval2=None&compval=&Graphval=yes&cat=5&loc=XX&year=2005&quest=Q67&Byvar=Q2&ByResvar=CI (accessed October 11, 2007)

FIGURE 4.14

**Percentage of students who took diet pills, powders, or liquids without a doctor's advice to lose or keep from gaining weight, 2005**

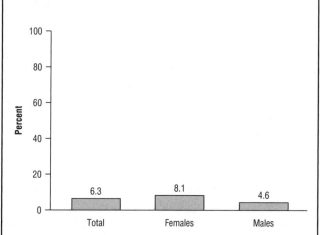

SOURCE: "Percentage of Students Who Took Diet Pills, Powders, or Liquids without a Doctor's Advice to Lose Weight or to Keep from Gaining Weight during the Past 30 Days," in *YRBSS Youth Online: Comprehensive Results*, Centers for Disease Control and Prevention, National Center for Chronic Disease Prevention and Health Promotion, Division of Adolescent and School Health, March 19, 2007, http://apps.nccd.cdc.gov/yrbss/displaygraphV.asp?path=byHT&colval=2005&rowval1=Sex&rowval2=None&compval=&Graphval=yes&cat=5&loc=XX&year=2005&quest=Q69&Byvar=Q2&ByResvar=CI (accessed October 11, 2007)

California teen drinks 1.4 sodas per day, and consumption of soda and other sugary beverages increases with age. Seventeen-year-olds reported drinking 40% more soda (1.7 per day) than twelve-year-olds (1.2 per day). Teenaged boys drink about 25% more soda and sweet drinks than do teenaged girls, and African-American teens drink the most—averaging two sodas per day. Soda consumption declines with increasing household income. Teens with household income below 300% of the federal poverty limit drink more soda (1.5 to 1.6 per day) than teens from more affluent homes. Soda consumption was 25% higher among teens who said that sodas were available in school vending machines.

Nearly half (48%) of the state's teens eat fast food every day, and many eat fast food more than once a day. Almost 10%—more than three hundred thousand California teens—have fast food twice a day, and ninety thousand (2.7%) eat fast food three or four times a day. As with soda consumption, more teens from low- and moderate-income homes eat fast food every day, and daily fast food consumption increases with age from 43.7% of twelve-year-olds to more than half (51.9%) of seventeen-year-olds. Hastert et al. find that 20.9% of California teens eat the recommended five servings of fruit and vegetables each day. Not surprisingly, they also note a relationship between fast-food consumption and eating the recommended servings of fruits and vegetables. The more often teens eat fast food, the less likely they are to eat fruits and vegetables. Significantly more teens who do not eat fast food eat five or more servings of fruit and vegetables per day.

**The Role of the Media**

Despite recent television and print media antiobesity campaigns, many industry observers condemn corporate marketing efforts and media for continuing to assault children with unhealthy messages that encourage them to eat junk foods. The CDC defines junk foods as those that provide calories primarily through fats or added sugars and have minimal amounts of vitamins and minerals. According to the article "Beware of Junk Food Marketeers" (CBSNews.com, November 11, 2003), Michael F. Jacobson, the executive director of the Center for Science in the Public Interest (CSPI), a nonprofit nutrition advocacy group based in Washington, D.C., believes that the United States has permitted junk-food marketers—not only fast-food companies but also makers of sugary cereals and high-fat, high-calorie chips—to target children. He charges that the marketing of fatty, sugary, and low-nutrient foods has reached an all-time high and is fueling childhood obesity, and he calls for restricting promotions targeted at the young.

Jacobson observes that even if parents lead by example in terms of healthy eating habits, it is still unfair to allow companies with slick, aggressive, and sophisticated advertising campaigns to bypass parents, undermine

parental authority, and directly influence children's food choices. Jacobson believes that parents must assume responsibility for ensuring that their children eat healthy meals and snacks; however, he says the marketers and media have an unfair advantage, "Companies are going directly to kids and saying, 'Eat this, eat this, drink this, drink this, it's yummy—you'll love it.' Parents have to say 'No, no, no,' and how many parents say no a thousand times?"

The CSPI called on the U.S. Department of Health and Human Services to work with Congress and the Federal Trade Commission to limit junk-food advertising aimed at children. Currently, federal rules do not restrict advertising content to children, only how much time ads can interrupt children's programming—10.5 minutes per hour on weekends and 12 minutes per hour during the week. The CSPI also advocates government-sponsored media campaigns that encourage healthy eating and physical activity.

According to the press release "BBB Announces Burger King Corp. Joins Children's Food and Beverage Advertising Initiative" (September 12, 2007, http://www.bbb.org/Alerts/article.asp?ID=798), the Children's Food and Beverage Advertising Initiative announced in September 2007 that the companies that accounted for more than two-thirds of children's food and beverage advertising expenditures in 2004—Burger King Corp.; Cadbury Adams, USA, LLC; Campbell Soup Company; the Coca-Cola Company; General Mills, Inc.; the Hershey Company; Kellogg Company; Kraft Foods Inc.; Mars, Inc.; McDonald's USA, LLC; PepsiCo, Inc.; and Unilever United States—had voluntarily agreed to reduce or stop marketing unhealthy foods to children. Participating companies have also resolved to:

- Limit products shown in interactive games to healthier dietary choices, or incorporate healthy lifestyle messages into the games

- Not engage in food and beverage product placement in editorial and entertainment content

- Reduce the use of popular cartoon and other characters in advertising

- Not advertise food or beverage products in elementary schools

Educators and marketers observe that even though corporations may agree not to advertise in schools, they remain eager to maintain a high-profile presence in schools, which enables them to remain highly visible to students. In 2005 McDonald's launched its "Passport to Play" program, which provides free lesson plans and materials to third- through fifth-grade physical education teachers. In the fact sheet "McDonald's® Commitment to Balanced, Active Lifestyles" (2006,

McDonald's describes the program, which has been distributed to elementary schools nationwide, as reflecting the company's "commitment to balanced, active lifestyles today." Some critics object to commercialism of any kind in the schools, even if the message encourages healthy choices. Others believe that it is hypocritical for purveyors of low-nutrient foods to link these foods to physical fitness or athletic prowess.

Lloyd D. Johnston, Jorge Delva, and Patrick M. O'Malley note in "Soft Drink Availability, Contracts, and Revenues in American Secondary Schools" (*American Journal of Preventive Medicine*, vol. 33, no. 4, supplement, October 2007) that many schools still offer students ready access to junk food and beverages via vending machines (88%) and in the cafeteria at lunch (59%). In 2005, 83% of high schools and 67% of middle schools had contracts with a soft drink manufacturer.

However, students are beginning to have some healthier beverage options in school. In September 2007 the Alliance for a Healthier Generation, a collaboration of the American Heart Association and the William J. Clinton Foundation, announced in "Alliance for a Healthier Generation Statement of Support for American Beverage Association School Beverage Guidelines Progress Report" (http://www.healthiergeneration.org/uploadedFiles/ For_Media/Alliance%20Statement-ABA.pdf) that the three largest beverage companies—Coca-Cola, PepsiCo, and Cadbury Schweppes—agreed to remove high-calorie, sugar-laden sodas from schools. Elementary and middle schools will offer students water, low-fat milk, and 100% juice. High school students will be able to purchase sports drinks, light juices, and diet drinks.

Even though there is widespread agreement that removing soda from schools is a healthy move, some nutritionists feel the that the proposed beverage choices, which include vitamin-enhanced water that also contains sugar, still favor beverage company interests rather than students' health. According to Andrew Martin, in "Sugar Finds Its Way Back to the School Cafeteria" (*New York Times*, September 16, 2007), Margo Wootan, the CSPI director of nutrition policy, opined that the terms of the agreement left "a huge loophole that will bring lots more sugar and calories into kids' diets."

### The Media Can Deliver Powerful Nutrition and Health Education

Greater emphasis on children's diets has inspired the media to offer nutrition education. Rather than subsisting on a diet of cookies alone, *Sesame Street*'s Cookie Monster now champions healthy food choices. The beloved character is singing a new tune, "A Cookie Is a Sometimes Food." SpongeBob SquarePants, who in the past appeared on Breyer's ice cream cartons and Kellogg's sweetened cereals, has relocated to the produce section and is advocating fresh produce consumption. Along with SpongeBob SquarePants, Dora the Explorer and other Nickelodeon characters appear on packages of fruit and vegetables, under licensing agreements with produce companies. Clifford the Big Red Dog promotes an organic cereal with his name and likeness, and Arthur the aardvark has loaned his name and likeness to Arthur's Loops, another organic cereal.

Children's television programming such as *Jo Jo's Circus* on Disney and Nickelodeon's *Lazy Town* aim to inspire young viewers to be physically active. Blending fitness and entertainment, videogame makers have developed a genre of active rhythm games including *Dance Dance Revolution*, which features a workout mode that can track how many calories the user burns while playing. *In the Groove* and *Pump It Up: Exceed* are videogames in which players try to match the onscreen action by stepping on different sections of a floor pad, and *Yourself!Fitness* and *Kinetic* offer teens exercise routines in videogame formats. *Escape from Obeez City* is an interactive DVD game that teaches children about the dangers of poor nutrition and inactivity, motivating them to change their behavior. As part of its antiobesity efforts, the National Institutes of Health is funding videogame research projects in the United States.

### Many Schools Offer and Promote Unhealthy Food Choices

Food manufacturers and marketers know that schools are ideal sites to promote their products to children and teens. Nearly all youth attend school and spend many of their waking hours at school. Furthermore, the presence of foods in schools allows food companies to benefit from the implied endorsement of the schools and teachers. According to the CDC, in "Competitive Foods and Beverages Available for Purchase in Secondary Schools—Selected Sites, United States, 2004" (*Morbidity and Mortality Weekly Report*, vol. 54, no. 37, September 23, 2005), 98% of high schools, 74.5% of middle schools, and 43.1% of elementary schools have vending machines, stores, or snack bars on campus that sell "competitive foods"—foods that are not part of federally reimbursable school meals. Figure 4.15 shows the types and sources of competitive foods in schools. The nutritional value of competitive foods is essentially unregulated, and students often purchase these foods instead of, or besides, school meals.

Besides selling food in schools, food manufacturers advertise on vending machines, posters, book covers, scoreboards, and banners and offer schools educational materials, contests in which children receive prizes or food rewards for achievement, and fund-raising opportunities. Some critics, including the CSPI, assert that

# FIGURE 4.15

Categories and sources of competitive foods in schools

Other competitive foods

Foods of minimal nutritional value

| Sources of competitive foods | Cafeteria a la carte lines | School stores, canteens, and snack bars | Vending machines | Other locations: teachers, fund-raising, parties, etc |

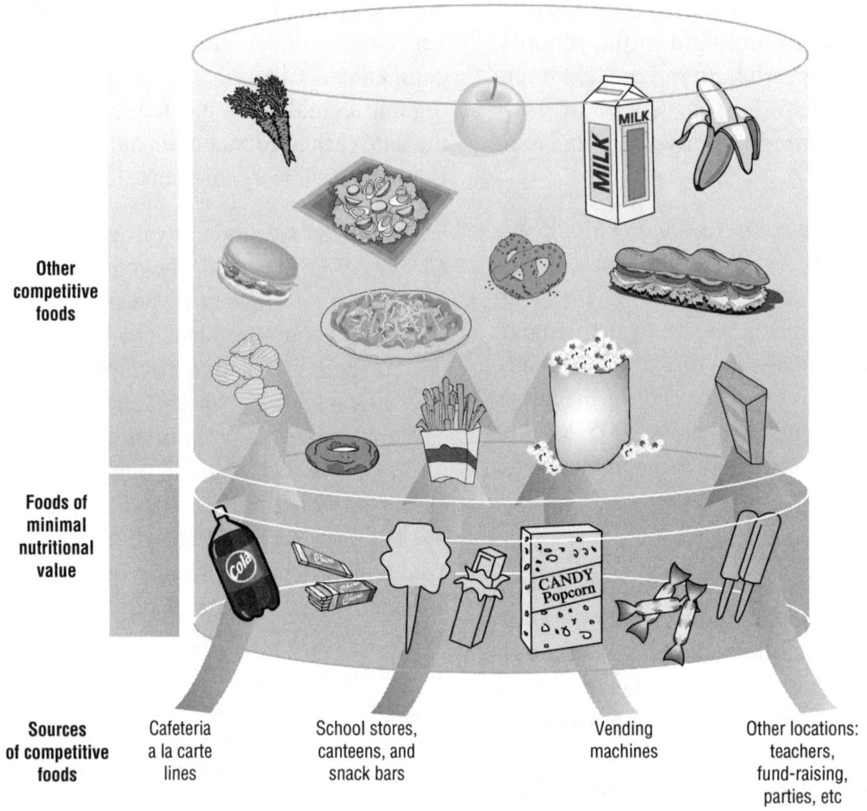

SOURCE: "Figure 1. Categories and Sources of Competitive Foods in Schools," in *School Meal Programs: Competitive Foods Are Widely Available and Generate Substantial Revenues for Schools*, U.S. Government Accountability Office, August 2005, http://www.gao.gov/new.items/d05563.pdf (accessed October 12, 2007)

the manufacturers are taking unfair advantage of cash-strapped school districts.

**SCHOOLS SELL COMPETITIVE FOODS AND OBTAIN SUBSTANTIAL REVENUES FROM THEIR SALE.** In *School Meal Programs: Competitive Foods Are Available in Many Schools; Actions Taken to Restrict Them Differ by State and Locality* (April 2004, http://www.gao.gov/new.items/d04673.pdf), the U.S. General Accounting Office (GAO; now the U.S. Government Accountability Office) finds that by 2004 several states had enacted competitive food policies that were more stringent than those required by federal regulations. However, the policies and practices varied widely. In 2005 the GAO analyzed data from two nationally representative surveys to determine the prevalence of competitive foods in schools, the groups involved in their sale, restrictions on competitive foods, and the amounts and use of revenue generated by their sale. The GAO issued its findings in *School Meal Programs: Competitive Foods Are Widely Available and Generate Substantial Revenues for Schools* (August 2005, http://www.gao.gov/new.items/d05563.pdf).

The GAO analysis finds that nearly all schools sold competitive foods during the 2003–04 school year, with middle schools and high schools more likely than elementary schools to offer competitive foods. In one-third of schools, sweet baked goods, salty snacks, and other less-nutritious foods were available in cafeteria snack lines. Schools often sold competitive foods at lunchtime, in the cafeteria or nearby, allowing kids to buy them for lunch or to supplement their lunch. Table 4.2 categorizes the types of competitive foods that are frequently available through all venues in elementary, middle, and high schools.

The GAO analysis reveals that the between 1998–99 and 2003–04 the availability of competitive food venues in middle schools rose from 83% to 97%. During this same period, the number of middle schools with exclusive beverage contracts and the number of vending machines per school also increased. Nearly 75% of high schools, 65% of middle schools, and 30% of grade schools had exclusive beverage contracts in 2004. Similarly, the volume and variety of competitive foods sold increased in more than two-thirds of high schools, more

## TABLE 4.2

**Types of competitive foods often or always available through any venue in schools, by school level and nutrition category**

| | Elementary schools | Middle schools | High schools |
|---|---|---|---|
| Water | m | m | m |
| Milk, 1% or skim | m | m | m |
| Milk, whole or 2% | m | m | m |
| 100% juice | m | m | m |
| Fruit | m | m | m |
| Vegetables and/or salad | m | m | m |
| Yogurt | | O | m |
| Less than 100% juice | O | m | m |
| Sports drinks | | m | m |
| Low-fat salty snacks | | m | m |
| Low-fat sweet baked goods | | O | m |
| Low-fat frozen desserts | | | O |
| Sandwiches | O | m | m |
| Pizza | | m | m |
| Fried vegetables | | | O |
| Frozen desserts (not low-fat) | | O | m |
| Salty snacks (not low-fat) | | m | m |
| Sweet baked goods (not low-fat) | | m | m |
| Candy | | O | m |
| Soda | | O | m |

☐ Nutritious
▨ Neither clearly nutritious nor less nutritious
▨ Less nutritious

m Item is estimated to be available in approximately half or more schools with any venue

O Item is estimated to be available in approximately one-third or more schools with any venue

Note: The nutrition categories, as signified by the shading, are general descriptions of the foods in each category. Nutritional content can vary depending on the ingredients and the methods used to prepare foods.

SOURCE: "Table 4. Types of Competitive Foods Often or Always Available through Any Venue in Schools, by School Level and Nutrition Category," in *School Meal Programs: Competitive Foods Are Widely Available and Generate Substantial Revenues for Schools*, U.S. Government Accountability Office, August 2005, http://www.gao.gov/new.items/d05563.pdf (accessed October 12, 2007)

---

than half of middle schools, and nearly one-third of elementary schools. School administrators attributed the increases to student demand, providing more nutritious and appealing food choices, and generating additional revenues for the school food service.

In 2003–04 schools, particularly middle and high schools, generated considerable revenues through competitive food sales. In terms of sales, the top 29% of high schools generated more than $125,000 per school. The GAO also finds that all the school districts it examined had taken action to substitute healthy foods for less-nutritious competitive foods. The districts acknowledged that chief among the obstacles to enacting these changes was concern about revenue losses.

### Food for Thought Has New Meaning At Many Schools

The 2004 reauthorization of the federal Child Nutrition Act required every school district that receives federal funds to establish a local wellness policy by June 30, 2006, and U.S. Department of Agriculture (USDA) diet-

ary guidelines released in January 2005 prompted many schools' food services to offer more whole grains and fresh fruits and vegetables. Even before legislation mandated changes, and certainly afterward, many school districts replaced some of the food and beverages available in their schools. For example, Andrew Martin states in "The School Cafeteria, on a Diet" (*New York Times*, September 5, 2007) that in 2007 California schools eliminated deep-fried foods, substituting baked chicken nuggets and fries. An Alabama school stopped serving sweetened tea and substituted baked potato chips for regular chips.

Connecticut was the first state to pass a ban on selling sugar-sweetened sodas in schools. Similar bills have been introduced in many other states. In 2004 schools in Philadelphia, Pennsylvania, instituted a no-soda policy, and in 2005 California banned the sale of soda in state high schools. According to the article "Schools Serving Healthier Drinks—Report" (CNN.com, September 17, 2007), in 2007 twenty-two states limited the sale of soda and other sugary drinks in some grades.

New Jersey schools have adopted what may be the most ambitious statewide school nutrition policy in the nation. Since 2007 all the state's public schools adhere to a policy stipulating that soda, any food item listing sugar as its first ingredient, all forms of candy, and foods of minimal nutritional value (per the USDA definition) cannot be served, sold, or given for free anytime during the school day. Snacks and drinks sold anywhere on a school campus must have no more than eight grams of fat and two grams of saturated fat per serving, and drinks cannot exceed more than twelve ounces, except bottled water. This policy applies to vending machines, cafeterias, à la carte items, school stores, school fund-raisers, and the after-school snack program. The policy also makes nutrition education a requirement in school curricula.

### High School Physical Education Programs

School physical education programs, especially high school programs, have been found as lacking as school nutrition programs. The CDC finds in "Youth Risk Behavior Surveillance System" that, nationwide, 54.2% of students went to physical education (PE) classes on one or more days in an average week when they were in school. (See Figure 4.16.) Just one-third (33%) of students attended daily PE classes. (See Figure 4.17.) Among students enrolled in PE classes, 84% actually exercised or played sports for more than twenty minutes during an average PE class. (See Figure 4.18.)

The importance of school PE programs cannot be underestimated, especially in view of the survey finding that only a little more than one-third (35.8%) of students had participated in sufficient physical activity—sixty minutes or more on at least five days—during the week

**FIGURE 4.16**

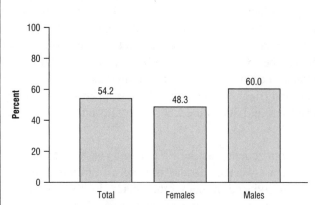

**Percentage of students who attended physical education classes one or more days per week, 2005**

SOURCE: "Percentage of Students Who Attended Physical Education (PE) Classes on One or More Days in an Average Week When They Were in School," *YRBSS Youth Online: Comprehensive Results*, Centers for Disease Control and Prevention, National Center for Chronic Disease Prevention and Health Promotion, Division of Adolescent and School Health, March 19, 2007, http://apps.nccd.cdc.gov/yrbss/displaygraphV.asp?path=byHT&colval=2005&rowval1=Sex&rowval2=None&compval=&Graphval=yes&cat=6&loc=XX&year=2005&quest=Q82&Byvar=Q2&ByResvar=CI (accessed October 13, 2007)

**FIGURE 4.18**

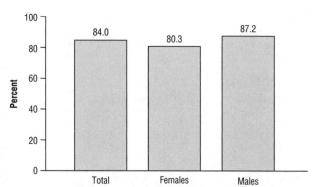

**Percentage of students who actually exercised or played sports more than 20 minutes during an average physical education class, 2005**

SOURCE: "Among Students Enrolled in Physical Education (PE) Class, the Percentage Who Actually Exercised or Played Sports More than 20 Minutes during an Average PE Class," in *YRBSS Youth Online: Comprehensive Results*, Centers for Disease Control and Prevention, National Center for Chronic Disease Prevention and Health Promotion, Division of Adolescent and School Health, March 19, 2007, http://apps.nccd.cdc.gov/yrbss/displaygraphV.asp?path=byHT&colval=2005&rowval1=Sex&rowval2=None&compval=&Graphval=yes&cat=6&loc=XX&year=2005&quest=Q83&Byvar=Q2&ByResvar=CI (accessed October 13, 2007)

**FIGURE 4.17**

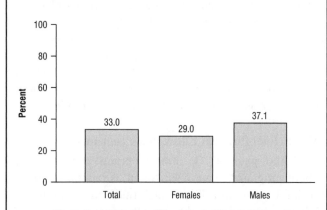

**Percentage of students who attended daily physical education classes, 2005**

SOURCE: "Percentage of Students Who Attended Physical Education (PE) Classes Daily in an Average Week When They Were in School," in *YRBSS Youth Online: Comprehensive Results*, Centers for Disease Control and Prevention, National Center for Chronic Disease Prevention and Health Promotion, Division of Adolescent and School Health, March 19, 2007, http://apps.nccd.cdc.gov/yrbss/displaygraphV.asp?path=byHT&colval=2005&rowval1=Sex&rowval2=None&compval=&Graphval=yes&cat=6&loc=XX&year=2005&quest=511&Byvar=Q2&ByResvar=CI (accessed October 13, 2007)

**FIGURE 4.19**

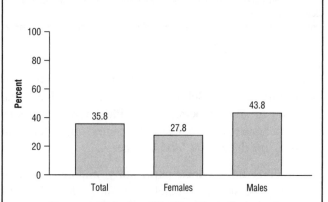

**Percentage of students who were physically active for 60 minutes or more per day on five of the past seven days, 2005**

SOURCE: "Percentage of Students Who Were Physically Active for a Total of 60 Minutes or More per Day on Five or More of the Past Seven Days," *YRBSS Youth Online: Comprehensive Results*, Centers for Disease Control and Prevention, National Center for Chronic Disease Prevention and Health Promotion, Division of Adolescent and School Health, March 19, 2007, http://apps.nccd.cdc.gov/yrbss/displaygraphV.asp?path=byHT&colval=2005&rowval1=Sex&rowval2=None&compval=&Graphval=yes&cat=6&loc=XX&year=2005&quest=Q80&Byvar=Q2&ByResvar=CI (accessed October 13, 2007)

preceding the survey. (See Figure 4.19.) Many more male teens (43.8%) than female teens (27.8%) said they had been active for sixty minutes or more on at least five of the past seven days.

More than two-thirds (68.7%) of students had exercised vigorously for twenty or more minutes, three or more times per week in the seven days preceding the survey. (See Figure 4.20.) Once again, more male

FIGURE 4.20

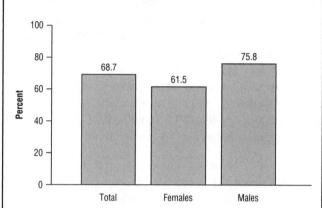

**Percentage of students who had participated in at least 20 minutes of vigorous physical activity on three or more of seven days and/or at least 30 minutes of moderate physical activity on five or more of seven days, 2005**

SOURCE: "Percentage of Students Who Had Participated in at least 20 Minutes of Vigorous Physical Activity on Three or More of the Past Seven Days and/or at least 30 Minutes of Moderate Physical Activity on Five or More of the Past Seven Days," in *YRBSS Youth Online: Comprehensive Results*, Centers for Disease Control and Prevention, National Center for Chronic Disease Prevention and Health Promotion, Division of Adolescent and School Health, March 19, 2007, http://apps .nccd.cdc.gov/yrbss/displaygraphV.asp?path=byHT&colval=2005& rowval1=Sex&rowval2=None&compval=&Graphval=yes&cat=6& loc=XX&year=005&quest=509&Byvar=Q2&ByResvar=CI (accessed October 13, 2007)

FIGURE 4.21

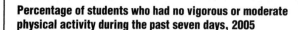

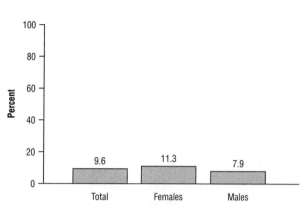

**Percentage of students who had no vigorous or moderate physical activity during the past seven days, 2005**

SOURCE: "Percentage of Students Who Had Not Participated in any Vigorous or Moderate Physical Activity during the Past Seven Days," in *YRBSS Youth Online: Comprehensive Results*, Centers for Disease Control and Prevention, National Center for Chronic Disease Prevention and Health Promotion, Division of Adolescent and School Health, March 19, 2007, http://apps.nccd.cdc.gov/yrbss/displaygraphV .asp?path=byHT&colval=2005&rowval1=Sex&rowval2=None& compval=&Graphval=yes&cat=6&loc=XX&year=2005&quest= 510&Byvar=Q2&ByResvar=CI (accessed October 13, 2007)

FIGURE 4.22

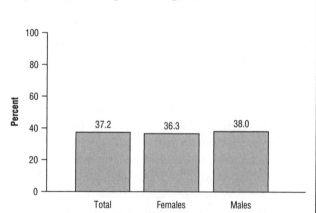

**Percentage of students who watched three or more hours per day of TV on an average school day, 2005**

SOURCE: "Percentage of Students Who Watched Three or More Hours per Day of TV on an Average School Day," in *YRBSS Youth Online: Comprehensive Results*, Centers for Disease Control and Prevention, National Center for Chronic Disease Prevention and Health Promotion, Division of Adolescent and School Health, March 19, 2007, http://apps .nccd.cdc.gov/yrbss/displaygraphV.asp?path=byHT&colval=2005& rowval1=Sex&rowval2=None&compval=&Graphval=yes&cat=6& loc=XX&year=2005&quest=Q81&Byvar=Q2&ByResvar=CI (accessed October 13, 2007)

(75.8%) than female (61.5%) students reported exercising vigorously. Figure 4.21 shows that 11.3% of female students and 7.9% of males had not participated in any vigorous or moderate physical activity in the seven days preceding the survey. Further reinforcing the notion that American teens lead relatively sedentary lives, 38% of male high school students and 36.3% of female students reported watching three or more hours of television on an average school day. (See Figure 4.22.)

## HEALTH RISKS AND CONSEQUENCES

The harmful health consequences of overweight and obesity can begin during childhood and adolescence. According to John J. Reilly et al., in "Health Consequences of Obesity" (*Archives of Disease in Childhood*, vol. 88, no. 9, 2003), nearly 60% of overweight children have at least one cardiovascular risk factor, compared to 10% of those with a BMI-for-age less than the eighty-fifth percentile, and 25% of overweight children have two or more risk factors. The most frequently occurring medical consequences of overweight among children and adolescents are:

- Elevated blood lipids—overweight children and adolescents display the same elevated levels of cholesterol, triglycerides, and/or low-density lipoproteins as overweight adults. These hyperlipidemias are linked

to an increased risk for cardiovascular disease and premature mortality (death) in adulthood.

- Glucose intolerance and Type 2 diabetes—glucose intolerance, a carbohydrate intolerance that varies in severity, is a forerunner of diabetes. The incidence of Type 2 diabetes (also called noninsulin-dependent diabetes mellitus) among adolescents is increasing in response to the national rise in overweight among teens. A skin condition known as acanthosis nigricans—velvety thickening and darkening of skinfold areas at the neck, elbow, and behind the knee—often coexists with glucose intolerance in youth.

- Fatty liver disease—high concentrations of liver enzymes are associated with fatty degeneration of the liver (also called hepatic steatosis) and have been found in overweight children and adolescents. Excessively high blood insulin levels (hyperinsulinemia) may contribute to the genesis of this disease.

- Gallstones—even though gallstones occur less frequently among children and adolescents who are overweight than in obese adults, nearly half of the cases of inflammation of the gallbladder (also called cholecystitis) in adolescents may be associated with overweight. Like adults, the risk for cholecystitis and gallstones in adolescents may decrease with weight reduction.

Another common health consequence of overweight is early maturation, a condition in which measurement of skeletal age is more than three months greater than chronological age. Early maturation is linked to overweight in adulthood and is also associated with the distribution of fat—it predicts the fat predominantly located on the abdomen and trunk that is in turn predictive of increased disease risk.

Less frequently occurring health consequences include hypertension (high blood pressure), a condition that is nine times more frequent among children who are overweight, compared to other children; obstructive sleep apnea (breathing becomes shallow or stops completely for short periods during sleep), a condition that afflicts an estimated 7% of overweight children; and orthopedic problems resulting from excessive stress on the feet, legs, and hips. Hypertension for children and adolescents one to seventeen years old is defined as average blood pressure readings at or above the ninety-fifth percentile (based on age, sex, and height) on at least three separate occasions. (See Table 4.3 and Table 4.4 for blood pressures by age and gender that are considered indicative of hypertension or at risk for hypertension. Children and adolescents between the ninetieth and ninety-fifth percentiles for their age, sex, and height are at risk for developing hypertension.) According to the CDC, in *Overweight Children and Adolescents: Recommendations to Screen, Assess, and Manage* (2002, http://www.cdc.gov/NCCdphp/dnpa/growthcharts/training/modules/module3/text/module3print.pdf), several studies confirm that blood pressure and change in BMI during childhood are the two most powerful predictors of adult blood pressure across all ages and both genders.

## Metabolic Syndrome

The metabolic syndrome is a group of risk factors for atherosclerotic cardiovascular disease and Type 2 diabetes mellitus in adults that includes insulin resistance, obesity, hypertension, and hyperlipidemia. (Atherosclerosis is a hardening of the walls of the arteries caused by the buildup of fatty deposits on the inner walls of the arteries that interferes with blood flow.) Atherosclerotic cardiovascular disease is the leading cause of death among adults, but rarely occurs in young people. Recently, however, the risk factors—high blood pressure, elevated triglycerides (a fatty substance found in the blood), obesity, and low levels of the "good" high-density lipoprotein (HDL) cholesterol—associated with its development have been appearing during childhood.

Joanne S. Harrell, Ann Jessup, and Natasha Greene of the University of North Carolina at Chapel Hill find in "Changing Our Future: Obesity and the Metabolic Syndrome in Children and Adolescents" (*Journal of Cardiovascular Nursing*, vol. 21, no. 4, July–August 2006), a study of thirty-two hundred boys and girls aged eight to seventeen years old, that there was a much higher prevalence of risk factors for metabolic syndrome than in previous studies. More than half of the subjects had a least one of six risk factors—obesity, high blood pressure, high triglycerides, low levels of HDL cholesterol, glucose intolerance, and elevated insulin levels—for metabolic syndrome. The most common risk factor, found in more than 43% of the subjects, was a low HDL cholesterol level. More than 27% had two or more risk factors, and 13.5% had at least three risk factors. More girls (16.3%) than boys (10.7%) had at least three risk factors for metabolic syndrome. More than 8% of the children who had three or more factors were between eight and nine years old. Harrell, Jessup, and Greene hope that the results of their study will serve as a warning that without effective intervention many children and teenagers with these risk factors will develop Type 2 diabetes and heart disease.

In "Diagnosis of the Metabolic Syndrome in Children" (*Current Opinion in Lipidology*, vol. 14, no. 6, December 2003), a review of recent research, Julia Steinberger of the University of Minnesota Medical School finds that the process of atherosclerosis starts at an early age and is linked to obesity in childhood. Obesity beginning in childhood often precedes the hyperinsulinemia, and other components of the metabolic syndrome are also present in children and adolescents. Being overweight during childhood and adolescence is significantly associated with insulin resistance, dyslipidemia (high low-density lipoprotein and triglycerides, and low HDL), and high blood pressure in young adulthood. In view of the increasing prevalence of metabolic syndrome in children and adolescents, Steinberger recommends that "the first approach should focus on prevention of obesity in

TABLE 4.3

**Blood pressure levels for the 90th and 95th percentiles of blood pressure for boys ages 1 to 17 years**

| Age | BP percentile* | Systolic BP (mm Hg), by height percentile from standard growth curves | | | | | | | Diastolic BP (mm Hg), by height percentile from standard growth curves | | | | | | |
|---|---|---|---|---|---|---|---|---|---|---|---|---|---|---|---|
| | | 5% | 10% | 25% | 50% | 75% | 90% | 95% | 5% | 10% | 25% | 50% | 75% | 90% | 95% |
| 1 | 90th | 94 | 95 | 97 | 98 | 100 | 102 | 102 | 50 | 51 | 52 | 53 | 54 | 54 | 55 |
| | 95th | 98 | 99 | 101 | 102 | 104 | 106 | 106 | 55 | 55 | 56 | 57 | 58 | 59 | 59 |
| 2 | 90th | 98 | 99 | 100 | 102 | 104 | 105 | 106 | 55 | 55 | 56 | 57 | 58 | 59 | 59 |
| | 95th | 101 | 102 | 104 | 106 | 108 | 109 | 110 | 59 | 59 | 60 | 61 | 62 | 63 | 63 |
| 3 | 90th | 100 | 101 | 103 | 105 | 107 | 108 | 109 | 59 | 59 | 60 | 61 | 62 | 63 | 63 |
| | 95th | 104 | 105 | 107 | 109 | 111 | 112 | 113 | 63 | 63 | 64 | 65 | 66 | 67 | 67 |
| 4 | 90th | 102 | 103 | 105 | 107 | 109 | 110 | 111 | 62 | 62 | 63 | 64 | 65 | 66 | 66 |
| | 95th | 106 | 107 | 109 | 111 | 113 | 114 | 115 | 66 | 67 | 67 | 68 | 69 | 70 | 71 |
| 5 | 90th | 104 | 105 | 106 | 108 | 110 | 111 | 112 | 65 | 65 | 66 | 67 | 68 | 69 | 69 |
| | 95th | 108 | 109 | 110 | 112 | 114 | 115 | 116 | 69 | 70 | 70 | 71 | 72 | 73 | 74 |
| 6 | 90th | 105 | 106 | 108 | 110 | 111 | 113 | 114 | 67 | 68 | 69 | 70 | 70 | 71 | 72 |
| | 95th | 109 | 110 | 112 | 114 | 115 | 117 | 117 | 72 | 72 | 73 | 74 | 75 | 76 | 76 |
| 7 | 90th | 106 | 107 | 109 | 111 | 113 | 114 | 115 | 69 | 70 | 71 | 72 | 72 | 73 | 74 |
| | 95th | 110 | 111 | 113 | 115 | 116 | 118 | 119 | 74 | 74 | 75 | 76 | 77 | 78 | 78 |
| 8 | 90th | 107 | 108 | 110 | 112 | 114 | 115 | 116 | 71 | 71 | 72 | 73 | 74 | 75 | 75 |
| | 95th | 111 | 112 | 114 | 116 | 118 | 119 | 120 | 75 | 76 | 76 | 77 | 78 | 79 | 80 |
| 9 | 90th | 109 | 110 | 112 | 113 | 115 | 117 | 117 | 72 | 73 | 73 | 74 | 75 | 76 | 76 |
| | 95th | 113 | 114 | 116 | 117 | 119 | 121 | 121 | 76 | 77 | 78 | 79 | 80 | 81 | 81 |
| 10 | 90th | 110 | 112 | 113 | 115 | 117 | 118 | 119 | 73 | 74 | 74 | 75 | 76 | 77 | 78 |
| | 95th | 114 | 115 | 117 | 119 | 121 | 122 | 123 | 77 | 78 | 79 | 80 | 80 | 81 | 82 |
| 11 | 90th | 112 | 113 | 115 | 117 | 119 | 120 | 121 | 74 | 74 | 75 | 76 | 77 | 78 | 78 |
| | 95th | 116 | 117 | 119 | 121 | 123 | 124 | 125 | 78 | 78 | 79 | 80 | 81 | 82 | 82 |
| 12 | 90th | 115 | 116 | 117 | 119 | 121 | 123 | 123 | 75 | 75 | 76 | 77 | 78 | 78 | 79 |
| | 95th | 119 | 120 | 121 | 123 | 125 | 126 | 127 | 79 | 79 | 80 | 81 | 82 | 82 | 83 |
| 13 | 90th | 117 | 118 | 120 | 122 | 124 | 125 | 126 | 75 | 76 | 76 | 77 | 78 | 79 | 80 |
| | 95th | 121 | 122 | 124 | 126 | 128 | 129 | 130 | 79 | 80 | 81 | 82 | 83 | 83 | 84 |
| 14 | 90th | 120 | 121 | 123 | 125 | 126 | 128 | 128 | 76 | 76 | 77 | 78 | 79 | 80 | 80 |
| | 95th | 124 | 125 | 127 | 128 | 130 | 132 | 132 | 80 | 81 | 81 | 82 | 83 | 84 | 85 |
| 15 | 90th | 123 | 124 | 125 | 127 | 129 | 131 | 131 | 77 | 77 | 78 | 79 | 80 | 81 | 81 |
| | 95th | 127 | 128 | 129 | 131 | 133 | 134 | 135 | 81 | 82 | 83 | 83 | 84 | 85 | 85 |
| 16 | 90th | 125 | 126 | 128 | 130 | 132 | 133 | 134 | 79 | 79 | 80 | 81 | 82 | 82 | 83 |
| | 95th | 129 | 130 | 132 | 134 | 136 | 137 | 138 | 83 | 83 | 84 | 85 | 86 | 87 | 87 |
| 17 | 90th | 128 | 129 | 131 | 133 | 134 | 136 | 136 | 81 | 81 | 82 | 83 | 84 | 85 | 85 |
| | 95th | 132 | 133 | 135 | 136 | 138 | 140 | 140 | 85 | 85 | 86 | 87 | 88 | 89 | 89 |

*Blood pressure percentile determined by a single measurement.

SOURCE: "Table 16. Blood Pressure Levels for the 90th and 95th Percentiles of Blood Pressure for Boys Ages 1 to 17 Years," in *Overweight Children and Adolescents: Screen, Access, and Manage*, Centers for Disease Control and Prevention, Division of Nutrition, Physical Activity and Obesity, National Center for Chronic Disease Prevention, May 2000, http://www.cdc.gov/nccdphp/dnpa/growthcharts/training/modules/module3/text/hypertension_tables.htm (accessed October 13, 2007)

## TABLE 4.4

### Blood pressure levels for the 90th and 95th percentiles of blood pressure for girls ages 1 to 17 years

| Age | BP percentile* | Systolic BP (mm Hg), by height percentile from standard growth curves | | | | | | | Diastolic BP (mm Hg), by height percentile from standard growth curves | | | | | | |
|---|---|---|---|---|---|---|---|---|---|---|---|---|---|---|---|
| | | 5% | 10% | 25% | 50% | 75% | 90% | 95% | 5% | 10% | 25% | 50% | 75% | 90% | 95% |
| 1 | 90th | 97 | 98 | 99 | 100 | 102 | 103 | 104 | 53 | 53 | 53 | 54 | 55 | 56 | 56 |
| | 95th | 101 | 102 | 103 | 104 | 105 | 107 | 107 | 57 | 57 | 57 | 58 | 59 | 60 | 60 |
| 2 | 90th | 99 | 99 | 100 | 102 | 103 | 104 | 105 | 57 | 57 | 58 | 58 | 59 | 60 | 61 |
| | 95th | 102 | 103 | 104 | 105 | 107 | 108 | 106 | 61 | 61 | 62 | 62 | 63 | 64 | 65 |
| 3 | 90th | 100 | 100 | 102 | 103 | 104 | 105 | 110 | 61 | 61 | 61 | 62 | 63 | 63 | 64 |
| | 95th | 104 | 104 | 105 | 107 | 108 | 109 | 108 | 65 | 65 | 65 | 66 | 67 | 67 | 68 |
| 4 | 90th | 101 | 102 | 103 | 104 | 106 | 107 | 111 | 63 | 63 | 64 | 65 | 65 | 66 | 67 |
| | 95th | 105 | 106 | 107 | 108 | 109 | 111 | 109 | 67 | 67 | 68 | 69 | 69 | 70 | 71 |
| 5 | 90th | 103 | 103 | 104 | 106 | 107 | 108 | 113 | 65 | 66 | 66 | 67 | 68 | 68 | 69 |
| | 95th | 107 | 107 | 108 | 110 | 111 | 112 | 111 | 69 | 70 | 70 | 71 | 72 | 72 | 73 |
| 6 | 90th | 104 | 105 | 106 | 107 | 109 | 110 | 114 | 67 | 67 | 68 | 69 | 69 | 70 | 71 |
| | 95th | 108 | 109 | 110 | 111 | 112 | 114 | 112 | 71 | 71 | 72 | 73 | 73 | 74 | 75 |
| 7 | 90th | 106 | 107 | 108 | 109 | 110 | 112 | 116 | 69 | 69 | 69 | 70 | 71 | 72 | 72 |
| | 95th | 110 | 110 | 112 | 113 | 114 | 115 | 114 | 73 | 73 | 73 | 74 | 75 | 76 | 76 |
| 8 | 90th | 108 | 109 | 110 | 111 | 112 | 113 | 118 | 70 | 70 | 71 | 71 | 72 | 73 | 74 |
| | 95th | 112 | 112 | 113 | 115 | 116 | 118 | 116 | 74 | 74 | 75 | 75 | 76 | 77 | 78 |
| 9 | 90th | 110 | 110 | 112 | 113 | 114 | 115 | 120 | 71 | 72 | 72 | 73 | 74 | 74 | 75 |
| | 95th | 114 | 114 | 115 | 117 | 118 | 119 | 118 | 75 | 76 | 76 | 77 | 78 | 78 | 79 |
| 10 | 90th | 112 | 112 | 114 | 115 | 116 | 117 | 122 | 73 | 73 | 73 | 74 | 75 | 76 | 76 |
| | 95th | 116 | 116 | 117 | 119 | 120 | 121 | 120 | 77 | 77 | 77 | 78 | 79 | 80 | 80 |
| 11 | 90th | 114 | 114 | 116 | 117 | 118 | 119 | 124 | 74 | 74 | 75 | 75 | 76 | 77 | 77 |
| | 95th | 118 | 118 | 119 | 121 | 122 | 123 | 122 | 78 | 78 | 79 | 79 | 80 | 81 | 81 |
| 12 | 90th | 116 | 116 | 118 | 119 | 120 | 121 | 126 | 75 | 75 | 76 | 76 | 77 | 78 | 78 |
| | 95th | 120 | 120 | 121 | 123 | 124 | 125 | 124 | 79 | 79 | 80 | 80 | 81 | 82 | 82 |
| 13 | 90th | 118 | 118 | 119 | 121 | 122 | 123 | 128 | 76 | 76 | 77 | 78 | 78 | 79 | 80 |
| | 95th | 121 | 122 | 123 | 125 | 126 | 127 | 126 | 80 | 80 | 81 | 82 | 82 | 83 | 84 |
| 14 | 90th | 119 | 120 | 121 | 122 | 124 | 125 | 130 | 77 | 77 | 78 | 79 | 79 | 80 | 81 |
| | 95th | 123 | 124 | 125 | 126 | 128 | 129 | 127 | 81 | 81 | 82 | 83 | 83 | 84 | 85 |
| 15 | 90th | 121 | 121 | 122 | 124 | 125 | 126 | 131 | 78 | 78 | 79 | 79 | 80 | 81 | 82 |
| | 95th | 124 | 125 | 126 | 128 | 129 | 130 | 128 | 82 | 82 | 83 | 83 | 84 | 85 | 86 |
| 16 | 90th | 122 | 122 | 123 | 125 | 126 | 127 | 132 | 79 | 79 | 79 | 80 | 81 | 82 | 82 |
| | 95th | 125 | 126 | 127 | 128 | 130 | 131 | 128 | 83 | 83 | 83 | 84 | 85 | 86 | 86 |
| 17 | 90th | 122 | 123 | 124 | 125 | 126 | 128 | 132 | 79 | 79 | 79 | 80 | 81 | 82 | 82 |
| | 95th | 126 | 126 | 127 | 129 | 130 | 131 | | 83 | 83 | 83 | 84 | 85 | 86 | 86 |

*Blood pressure percentile determined by a single measurement.

SOURCE: "Table 17. Blood Pressure Levels for the 90th and 95th Percentiles of Blood Pressure for Girls Ages 1 to 17 Years," in *Overweight Children and Adolescents: Screen, Access, and Manage,* Centers for Disease Control and Prevention, Division of Nutrition, Physical Activity and Obesity, National Center for Chronic Disease Prevention, May 2000, http://www.cdc.gov/nccdphp/dnpa/growthcharts/training/modules/module3/text/hypertension_tables.htm (accessed October 13, 2007) childhood. More attention should be paid to increasing physical activity and decreasing calorie consumption in this age group. Once obesity is established in a child or adolescent, vigorous clinical efforts should be directed at treating it."

Sarah P. Garnett et al. of the Institute of Endocrinology and Diabetes in Westmead, Australia, observe in "Body Mass Index and Waist Circumference in Midchildhood and Adverse Cardiovascular Disease Risk Clustering in Adolescence" (*American Journal of Clinical Nutrition*, vol. 86, no. 3, September 1, 2007), that overweight and obese eight-year-olds were seven times more likely than their healthy-weight peers to have heart disease risk factors such as high blood pressure, unhealthy cholesterol levels, and elevations in blood sugar and insulin, by age fifteen. Garnett et al. opine that their findings, that these risk factors persist and increase over time, underscore the importance of preventing childhood obesity.

## Mental Health Consequences

One of the most immediate, distressing, and widespread consequences of being overweight as described by children themselves is social discrimination and low self-esteem. Overweight and obese children and adolescents are at risk for psychological and social adjustment problems such as considering themselves less competent than normal-weight youth in social, athletic, and appearance arenas, as well as suffering from overall diminished self-worth. In "Health-Related Quality of Life of Severely Obese Children and Adolescents" (*Journal of the American Medical Association*, vol. 289, no. 14, April 9, 2003), Jeffrey B. Schwimmer, Tasha M. Burwinkle, and James W. Varni find that obese children rated their quality of life with scores as low as those of young cancer patients undergoing chemotherapy (medical treatment to combat cancer). The researchers analyzed the responses of 106 children aged five to eighteen to a questionnaire used by pediatricians to evaluate quality-of-life issues. Study participants were asked to rate attributes such as their ability to walk more than one block, play sports, sleep well, get along with others, and keep up in school.

The results indicated that teasing at school, difficulties playing sports, fatigue, sleep apnea, and other obesity-linked problems severely affected obese children's well-being. The obese subjects were five times more likely than healthy children and adolescents to have impaired physical functioning and nearly six times more likely to suffer impaired psychosocial functioning. They were four times more likely than healthy children and adolescents to report impaired school function and had missed a mean of 4.2 days of school in the month before the study, compared to less than one day of school missed for children who were not overweight. When the parents of the subjects completed the same questionnaire, their ratings of their children's abilities and well-being were even lower than the children's self-reported ratings.

Lucy Jane Griffiths et al. report in "Obesity and Bullying: Different Effects for Boys and Girls" (*Archives of Disease in Childhood*, vol. 91, 2006) that obese children engage in more bullying behavior, at least in part because they deviate from appearance ideals. Obese boys were more than one and a half times more likely to use their physical dominance to bully other children or to be victims of bullying than their normal-weight or overweight peers. Obese girls were more likely to be victims of bullying than their normal-weight peers.

In "Obesity, Shame, and Depression in School-Aged Children: A Population Study" (*Pediatrics*, vol. 116, no. 3, September 2005), Rickard L. Sjöberg, Kent W. Nilsson, and Jerzy Leppert indicate that depression is common among obese teenagers and largely results from teens' experiences of being shamed. The researchers analyzed data from 4,703 teens aged fifteen and seventeen years and found that obese teens reported experiencing more symptoms of depression than their normal-weight or overweight peers and had a higher risk of depression. Obese teens were more likely than their normal-weight or overweight peers to say they had been treated in a degrading manner, had been ignored, or otherwise had shaming experiences within the past three months. Furthermore, adolescents who reported the highest number of shame experiences were more than eleven times more likely to be depressed than those who reported the lowest number of shame experiences. Sjöberg, Nilsson, and Leppert conclude, "These results suggest that clinical treatment of obesity may sometimes not just be a matter of diet and exercise but also of dealing with issues of shame and social isolation."

Obese children are also absent from school more frequently than their healthy-weight peers. Andrew B. Geier et al. find in "The Relationship between Relative Weight and School Attendance among Elementary Schoolchildren" (*Obesity*, vol. 15, 2007) that on average the healthy-weight students were absent 10.1 days, overweight children missed 10.9 days, and obese children missed 12.2 days. The researchers note that being overweight or obese is a better predictor of absenteeism than the factors (race, socioeconomic status, age, and gender) previously thought to have the best predictive value for school attendance. Even though some of the increased absenteeism may be due to health problems, Geier et al. think that social problems such as fear of being bullied, embarrassed, or excluded account for a considerable amount of the observed absences.

## SCREENING AND ASSESSMENT OF OVERWEIGHT CHILDREN AND ADOLESCENTS

In view of the rising prevalence of overweight youth, screening children and adolescents for overweight and risk for overweight has assumed a prominent place in pediatric practice (the medical specialty devoted to diagnosis and treatment of children) and public health programs. The Recommendations for Preventive Pediatric Health Care by the American Academy of Pediatrics advise a frequent schedule of accurate weight and height measurements to determine whether children require further assessment or treatment for overweight. Screening distinguishes between youths who are not at risk of overweight, at risk of overweight, and overweight. Those deemed overweight receive an in-depth medical assessment; those considered at risk are assessed for changes in BMI, blood pressure, and cholesterol levels; and annual screening is advised for those who are not at risk of overweight.

The comprehensive assessment performed on overweight children and adolescents generally includes obtaining a detailed medical history to identify any underlying medical conditions that may contribute to overweight and analyzing family history for the presence of familial risks for overweight or obesity. Relevant familial factors include the occurrence of obesity, eating disorders, Type 2 diabetes, heart disease, high blood pressure, and abnormal lipid profiles such as high cholesterol among immediate family members. The assessment may also involve:

- Dietary evaluation to consider the quantity, quality, and timing of food consumed to identify foods and patterns of eating that may lead to excessive calorie intake. A food record or food diary may be used to assess eating habits.

- An evaluation of daily activities. This assessment involves an estimate of time devoted to exercise and activity as well as time spent on sedentary behaviors such as television, videogames, and computer use.

- A physical examination to provide information about the extent of overweight and any complications of overweight, including high blood pressure. Children and adolescents with a BMI-for-age at or above the ninety-fifth percentile and who are athletic and muscular may be further assessed using triceps skinfold measurement to assess body fat. A measurement of greater than the ninety-fifth percentile indicates that the child has excess fat rather than increased lean body mass or a large frame.

- Laboratory tests, such as cholesterol screening, dictated by the degree of overweight, family history, and the results of the physical examination. Table 4.5 shows the range of values for total blood cholesterol and low-density lipoprotein cholesterol that are considered acceptable, borderline, and high.

**TABLE 4.5**

**Classification of cholesterol levels in high-risk children and adolescents**

|  | Total cholesterol, ng/dL | L DL cholesterol, ng/dL |
|---|---|---|
| Acceptable | <170 | <110 |
| Borderline | 170–199 | 110–129 |
| High | Greater than or equal to 200 | Greater than or equal to 130 |

Note: High-risk children are defined as those from families with hypercholesterolemia or premature cardiovascular disease.

SOURCE: "Table 15. Classification of Cholesterol Levels in High-Risk Children and Adolescents," in *Overweight Children and Adolescents: Screen, Assess, and Manage*, Centers for Disease Control and Prevention, 2001, http://www.cdc.gov/nccdphp/dnpa/growthcharts/training/modules/module3/text/chole sterol.htm (accessed October 13, 2007)

- A mental health evaluation to determine the readiness of children and adolescents to change behaviors and to identify a history of eating disorders or depression that may require treatment. An assessment of the family's ability to support a child's weight-loss or weight-management efforts may also be performed.

## INTERVENTION AND TREATMENT OF OVERWEIGHT AND OBESITY

In the absence of acute medical necessity, such as with children who are dangerously obese, most health professionals concur that drastic caloric restriction is an inappropriate weight-loss strategy for children who are still growing. Instead, they advise efforts to stabilize body weight with a healthy, balanced diet, increased physical activity, and education about nutrition, food choices, and preparation. This approach is especially effective for children who are just slightly overweight, because maintaining body weight often allows them to outgrow overweight and become normal-weight adults.

When active weight loss is indicated, it is generally for children with a BMI greater than the ninety-fifth percentile or those experiencing complications of overweight or obesity. Among children aged two to seven, gradual weight loss of about one pound per month is advised. Older children with serious health risks who are severely overweight (BMI greater than 35) may be advised to lose between one and two pounds per week.

Many studies confirm that dietary interventions with children and teens are as ineffective long term as they are with adults. In "Treatment of Pediatric and Adolescent Obesity" (*Journal of the American Medical Association*, vol. 289, no. 14, April 9, 2003), Jack A. Yanovski and Susan Z. Yanovski of the National Institutes of Health observe that studies find that long-term weight reduction is maintained in only about half of children and adolescents treated with intensive behavioral-modification therapy. Furthermore, they characterize effective behavior-modification programs lacking widespread applicability because they are labor-intensive, not easily conducted by primary care physicians (pediatricians and family medicine physicians), and require intensive involvement from parents. Many practitioners believe that behavior modification alone is insufficient for severely obese children and adolescents. For this population, researchers and practitioners have had success with pharmacotherapy—drug treatment with medications known as anorexiants, which reduce appetite by blocking the reuptake of the neurotransmitters norepinephrine and serotonin. The most serious adverse effects of these medications are an increase in blood pressure and pulse rate sufficient to warrant reducing the drug dose or discontinuing it altogether. Like many other researchers and clinicians, Yanovski and Yanovski conclude that it "remains exceedingly difficult for overweight children and adolescents to lose weight, and even more difficult for them to sustain that weight loss long term. The ultimate goal must be prevention of the development of overweight in children and adolescents."

Robert I. Berkowitz et al. compared the efficacy of family-based behavioral treatment alone to a combined regimen of family-based behavioral therapy and weight-loss medication among adolescents. The researchers reported the results of their study in "Behavior Therapy and Sibutramine for the Treatment of Adolescent Obesity" (*Journal of the American Medical Association*, vol. 289, no. 14, April 9, 2003). For the first six months of the study, eighty-two participants aged thirteen to seventeen with BMIs ranging from 32 to 44 received behavior therapy and sibutramine (an anorexiant medication) or behavior therapy and a placebo (an inactive compound). During the second six months, all participants received behavioral treatment and sibutramine.

During the first phase, behavioral treatment called for participants to attend thirteen weekly group sessions followed up by six biweekly group sessions. In the second phase, the group sessions were conducted biweekly from months seven to nine and monthly from months ten to twelve. Parents met in separate group sessions held on the same schedule as the adolescents' meetings. Dietitians, psychologists, or psychiatrists conducted the groups. Participants in both treatment groups were instructed to consume a twelve-hundred- to fifteen-hundred-calorie diet of conventional foods, with approximately 30% of their calories derived from fat, 15% from protein, and the remainder from carbohydrates. They were advised to incrementally increase their physical activity with the goal of walking or participating in aerobic activity for 120 minutes per week or more. Participants kept daily eating and activity logs that they submitted at each session.

At the end of the first six months, participants in the behavioral treatment and sibutramine group lost a mean of 17.2 pounds and had an 8.5% reduction in BMI, which was significantly more than the weight loss of 7.1 pounds and a reduction in BMI of 4% in the behavioral treatment and placebo group. Participants who received behavioral treatment and sibutramine also reported significantly less hunger. From months seven to twelve, participants initially treated with sibutramine maintained their weight loss with continued use of the medication, whereas those who switched from placebo to sibutramine lost an additional 2.9 pounds. Berkowitz et al. explain the behavioral treatment and sibutramine participants' failure to lose further weight during the second phase of the study as consistent with the observation that weight loss tends to plateau in obese adults after six months of treatment with behavior therapy or pharmacotherapy.

Berkowitz et al. conclude that weight-loss medications may be of benefit to adolescents. However, they caution that their use must be carefully monitored in adolescents, as in adults, to control increases in blood pressure and pulse rate. Absent the many large-scale studies necessary to confirm the safety and effectiveness of pharmacological treatment of obesity in adolescents, Berkowitz et al. advise that "medications for weight loss should be used only on an experimental basis in adolescents and children."

Because maintenance—keeping pounds lost off over time—is an issue for adults and children, Denise E. Wilfley et al. examine in "Efficacy of Maintenance Treatment Approaches for Childhood Overweight" (*Journal of the American Medical Association*, vol. 298, no. 14, October 10, 2007) how effectively children and parents who completed a five-month weight-loss program and lost an average of 11% of their bodyweight maintained their weight loss. Children who received the most intensive follow-up, which included socializing with more physically active peers, learning how to cope with teasing, and improving their body image, fared the best in terms of maintaining weight loss; however, at the one- and two-year follow-up, many had regained the weight they had lost.

## Educating Parents

Researchers agree that primary prevention is the strategy with the greatest potential for reversing the alarming rise in overweight and obesity among children and teens. Public health educators recommend counseling parents and caregivers about healthy eating habits for children. They advise offering children a variety of healthy foods, in reasonable quantities, to assist children to make wise food choices. Children should be encouraged, but not forced, to sample new foods and should not be pressured to clean their plates. No foods or food groups should be entirely off-limits, or children may become fixated on obtaining the forbidden foods.

Even though it is difficult to impress children with the future health risks associated with excess weight, parents should be informed that obese children are more likely to suffer from diabetes, heart, and joint diseases such as osteoarthritis, as well as breast and colon cancer. Adults should model healthy habits, consuming no more that 30% of calories from fat, exercising regularly, and limiting time spent in front of the television. Health educators are especially eager to reduce children's television viewing, with its destructive blend of junk-food advertising and enforced inactivity. Finally, health professionals caution that food should not be used to punish or reward behavior or as a way to comfort or console children. The undivided attention of a parent or caregiver or an expression of sympathy, reassurance, or encouragement may satisfy a child's need better than an ice cream cone or an order of French fries.

In "Factors Associated with Parental Readiness to Make Changes for Overweight Children" (*Pediatrics*, vol. 116, no. 1, July 2005), Kyung E. Rhee et al. find that parents are not always receptive to making lifestyle changes that could help their overweight children lose weight—particularly if the parents do not see their child's weight as a health issue. A study of 151 parents found that 44% of parents of children who were overweight or obese did not see their child's weight as a problem and as a result were not planning on instituting lifestyle changes soon. Another 17% of parents did recognize that their child had a problem and were considering making behavioral or lifestyle changes, but not soon. Rhee et al. find that parents of children who were eight years old or older were more likely to be ready to address their child's weight issue than parents of younger children. The same was true of parents who believed their child's weight was a health issue; they were nearly ten times more likely than other parents to say they were ready to take actions such as increasing their children's fruit and vegetable consumption, limiting television time, and encouraging exercise. Parents were also more open to change if they viewed themselves as overweight.

## EATING DISORDERS

Overweight and obesity are among the most stigmatizing and least socially acceptable conditions in childhood and adolescence. Society, culture, and the media send children powerful messages about body weight and shape ideals. For girls these include the "thin ideal" and encouragement to diet and exercise. Messages to boys emphasize a muscular body and pressure to body build and even use potentially harmful dietary supplements and steroids. Gender has not been identified as a specific risk factor for obesity in children, but the pressure placed on girls to be thin may put them at a greater risk for developing eating-disordered behaviors. Even though society presents boys with a wider range of acceptable body images, they are also at risk for developing disordered eating and body image disturbances.

Adolescence is a developmental period marked by great physical change, and it is a time when many teens subject themselves to painful scrutiny. Uneven growth, puberty, and sexual maturation may make teens feel awkward and self-conscious about their bodies. Teenaged girls are especially susceptible to developing negative body images—ignoring other qualities and focusing exclusively on appearance to measure their self-worth. This single-minded, and often distorted, destructive focus can result in lowered self-esteem and increased risk for mental health problems, including eating disorders.

## Who Is at Risk?

Even though there are biological, genetic, and familial factors that predispose to eating disorders such as anorexia nervosa (intense fear of becoming fat even when dangerously underweight) and bulimia (recurrent episodes of binge eating followed by purging to prevent weight gain), the emergence of these disorders is triggered by environmental factors. Chief among the environmental triggers is body image. Many researchers and health professionals believe that teenaged girls who identify with the idealized body images projected throughout American culture are at an increased risk for eating disorders.

Other risk factors are peer group pressures and sociocultural forces such as the fashion and entertainment industries and the media. The National Eating Disorders Association identifies media definitions of beauty, attractiveness, and health as among the myriad factors contributing to the rise of eating disorders. In the landmark survey *The Commonwealth Fund Survey of the Health of Adolescent Girls* (November 1997, http://www.commonwealthfund .org/usr_doc/Schoen_adolescentgirls.pdf?section=4039), Cathy Schoen et al. find that the media are girls' primary source of information about women's health issues. In another study, E. O. Guillen and Susan I. Barr of the University of British Columbia report in "Nutrition, Dieting, and Fitness Messages in a Magazine for Adolescent Women, 1970–1990" (*Journal of Adolescent Health*, vol. 15, no. 6, September 1994) that in the course of twenty years three-quarters of articles about fitness or exercise plans in one teen adolescent magazine named "to become more attractive" as the reason to start exercising, and 51% cited the need to lose weight or burn calories.

Historically, most adolescents with eating disorders have been first- or second-born white females from middle- to upper-class families. Girls who suffer from anorexia are often academically successful, with athletic prowess or training in dance. They tend to be perfectionists, well behaved, emotionally dependent, socially anxious, and intent on receiving approval from others. Adolescent girls with bulimia are generally more extroverted and socially involved. In "Statistics and Study Findings" (2008, http://www.eatingdisorderscoalition.org/reports/

statistics.html), the Eating Disorders Coalition for Research, Policy, and Action notes that in the early twenty-first century the occurrence of eating disorders is increasing among younger children and throughout diverse ethnic and sociocultural groups.

The National Eating Disorders Association notes in "Statistics: Eating Disorders and Their Precursors" (2006, http://www.nationaleatingdisorders.org/p.asp?WebPage _ID=286&Profile_ID=41138) that a preoccupation with thinness and dieting begins at an early age. One study reports that 42% of first- to third-grade girls said they wanted to be thinner, and another finds that 81% of ten-year-olds feared becoming fat. Between 30% and 40% of middle school girls are worried about their weight, and 40% to 60% of high school girls diet. A survey of female college students finds that 91% had attempted to control their weight by dieting, and 22% said they were "often" or "always" dieting.

## Which Variables Are Associated with Dieting, Overweight, and Eating Disorders?

Dianne Neumark-Sztainer and Peter J. Hannan of the University of Minnesota at Minneapolis analyzed a representative sample of 6,728 adolescents in grades five through twelve who completed the Commonwealth Fund survey about the health of adolescent girls and boys. The results of the research were detailed in "Weight-Related Behaviors among Adolescent Girls and Boys: Results from a National Survey" (*Archives of Pediatrics and Adolescent Medicine*, vol. 154, no. 6, June 2000). The research aimed to assess the prevalence of dieting and disordered eating among adolescents; the sociodemographic, psychosocial, and behavioral variables that were associated with dieting and disordered eating; and whether adolescents report having discussed weight-related issues with their health-care providers. (Neumark-Sztainer and Hannan defined disordered eating as weight-related behaviors such as anorexia and bulimia nervosa, self-induced vomiting, binge eating, inappropriate or extreme dieting, and obesity.)

Subjects were assessed by calculating their BMI and eliciting weight-related attitudes and behaviors. For example, dieting was assessed by asking questions such as "Have you ever been on a diet?" and "Why were you dieting?" Behaviors were assessed by posing a question such as "Have you ever binged and purged (which is when you eat a lot of food and then make yourself throw up, vomit, or take something that makes you have diarrhea) or not?" Subjects were also asked "Right now, how would you describe yourself?" to gain an understanding of their perceptions of their weight. Psychosocial and behavioral variables including self-esteem, stress, depression, substance use (of tobacco, alcohol, or illegal drugs), and level of physical activity were also measured and scored using standardized questionnaires and inventories.

Neumark-Sztainer and Hannan reveal that 24% of the population was overweight, with 45% of the girls and 20% of the boys reporting a history of dieting. Twenty percent (13% of girls and 7% of boys) of the population reported disordered eating, which was associated with a range of behavioral variables, including overweight, low self-esteem, depression, suicidal ideation (thoughts, intent, or plans to take one's own life), and substance use. Nearly half of the adolescents recalled discussions about nutrition with a health-care provider, but just 24% of girls and 15% of boys said they had discussed eating disorders with a health-care provider.

Younger girls (grades five through eight) were significantly less likely to engage in dieting and disordered eating than older girls (grades nine through twelve), and dieting was reported by 31.1% of the fifth-grade girls and increased to 62.1% among twelfth-grade girls. The prevalence of disordered eating was highest among Hispanic girls and lowest among non-Hispanic African-American girls, and the prevalence of dieting was highest among white non-Hispanic girls and lowest among non-Hispanic African-American girls. Neumark-Sztainer and Hannan observe that the prevalence rates of dieting behaviors were lowest among African-American girls, suggesting that African-American girls may experience lower levels of body dissatisfaction than white girls.

Alcohol and drug use were directly associated with dieting and disordered eating among girls and boys; however, the association between substance use and disordered eating was stronger than the association between substance use and dieting. Tobacco use was associated with dieting and disordered eating among girls, but not among boys.

Neumark-Sztainer and Hannan note that "about half of the youth reported that a health care provider had discussed nutrition and weight issues with them" and observe that even though the content of such discussions was unclear, "at least the youth remembered that these issues had been discussed." They conclude that "the high rates of dieting and disordered eating behaviors, coupled with the high prevalence of obesity found in this and previous studies, indicate a clear need for interventions aimed at the primary and secondary prevention of weight-related disorders. The large scope of the problem and the complexity of the issues at hand indicate that there is a need for multiple interventions at the individual and familial level (e.g., within clinical practices), at the group level (e.g., within school settings), and at the community or larger societal level (e.g., changes in the physical and social environment)."

# CHAPTER 5
# DIETARY TREATMENT FOR OVERWEIGHT
# AND OBESITY

*We rarely repent of having eaten too little.*

—Thomas Jefferson

Americans have long been consumed with losing weight, seemingly willing to suffer deprivation and to embrace each new diet that debuts—even if the "new diet" is simply a twist on a previous weight-loss plan. The fixation with weight loss is so long-standing that even the word *diet* has assumed a new meaning. As a verb, diet means to eat and drink a prescribed selection of foods; however, since the latter part of the twentieth century dieting became synonymous with an effort to lose weight.

During the nineteenth century, fashionable body shapes and sizes varied from decade to decade, but most periods celebrated plumpness as a sign of health and prosperity and considered being thin a sign of poverty and ill health. At the turn of the twentieth century, rising interest in dieting seemingly coincided with some of the social and cultural changes that would make it necessary: food became increasingly plentiful, and sedentary work and public transportation reduced Americans' level of physical activity. In *Fat History: Bodies and Beauty in the Modern West* (2002), Peter N. Stearns explains how fat became "a turn-of-the-century target" with anti-fat sentiments intensifying from the 1920s to the present.

Stearns asserts that the contemporary obsession with fat arose in tandem with the dramatic growth in consumer culture, women's increasing equality, and changes in women's sexual and maternal roles. Dieting, with its emphasis on deprivation, self-control, and moral discipline, seemed the perfect antidote to the indulgence of consumer culture, and Stearns contends that "weight morality bore disproportionately on women precisely because of their growing independence, or seeming independence, from other standards."

Fashion trends fueled anti-fat sentiments as women shed the corsets that had created the illusion of narrow waists and aspired to duplicate the wasp-waisted silhouettes by becoming slimmer. The shorter, close-fitting "flapper" dresses of the 1920s revealed women's legs and rekindled their desires to be slender. The emergence of the first actuarial tables (data compiled to assess insurance risk and formulate life insurance premiums), which showed the relationship between overweight and premature mortality (death), reinforced the growing sentiment that thinness was the key to health and longevity. Capitalizing on the increasing interest in monitoring and reducing body weight, the new Detecto and Health-o-Meter bathroom scales enabled people to weigh themselves regularly in the privacy of their own homes, as opposed to relying on periodic visits to the physician's office or pharmacy to use the balance scale.

## SELECTED MILESTONES IN THE HISTORY OF DIETING

Not unlike fashion trends, the history of dieting reveals the emergence and popularity of specific diets, which over time are cast aside in favor of different approaches but then are recycled and resurface as "new and miraculous." The first low-carbohydrate diet to earn popular acclaim was described by William Banting (1797–1878) in the 1860s. In *Letter on Corpulence, Addressed to the Public* (1863), Banting, then sixty-six years old, claimed that by adhering to his low-carbohydrate regimen he was never hungry and had lost 46 of his initial 202 pounds in one year.

The early 1900s marked the beginning of diets that restricted calories. *Diet and Health, with Key to the Calories* (1918) by Lulu Hunt Peters (1873–1930) advised readers to think in terms of consuming calories rather than food items and remained in print for twenty years. Peters wrote, "You should know and use the word calorie as frequently, or more frequently, than you use the foot, yard, quart, gallon and so forth ... hereafter you are going to eat calories of food. Instead of saying one slice of bread, or a piece of pie, you will say 100 calories of bread, 350 calories of pie." The 1920s also saw the rise of very-low-calorie diets to promote weight loss. For example, the Hollywood eighteen-day diet advised just 585 calories per day, which required the dieter to eat mostly citrus fruit.

Throughout the 1920s and 1930s the low-calorie diet remained a popular weight-loss strategy; however, other approaches, such as food-limiting plans that restricted dieters to just one or two foods (e.g., lamb chops, pineapples, grapefruits, or cabbage), were introduced, as were diets that prescribed combinations of certain foods and forbid others. For example, some diets prohibited eating protein and carbohydrates together; others were more specific, advising which vegetables could be served together. The 1930s also saw the first condemnations of carbohydrates as causes of overweight. A high-fat, low-fiber diet consisting primarily of milk and meat was thought to be protective against disease. The Italian poet Filippo Tommaso Marinetti (1876–1944) exhorted Italians to forgo their pasta because he claimed it made them sluggish, pessimistic, and fat.

In 1943 the U.S. Department of Agriculture (USDA) released the "Basic Seven" food guide in the *National Wartime Nutrition Guide*. It emphasized a patriotic wartime austerity diet that included between two to four servings of protein-rich meat and milk products, three servings of fruits or vegetables, and the rather vague recommendations of "bread, flour, and cereals every day and butter, fortified margarine—some daily."

In 1948 Esther Manz (1908–1996), a 5-foot, 2-inch, 208-pound homemaker established Take Off Pounds Sensibly (TOPS; http://www.tops.org/), the first support-group program for weight loss. Manz was inspired to start the program after she attended childbirth preparation classes, where women benefited from mutual support and encouragement. As of 2007, the annual membership of $24 supported the international nonprofit organization, which was based in Milwaukee. Along with weekly meetings and private weigh-ins, TOPS participants are encouraged to adhere to a calorie-counting meal plan based on a program developed by the American Dietetics Association. In 2007 TOPS boasted about two hundred thousand members in ten thousand chapters worldwide. Members who achieve their weight goals become KOPS (Keep Off Pounds Sensibly), and often keep attending meetings to maintain their weight and serve as role models for others.

In 1950 the American physician and biophysicist John Gofman (1918–2007) of the University of California at Berkeley hypothesized that blood cholesterol was involved in the rise in coronary heart disease. Gofman found not only that heart attacks correlated with elevated levels of cholesterol but also that the cholesterol was contained in one lipoprotein particle: low-density lipo-

protein (LDL). Early reports of the connection between overweight and elevated blood cholesterol intensified interest in weight loss, which was now promoted as a strategy for preventing heart disease. During the late 1950s, injections of human chorionic gonadotropin, which was derived from the urine of pregnant women or animals, enjoyed fleeting popularity as a weight-loss agent; however, it was quickly proven entirely ineffective. Fad diets, such as a diet advocating the consumption of several bananas to satisfy sugar cravings, and another that involved ingesting a blend of oils to boost metabolism, continued to lure Americans seeking quick weight loss. In 1959 the American Medical Association called dieting a "national neurosis."

In 1960 Metrecal, the first high-protein beverage, was widely advertised by the Mead Johnson Company as a weight-reducing aid. It was originally sold as a powder, which when mixed with a quart of water yielded four eight-ounce glasses intended to serve as four meals per day, totaling nine hundred calories. The powder was made from milk, soy flour, starch, corn oil, yeast, vitamins, coconut oil, and vanilla, chocolate, or butterscotch flavoring. The low-calorie regimen enabled a dieter to lose ten pounds in a few weeks, without the trouble of meal preparation or counting calories. Later, Metrecal was sold in a premixed, liquid form that could be consumed right from the can. Mead Johnson made over $10 million selling Metrecal in the first two years. It was the forerunner of liquid diet products such as Slim-Fast.

The 1960s also witnessed the birth of Overeaters Anonymous (OA) and Weight Watchers. OA began as a support group modeled on the twelve-step physical, emotional, and spiritual recovery program used by Alcoholics Anonymous. In "About Overeaters Anonymous" (2008, http://www.oa.org/about_oa.html), the OA notes that about sixty-five hundred OA groups meet each week in sixty-five countries. In 1961 Jean Nidetch (1923–), an overweight housewife in New York City, invited a few friends to her home to gain support for her efforts to diet and overcome an "obsession for cookies." From this first meeting, the friends gathered weekly, offering one another encouragement and sharing advice and ideas. The weekly support meetings proved successful, providing motivation and encouragement for long-term weight loss. In 1963 Nidetch incorporated Weight Watchers, and hundreds of people turned out for its first meeting. Weight Watchers grew in both size and popularity, developing nutritious and convenient eating plans, promoting exercise, cookbooks, healthy prepared food, and a magazine. The company became so successful that in 1978 it was acquired by the Heinz company. Weight Watchers states in "About Us: History and Philosophy" (2008, http://www.weightwatchers.com/about/his/hello.aspx) that about fifty thousand Weight Watcher groups meet weekly.

Two best-selling diet books also debuted during the 1960s. The first was Herman Taller's *Calories Don't*

*Count* (1961), which told dieters to avoid carbohydrates and refined sugars and to eat a high-protein diet that included large quantities of unsaturated fat. The second was Irwin Maxwell Stillman with Samm Sinclair Baker's *The Doctor's Quick Weight Loss Diet* (1967), which instructed dieters to avoid carbohydrates altogether and to consume just meat, poultry, fish, cheese, eggs, and water. Even though Taller and Stillman and Baker were not the first to tout low-carbohydrate diets, they introduced the first modern high-protein weight-loss diets. Taller's career as a diet guru ended abruptly in 1967, when he was convicted of mail fraud for the sale of safflower capsules as weight-loss aids. Stillman and Baker, however, followed up their wildly successful first book with several other additional weight-loss titles, including *The Doctor's Quick Teenage Diet* (1971), one of the first diet books to address the needs of overweight adolescents. High-protein, low-carbohydrate diets washed down by liberal amounts of alcohol were also advocated by other books from the 1960s, including Gardener Jameson's *The Drinking Man's Diet* (1965) and Sidney Petrie's *Martinis and Whipped Cream: The New Carbo-Cal Way to Lose Weight and Stay Slim* (1966) and *The Lazy Lady's Easy Diet: A Fast-Action Plan to Lose Weight Quickly for Sustained Slenderness and Youthful Attractiveness* (1969).

During this same decade, chemically processed, non-nutritive sweeteners were marketed as calorie- and guilt-free substitutes that enabled dieters to enjoy many of their favorite sweet treats. Saccharin, which is three hundred times sweeter than sugar, was the first artificial sweetener to be widely used in diet foods and beverages. Other chemically processed, artificial, and nonnutritive sweeteners followed, including cyclamate, which was withdrawn from the U.S. market in 1969 because research findings in animals suggested that it might increase the risk of bladder cancer in humans. According to the National Cancer Institute, more recent animal studies fail to demonstrate that cyclamate is a carcinogen (a substance known to cause cancer) or a cocarcinogen (a substance that enhances the effect of a cancer-causing substance); regardless, cyclamate is not approved for commercial use as a food additive in the United States. Aspartame and acesulfame K were approved by the U.S. Food and Drug Administration (FDA) in 1981 and 1988, respectively. Sucralose, a noncaloric sweetener, was approved by the FDA for general use in 1999. Sucralose has gained popularity because it is derived from and tastes like sugar, has no aftertaste, does not promote tooth decay, and is deemed safe for use by pregnant women and diabetics, as well as by those in the general population who are trying to cut down on their sugar intake. In 2002 the FDA approved neotame, another nonnutritive sweetener, for use as a general-purpose sweetener. Neotame is approximately seven thousand to thirteen thousand times sweeter than sugar and has been approved for

use in food products including baked goods, nonalcoholic beverages (including soft drinks), chewing gum, confections and frostings, frozen desserts, gelatins and puddings, jams and jellies, processed fruits and fruit juices, toppings, and syrups.

However, some researchers think sugar substitutes may sabotage dieters by interfering with the body's own innate ability to monitor calorie consumption based on a food's flavor—sweet or savory. Susan E. Swithers, Alicia Doerflinger, and Terry L. Davidson indicate in "Consistent Relationships between Sensory Properties of Savory Snack Foods and Calories Influence Food Intake in Rats" (*International Journal of Obesity*, vol. 30, 2006) that absent sensory clues about the relative caloric value of a food, both animals and humans may overeat and as a result become overweight.

In 1972 the cardiologist Robert C. Atkins (1930–2003) published *Dr. Atkins' Diet Revolution*, which provided a new explanation about how an extremely low-carbohydrate diet targets insulin to promote weight loss. Atkins called insulin, the hormone that regulates blood sugar levels, a "fat-producing hormone." He asserted that most overeaters are continually in a state of hyperinsulinism primed and ever-ready to convert excess carbohydrates to fat. As a result, they have excess circulating insulin, which primes the body to store fat. Atkins contended that when people with hyperinsulinism dieted to lose weight—especially when they reduced their fat intake and increased carbohydrate consumption—their efforts were doomed to fail. He claimed that dieters could alter their metabolisms and burn fat by inducing a state of ketosis (the accumulation of ketones from partly digested fats due to inadequate carbohydrate intake) that they monitored by testing their urine for the presence of ketones. Dieters who were tired of limiting portion size, weighing and measuring their foods, counting calories, and assiduously avoiding fatty foods such as steak, bacon, butter, cheese, and heavy cream embraced the low-carbohydrate diet with religious fervor.

The high-protein, low-carbohydrate diet not only was satisfying but also produced the immediate benefit of weight loss through water loss because the body flushes the waste products of protein digestion in the form of urine. Especially during the early weeks of dieting this additional weight loss delivered a psychological boost to dieters and provided the motivation to continue. As of 2008 many researchers and health professionals agreed with Atkins's premise that sharply limiting carbohydrate intake can help curb the appetite by maintaining even levels of insulin and preventing the insulin surges and blood sugar drops that may trigger hunger.

Even though Atkins and his devotees were celebrating weight loss, good health, and improved mood as a result of the low-carbohydrate diet, nutritionists and health professionals were countering by trumpeting the benefits of low-fat diets that were high in complex carbohydrates and fiber. Fat was demonized, and nutritionists pointed dieters to the USDA Food Guide Pyramid (http://www.mypyramid.gov/), which advised using fats sparingly. (The updated 2005 USDA Food Guide Pyramid continued to promote a low-fat diet and minimal use of fats and oils.) Critics of the low-carbohydrate regimen were concerned about the long-term health consequences of the high-fat diet and wondered if it might elevate cholesterol and triglyceride levels in people who by virtue of being overweight were already at increased risk for heart disease. There were also concerns that high-protein diets might cause kidney damage or bone loss over time. Rigorous research to compare the effectiveness and assess the health outcomes of low-carbohydrate and low-fat diets was not conducted until the late 1990s. Even though Atkins enjoyed tremendous celebrity, published a series of weight-loss books, and oversaw the sale of food products bearing his name, his significant contributions to the scientific understanding of nutrition and weight loss were not fully appreciated until the year preceding his death in 2003.

The 1970s also had its share of fad diets. Robert Linn's *The Last Chance Diet—When Everything Else Has Failed* (1976) advised a protein-sparing fast, which was so dangerously deficient in essential nutrients that several deaths were attributed to it. In *The Complete Scarsdale Medical Diet Plus Dr. Tarnower's Lifetime Keep-Slim Program* (1978), Herman Tarnower (1910–1980) advocated a fat-free, high-protein diet that allowed seven hundred calories per day.

At the close of the 1970s, Nathan Pritikin's (1915–1985) *The Pritikin Program for Diet and Exercise* (1979) championed a nearly fat-free diet that consisted of fresh and cooked fruits and vegetables, whole grains, breads and pasta, and small amounts of lean meat, fish, and poultry, in concert with daily aerobic exercise. Advocating heart health and fitness, in 1976 Pritikin opened the Pritikin Longevity Centers, where people could learn to modify not only their diets but also their lifestyles. Even though Pritikin's plan, which essentially eliminated fat from the diet, was considered by many health professionals too extreme to gain long-term adherents, Pritikin enjoyed as loyal a following as did Atkins.

During the 1980s Judy Mazel (1943–2007) resurrected the notion of specific food combinations as central to weight loss in *The Beverly Hills Diet* (1981). Mazel asserted that eating foods together, such as protein and carbohydrates, destroyed digestive enzymes and caused weight gain and poor digestion. Her diet featured an abundance of fruit, and some observers speculated that weight loss attributable to the diet resulted from the combined effects of caloric restriction and fluid loss

resulting from diarrhea. Celebrity endorsements and the glamorous author's frequent media interviews stimulated interest in the diet.

In 1983 Jenny Craig (1932–) launched a weight-loss program that would become one of the world's two largest diet companies (the other being Weight Watchers). With 660 centers in North America, Australia, Guam, New Zealand, and Puerto Rico, the company (2007, http://www.jennycraig.com/corporate/company/index.asp) that bears her name sells prepared foods, along with other weight-loss materials. The company offers telephone and online support and home delivery of food and support materials. In 2002 company founders Jenny Craig and Sid Craig sold their majority stake in the company to ACI Capital Co. and MidOcean Capital Partners, Inc., but retain 20% interest in the company.

The 1990s served up so-called new and revised versions of high-protein, high-fat, and low-carbohydrate diets and the low-fat diet as well as an update of Mazel's Beverly Hills diet. The cardiologist Dean Ornish (1953–) rekindled enthusiasm for low-fat eating with *Eat More, Weigh Less: Dr. Dean Ornish's Life Choice Program for Losing Weight Safely While Eating Abundantly* (1993). Atkins's 1999 update of *Dr. Atkins' New Diet Revolution*, which offered advice about how to achieve total wellness and weight loss, spent more than four years on the *New York Times* best-seller list and won over a new generation of dieters. Ornish's approach was directly opposed to Atkins's—he espoused the health benefits of vegetarianism and limiting dietary fat to just 10% of the total daily calories. However, both physicians took a holistic approach to health and weight loss, encouraging readers to engage in moderate exercise, foster social support, and reconnect with themselves to support their physical and emotional well-being.

Still, the diet that generated the most fanfare during the late 1990s was by the biochemist Barry Sears (1947–), who published *The Zone: A Dietary Road Map* (1995). Sears's high-protein, low-carbohydrate plan promised that by eating the correct ratio of protein, fat, and carbohydrates, dieters would lose weight permanently, avoid disease, enhance mental productivity, achieve maximum physical performance, balance and control insulin levels, and enter "that mysterious but very real state in which the body and mind work together at their ultimate best."

Since the turn of the twenty-first century, the fiery debate about the merits of low-carbohydrate and low-fat diets have intensified, with both sides citing scientific evidence to support the supremacy of one diet as the healthier and more effective weight-loss strategy. The cardiologist Arthur Agatston (1947–) offered a kind of compromise between the two regimens in *The South Beach Diet: The Delicious, Doctor-Designed, Foolproof*

*Plan for Fast and Healthy Weight Loss* (2003). Agatston condemned simple carbohydrates, such as white flour and white sugar, citing them as the source of the continuous cravings that sabotage dieters, but did not eliminate complex carbohydrates from the diet. (Carbohydrates are classified as simple or complex. The classification depends on the chemical structure of the particular food source and reflects how quickly the sugar is digested and absorbed. Simple carbohydrates have one or two sugars, whereas complex carbohydrates have three or more.) Agatston's diet program was a modified carbohydrate plan that recommended plenty of high-fiber foods, lean proteins, and healthy fats, while cutting back on, but not entirely banishing, bread, rice, pastas, and fruits.

Marian Burros notes in "Make That Steak a Bit Smaller, Atkins Advises Today's Dieters" (*New York Times*, January 18, 2004) that in 2004 the Atkins organization, which had previously advised dieters to satisfy their appetites with ample quantities of steak, bacon, eggs, heavy cream, and other saturated fats, modified its position. According to Burros, Colette Heimowitz, the director of research and education for Atkins Nutritionals, advised health professionals and dieters that just 20% of a dieter's calories should come from saturated fat. However, she and other Atkins representatives asserted that this did not represent a change in the diet itself, simply a revision in communicating how the diet should be followed. Diet industry observers maintained that the warning to reduce the consumption of saturated fat was in direct response to the debut of the South Beach Diet and other low-carbohydrate regimens that called for less saturated fat. Heimowitz asserted that the change was made because "we want physicians to feel comfortable with this diet, and we want people who are going to their physicians with this diet to feel comfortable."

Americans' enthusiasm for low-carbohydrate diets cooled during 2004, and Atkins Nutritionals Inc., the company that catapulted low-carbohydrate diets into a national obsession, filed for bankruptcy court protection in August 2005. Many dieters abandoned low-carbohydrate diets in favor of regimens focused on the glycemic index (GI)— a ranking system for carbohydrates according to their immediate effect on blood glucose levels, in which a numerical value is assigned to a carbohydrate-rich food based on its average increase in blood glucose. The GI measures how fast and how much a food raises blood glucose levels. Weight-loss diets based on the GI emphasize sharply restricting high-index foods and consuming primarily low-index foods.

Examples of foods with GI scores of seventy or above are cake, cookies, doughnuts, honey, French fries, rice, baked potato, and white bread. In contrast, lentils have a GI of twenty-nine, whereas broccoli, peanuts, and spinach have GIs of less than fifteen. Carbohydrates that break down slowly, such as whole-grain breads and cereals, beans, leafy greens, or cruciferous vegetables, generate slower glucose release into the blood stream and lower GI scores—fifty or less. Eating low GI foods supports weight loss by enhancing satiety (the feeling of fullness or satisfaction after eating) and thereby decreasing total food consumption.

In 2004 diet books that extolled the virtues of the low GI diet—including Michel Montignac's *Eat Yourself Slim* (1999), Rick Gallop's *The G.I. Diet: The Easy, Healthy Way to Permanent Weight Loss* (2003), and H. Leighton Steward et al.'s *The New Sugar Busters* (2003)—became quite popular. Proponents of low GI diets observed that the regimen not only produced weight loss but also improved overall health by reducing the risk for both Type 2 diabetes and cardiovascular disease.

Even though diet industry observers cannot predict the next craze, they are certain that a replacement for the low-carbohydrate diet will emerge. Contenders among the diets and diet books that debuted since 2004 include:

- *French Women Don't Get Fat* (2004) by Mireille Guiliano contends that the French are able to eat croissants and chocolate without becoming overweight because they take time to savor flavors and eat thoughtfully

- *The Fat Resistance Diet* (2005) by Leo Galland advises a diet rich in fish and other low-fat protein, vegetables, fruit, nuts, and green tea to help relieve inflammation and restore sensitivity to leptin, a hormone involved in fat metabolism that sends satiety signals to the brain

- *The Perricone Weight Loss Diet: A Simple 3-Part Plan to Lose the Fat, the Wrinkles, and the Years* (2005) by Nicholas Perricone recommends a diet composed of high- as opposed to low-GI foods and healthy (Omega-3-rich) versus unhealthy fats

- *The 3-Hour Diet: How Low Carb Makes You Fat and Timing Will Sculpt You Slim* (2005) by Jorge Cruise recommends eating frequently and timing meals and snacks to "stoke the metabolism"

- *The Diet Code: Revolutionary Weight-Loss Secrets from Da Vinci and the Golden Ratio* (2006) by Stephen Lanzalotta promotes Mediterranean-style eating and emphasizes bread, fish, cheese, vegetables, meat, nuts, and wine

- *The Total Wellbeing Diet* (2006) by Manny Noakes details a low-carbohydrate, high-protein diet developed by scientists at the Commonwealth Scientific and Industrial Research Organization to help Australians lose weight

- *The Rice Diet Cookbook: 150 Easy, Everyday Recipes and Inspirational Success Stories from the Rice Diet Community* (2007) by Kitty Gurkin Rosati counters the low-carbohydrate diet trend with a low-salt diet featuring rice, vegetables, and fruit; this diet was developed in 1939 by Walter Kempner (1903–1997) at Duke University.

TABLE 5.1

**Food energy and macronutrients per capita per day, selected years, 1909–2004**

| Year | Food energy (kcal) | Carbohydrate (g) | Fiber (g) | Protein (g) | Fat (g) | Saturated fatty acids (g) | Monounsaturated fatty acids (g) | Polyunsaturated fatty acids (g) | Cholesterol (mg) |
|---|---|---|---|---|---|---|---|---|---|
| 1909–19 | 3400 | 487 | 28 | 96 | 120 | 50 | 47 | 13 | 440 |
| 1920–29 | 3400 | 478 | 26 | 92 | 127 | 54 | 49 | 15 | 470 |
| 1930–39 | 3300 | 452 | 25 | 89 | 129 | 55 | 50 | 15 | 450 |
| 1940–49 | 3300 | 431 | 24 | 98 | 138 | 56 | 54 | 18 | 510 |
| 1950–59 | 3100 | 391 | 20 | 93 | 138 | 55 | 55 | 19 | 500 |
| 1960–69 | 3100 | 383 | 18 | 93 | 143 | 54 | 56 | 22 | 470 |
| 1970–79 | 3200 | 396 | 20 | 98 | 144 | 49 | 58 | 27 | 440 |
| 1980–89 | 3400 | 420 | 21 | 101 | 151 | 50 | 61 | 31 | 420 |
| 1990–99 | 3600 | 481 | 24 | 109 | 151 | 48 | 64 | 31 | 400 |
| 2000 | 3900 | 497 | 25 | 113 | 173 | 54 | 77 | 36 | 420 |
| 2001 | 4000 | 508 | 27 | 115 | 174 | 54 | 77 | 36 | 420 |
| 2002 | 3900 | 484 | 24 | 112 | 180 | 57 | 79 | 37 | 420 |
| 2003 | 3900 | 482 | 25 | 112 | 178 | 56 | 78 | 37 | 420 |
| 2004 | 3900 | 481 | 25 | 113 | 179 | 56 | 79 | 37 | 430 |

Note: kcal=kilo calorie. g=gram. mg=milligram.

SOURCE: H.A.B. Hiza and L. Bente, "Table 1. Food Energy and Macronutrients per Capita per Day in the U.S. Food Supply, Selected Years," in *Nutrient Content of the U.S. Food Supply, 1909–2004: A Summary Report*, U.S. Department of Agriculture, Center for Nutrition Policy and Promotion, February 2007, http://www.cnpp.usda.gov/publications/foodsupply/FoodSupply1909–2004Report.pdf (accessed October 23, 2007)

## AMERICANS' DIETS

H. A. B. Hiza and L. Bente of the Center for Nutrition Policy and Promotion offer in *Nutrient Content of the U.S. Food Supply, 1909–2004: A Summary Report* (February 2007, http://www.cnpp.usda.gov/publications/foodsupply/FoodSupply1909-2004Report.pdf) historical data about the nutrients in the U.S. food supply and trends in Americans' diets. Table 5.1 shows the consumption of macronutrients (nutrients that the body uses in relatively large amounts: carbohydrates, fats, and proteins) in selected years from 1909 to 2004. Trends include:

- An increase of eight hundred calories per day from 1960–69 to 2004

- An increase of twenty grams of protein per day from 1960–69 to 2004

- An increase of thirty-six grams of fat consumption from 1960–69 to 2004

Table 5.2 shows Americans' decreased consumption of whole milk in favor of low-fat milk; increased consumption of cheese, legumes, nuts, and soy; and decreased use of grain products. It also documents the shift from butter to margarine use, a decline in total consumption of vegetables, and a dramatic increase in consumption of salad, cooking, and other edible oils.

### Dietary Guidelines for Americans, 2005

Every five years the *Dietary Guidelines for Americans* are updated and revised to translate the most current scientific knowledge about individual nutrients and food components into dietary recommendations that may be adopted by the public. The recommendations are based on the preponderance of scientific evidence for reducing the risk of chronic disease and promoting health. Even though the recommendations focus on nutritional content, they recognize that a combination of poor diet and physical inactivity can lead to chronic diseases that include cardiovascular disease, Type 2 diabetes, hypertension (high blood pressure), osteoporosis, and certain cancers.

The U.S. Department of Health and Human Services and USDA's *Dietary Guidelines for Americans, 2005* (January 2005, http://www.health.gov/dietaryguidelines/dga2005/document/pdf/DGA2005.pdf) characterizes a healthy diet as one that includes plenty of fruits, vegetables, whole grains, and fat-free or low-fat milk and milk products, as well as lean meats, poultry, fish, beans, eggs, and nuts. A healthy diet is also low in saturated fats, trans fats (artificial fats created through the hydrogenation of oils, which solidifies the oil and limits the body's ability to regulate cholesterol), cholesterol, salt, and added sugars. Specific recommendations stipulate that fewer than 10% of calories should come from saturated fatty acids, and trans fatty acids, which are considered to be the most harmful to health, should be avoided. Cholesterol intake should be less than three hundred milligrams per day. Total fat intake should not exceed 20% to 35% of calories. Preferred fat sources are fish, nuts, and vegetable oils containing polyunsaturated and monounsaturated fatty acids. Lean, low-fat, or fat-free meats, poultry, dry beans, and milk or milk products are preferable to full-fat foods.

In general, the guidelines encourage most Americans to eat fewer calories, increase their physical activity, and choose nutrient-dense foods. They advocate increased consumption of fruits, vegetables, whole grains, and fat-free or low-fat milk and milk products. For example, two cups of fruit and two and a half cups of vegetables per day are recommended for a two-thousand-calorie diet,

**TABLE 5.2**

**Food energy contributed from major food groups, selected years, 1909–2004**

| Year | Meat, poultry, and fish | | | | Dairy products | | | | | Eggs | Legumes, nuts & soy | Grain products |
|---|---|---|---|---|---|---|---|---|---|---|---|---|
| | Meat | Poultry | Fish | Total | Whole milk | Lowfat milk | Cheese | Other | Total | | | |
| | | | | | | | Percent | | | | | |
| 1909–19 | 13.3 | 0.9 | 0.6 | 14.7 | 5.1 | 0.8 | 0.6 | 2.1 | 8.5 | 1.8 | 2.3 | 37.5 |
| 1920–29 | 12.9 | 0.9 | 0.5 | 14.3 | 5.6 | 0.7 | 0.7 | 2.8 | 9.7 | 1.9 | 2.4 | 32.0 |
| 1930–39 | 12.5 | 0.9 | 0.5 | 13.9 | 5.9 | 0.6 | 0.8 | 3.3 | 10.6 | 1.8 | 2.8 | 29.3 |
| 1940–49 | 14.6 | 1.2 | 0.5 | 16.3 | 7.2 | 0.5 | 1.0 | 3.7 | 12.4 | 2.1 | 3.1 | 26.4 |
| 1950–59 | 15.5 | 1.5 | 0.5 | 17.5 | 7.1 | 0.4 | 1.3 | 3.5 | 12.5 | 2.4 | 3.0 | 22.6 |
| 1960–69 | 16.2 | 2.2 | 0.5 | 18.9 | 6.1 | 0.7 | 1.6 | 3.1 | 11.6 | 2.1 | 3.0 | 21.1 |
| 1970–79 | 14.1 | 2.8 | 0.6 | 17.4 | 4.6 | 1.4 | 2.2 | 2.8 | 11.1 | 1.9 | 3.2 | 20.2 |
| 1980–89 | 11.9 | 3.4 | 0.6 | 15.9 | 2.9 | 1.9 | 2.9 | 2.7 | 10.3 | 1.6 | 3.2 | 21.9 |
| 1990–99 | 8.9 | 4.2 | 0.6 | 13.7 | 1.6 | 2.1 | 3.2 | 2.7 | 9.6 | 1.4 | 3.1 | 24.6 |
| 2000 | 8.3 | 4.4 | 0.6 | 13.2 | 1.4 | 1.9 | 3.3 | 2.5 | 9.0 | 1.3 | 3.0 | 23.8 |
| 2001 | 8.0 | 4.2 | 0.6 | 12.8 | 1.3 | 1.7 | 3.3 | 2.1 | 8.4 | 1.3 | 2.9 | 25.4 |
| 2002 | 8.3 | 4.5 | 0.6 | 13.3 | 1.3 | 1.7 | 3.3 | 2.1 | 8.5 | 1.4 | 3.0 | 23.3 |
| 2003 | 8.2 | 4.5 | 0.6 | 13.3 | 1.3 | 1.7 | 3.4 | 2.2 | 8.5 | 1.4 | 3.1 | 23.6 |
| 2004 | 8.2 | 4.6 | 0.6 | 13.4 | 1.2 | 1.7 | 3.4 | 2.2 | 8.6 | 1.4 | 3.1 | 23.5 |

| Year | Fruits | | | Vegetables | | | | | Fats and oils | | | | | | Sugars & sweeteners | Miscellaneous |
|---|---|---|---|---|---|---|---|---|---|---|---|---|---|---|---|---|
| | Citrus | Non-Citrus | Total | White potatoes | Dark green/ deep yellow | Tomatoes | Other | Total | Butter | Margarine | Shortening | Lard & beef tallow | Salad, cooking & other edible oils | Total | | |
| | | | | | | | | | Percent | | | | | | | |
| 1909–19 | 0.2 | 2.7 | 2.9 | 4.0 | 0.9 | 0.4 | 1.3 | 6.5 | 4.4 | 0.6 | 3.1 | 3.8 | 0.7 | 12.6 | 12.9 | 0.3 |
| 1920–29 | 0.3 | 2.8 | 3.1 | 3.5 | 0.9 | 0.4 | 1.4 | 6.1 | 4.6 | 0.7 | 2.7 | 4.2 | 1.4 | 13.5 | 16.4 | 0.5 |
| 1930–39 | 0.5 | 2.7 | 3.1 | 3.1 | 0.9 | 0.4 | 1.5 | 6.0 | 4.8 | 0.7 | 3.4 | 4.2 | 2.0 | 15.1 | 16.8 | 0.6 |
| 1940–49 | 0.7 | 2.5 | 3.2 | 2.9 | 0.8 | 0.5 | 1.6 | 5.8 | 3.4 | 1.1 | 3.2 | 4.3 | 2.3 | 14.3 | 15.7 | 0.6 |
| 1950–59 | 0.8 | 2.4 | 3.1 | 2.7 | 0.5 | 0.5 | 1.5 | 5.2 | 2.5 | 2.3 | 3.8 | 3.8 | 3.4 | 15.9 | 17.2 | 0.6 |
| 1960–69 | 0.7 | 2.1 | 2.8 | 2.8 | 0.4 | 0.5 | 1.4 | 5.1 | 1.8 | 2.8 | 5.1 | 2.2 | 4.8 | 16.8 | 17.8 | 0.7 |
| 1970–79 | 1.0 | 2.1 | 3.1 | 2.7 | 0.4 | 0.6 | 1.8 | 5.6 | 1.3 | 3.1 | 6.0 | 1.1 | 6.9 | 18.4 | 18.4 | 0.8 |
| 1980–89 | 1.0 | 2.4 | 3.4 | 2.6 | 0.4 | 0.6 | 1.6 | 5.2 | 1.2 | 2.9 | 6.7 | 1.0 | 8.0 | 19.8 | 17.8 | 0.9 |
| 1990–99 | 0.9 | 2.4 | 3.3 | 2.5 | 0.4 | 0.6 | 1.6 | 5.2 | 1.1 | 2.4 | 6.7 | 0.9 | 8.3 | 19.4 | 18.8 | 0.9 |
| 2000 | 1.0 | 2.2 | 3.1 | 2.4 | 0.4 | 0.6 | 1.4 | 4.8 | 1.0 | 1.7 | 8.9 | 1.4 | 9.6 | 22.6 | 18.2 | 0.9 |
| 2001 | 1.0 | 2.1 | 3.1 | 2.4 | 0.4 | 0.5 | 1.4 | 4.6 | 1.0 | 1.6 | 9.1 | 1.2 | 10.0 | 22.8 | 17.8 | 0.9 |
| 2002 | 0.8 | 2.1 | 3.0 | 2.3 | 0.3 | 0.6 | 1.4 | 4.6 | 1.0 | 1.5 | 9.6 | 1.3 | 10.9 | 24.3 | 17.8 | 0.8 |
| 2003 | 0.9 | 2.2 | 3.1 | 2.4 | 0.4 | 0.6 | 1.4 | 4.8 | 1.0 | 1.2 | 9.2 | 1.5 | 11.0 | 23.9 | 17.4 | 0.9 |
| 2004 | 0.9 | 2.2 | 3.1 | 2.3 | 0.4 | 0.6 | 1.4 | 4.7 | 1.0 | 1.2 | 9.2 | 1.4 | 11.1 | 23.9 | 17.3 | 0.9 |

SOURCE: H.A.B. Hiza and L. Bente, "Table 4. Food Energy Contributed from Major Food Groups to the U.S. Food Supply, Selected Years," in *Nutrient Content of the U.S. Food Supply, 1909–2004: A Summary Report*, U.S. Department of Agriculture, Center for Nutrition Policy and Promotion, February 2007, http://www.cnpp.usda.gov/publications/foodsupply/FoodSupply1909–2004Report.pdf (accessed October 23, 2007).

along with three or more servings of whole-grain products per day and three cups per day of fat-free or low-fat milk or equivalent milk products.

Two examples—the USDA Food Guide and the Dietary Approaches to Stop Hypertension (DASH) Eating Plan—offer instruction about how to allocate calories to various food groups. Table 5.3 shows the amounts of various food groups that are recommended each day or each week in the USDA Food Guide and in the DASH Eating Plan at the two-thousand-calorie level as well as the equivalent amounts for different food choices in each group. Acknowledging Americans' tendencies to eat out and eat while commuting, running errands, or working, the guidelines also offer tips for making healthy choices away from home. (See Table 5.4.)

The guidelines specifically address weight management, advising Americans "to maintain body weight in a healthy range, balance calories from foods and beverages with calories expanded," and "to prevent gradual weight gain over time, make small decreases in food and beverage calories and increase physical activity." For people who are overweight, the guidelines advise gradual, steady weight loss achieved by decreasing caloric consumption while maintaining sufficient nutrients and increasing physical activity. Parents of overweight children are counseled to reduce the rate of weight gain while children grow and develop and to consult a health-care provider before placing children on weight-reduction diets. Pregnant women are advised to gain weight as instructed by their health-care providers, and breastfeeding mothers are reassured that modest weight loss is safe and will not harm the development of nursing infants.

## HOW WEIGHT-LOSS DIETS WORK

Research demonstrates that weight loss is associated with the length of the diet, pre-diet weight (people who

**TABLE 5.3**

**Sample USDA Food Guide and the DASH Eating Plan at the 2,000-calorie level**

| Food groups and subgroups | USDA Food Guide amount[a] | DASH Eating Plan amount | Equivalent amounts |
|---|---|---|---|
| Fruit group | 2 cups (4 servings) | 2 to 2.5 cups (4 to 5 servings) | 1/2 cup equivalent is: 1/2 cup fresh, frozen, or canned fruit; 1 med fruit; 1/4 cup dried fruit; USDA: 1/2 cup fruit juice; DASH: 3/4 cup fruit juice |
| Vegetable group<br>Dark green vegetables<br>Orange vegetables<br>Legumes (dry beans)<br>Starchy vegetables<br>Other vegetables | 2.5 cups (5 servings)<br>3 cups/week<br>2 cups/week<br>3 cups/week<br>3 cups/week<br>6.5 cups/week | 2 to 2.5 cups (4 to 5 servings) | 1/2 cup equivalent is: 1/2 cup of cut-up raw or cooked vegetable; 1 cup raw leafy vegetable; USDA: 1/2 cup vegetable juice; DASH: 3/4 cup vegetable juice |
| Grain group<br>Whole grains<br>Other grains | 6 ounce-equivalents<br>3 ounce-equivalents<br>3 ounce-equivalents | 7 to 8 ounce-equivalents (7 to 8 servings) | 1 ounce-equivalent is: 1 slice bread; 1 cup dry cereal; 1/2 cup cooked rice, pasta, cereal; DASH: 1 oz dry cereal (1/2–1/4 cup depending on cereal type—check label) |
| Meat and beans group | 5.5 ounce-equivalents | 6 ounces or less meat, poultry, fish; 4 to 5 servings per week nuts, seeds, and dry beans[b] | 1 ounce-equivalent is: 1 ounce of cooked lean meats, poultry, fish; 1 egg; USDA: 1/4 cup cooked dry beans or tofu, 1 Tbsp peanut butter, 1/2 oz nuts or seeds; DASH: 1 1/2 oz nuts, 1/2 oz seeds, 2 Tbsp peanut butter, 1/2 cup cooked dry beans |
| Milk group | 3 cups | 2 to 3 cups | 1 cup equivalent is: 1 cup low-fat/fat-free milk, yogurt; 1 1/2 oz of low-fat or fat-free natural cheese 2 oz of low-fat or fat-free processed cheese |
| Oils | 27 grams (6 tsp) | 8 to 12 grams (2 to 3 tsp) | 1 tsp equivalent is: DASH: 1 tsp soft margarine; 1 Tbsp low-fat mayo; 2 Tbsp light salad dressing; 1 tsp vegetable oil |
| Discretionary calorie allowance<br>Example of distribution:<br>Solid fat[c]<br>Added sugars | 267 calories<br><br><br>18 grams<br>8 tsp | ~2 tsp of added sugar (5 Tbsp per week) | 1 Tbsp added sugar equivalent is: DASH: 1 Tbsp jelly or jam; 1/2 oz jelly beans; 8 oz lemonade |

Note: All servings are per day unless otherwise noted. USDA vegetable subgroup amounts and amounts of DASH (Dietary Approaches to Stop Hypertension) nuts, seeds, and dry beans are per week.

[a]The 2,000-calorie USDA Food Guide is appropriate for many sedentary males 51 to 70 years of age, sedentary females 19 to 30 years of age, and for some other gender/age groups who are more physically active.

[b]In the DASH Eating Plan, nuts, seeds, and dry beans are a separate food group from meat, poultry, and fish.

[c]The oils listed in this table are not considered to be part of discretionary calories because they are a major source of the vitamin E and polyunsaturated fatty acids, including the essential fatty acids, in the food pattern. In contrast, solid fats (i.e., saturated and trans fats) are listed separately as a source of discretionary calories.

SOURCE: "Table 1. Sample USDA Food Guide and the DASH Eating Plan at the 2,000-Calorie Level," in *Dietary Guidelines for Americans, 2005*, 6th ed., U.S. Department of Health and Human Services and U.S. Department of Agriculture, January 2005, http://www.health.gov/dietaryguidelines/dga2005/document/html/chapter2.htm (accessed October 22, 2007)

are more overweight tend to lose more weight, more quickly than those who are only mildly overweight), and the number of calories consumed. Any diet that restricts caloric intake such that calories consumed are less than those expended will promote short-term weight loss. The key to weight loss through diet is adherence—if people do not stick to their diets, then they will not lose weight. More than a century ago, Banting, in describing the benefits of his low-carbohydrate diet, wrote that "the great charms and comfort of this system are that its effects are palpable within a week of trial and creates a natural stimulus to persevere for a few weeks more."

The successes achieved using regimens that restrict dieters to a single food or food group such as grapefruit, pineapple, or cabbage are probably in part attributable to the human hankering for variety. When limited to just one food, most dieters experience boredom—there is just no appeal to eating the same food at every meal, for days on end, so naturally less food is consumed. In addition, these diets generally rely on low-calorie foods, so that even if dieters were inspired to consume fifteen grapefruits per day, their total daily caloric consumption would be about twelve hundred calories, which is sufficient to produce weight loss for most people who are overweight.

## TABLE 5.4

### Smart choices for eating out and on the go

It's important to make smart food choices and watch portion sizes wherever you are—at the grocery store, at work, in your favorite restaurant, or running errands. Try these tips:

- At the store, plan ahead by buying a variety of nutrient-rich foods for meals and snacks throughout the week.
- When grabbing lunch, have a sandwich on whole-grain bread and choose low-fat/fat-free milk, water, or other drinks without added sugars.
- In a restaurant, opt for steamed, grilled, or broiled dishes instead of those that are fried or sautéed.
- On a long commute or shopping trip, pack some fresh fruit, cut-up vegetables, string cheese sticks, or a handful of unsalted nuts—to help you avoid impulsive, less healthful snack choices.

SOURCE: "Don't Give in When You Eat Out and Are on the Go," in *Dietary Guidelines for Americans, 2005*, 6th ed., U.S. Department of Health and Human Services and U.S. Department of Agriculture, January 2005, http://www.health.gov/dietaryguidelines/dga2005/document/media/OnTheGo.pdf (accessed October 22, 2007)

## TABLE 5.5

### Low-calorie Step I diet

| Nutrient | Recommended intake |
|---|---|
| Calories[a] | Approximately 500 to 1,000 kcal/day reduction from usual intake |
| **Total fat[b]** | **30 percent or less of total calories** |
| Saturated fatty acids[c] | 8 to 10 percent of total calories |
| Monounsaturated fatty acids | Up to 15 percent of total calories |
| Polyunsaturated fatty acids | Up to 10 percent of total calories |
| Cholesterol[c] | <300 mg/day |
| Protein[d] | Approximately 15 percent of total calories |
| Carbohydrate[e] | 55 percent or more of total calories |
| Sodium chloride | No more than 100 mmol/day (approximately 2.4 g of sodium or approximately 6 g of sodium chloride) |
| Calcium[f] | 1,000 to 1,500 mg/day |
| Fiber[e] | 20 to 30 g/day |

[a]A reduction in calories of 500 to 1,000 kcal/day will help achieve a weight loss of 1 to 2 pounds/week. Alcohol provides unneeded calories and displaces more nutritious foods. Alcohol consumption not only increases the number of calories in a diet but has been associated with obesity in epidemiologic studies as well as in experimental studies. The impact of alcohol calories on a person's overall caloric intake needs to be assessed and appropriately controlled.
[b]Fat-modified foods may provide a helpful strategy for lowering total fat intake but will only be effective if they are also low in calories and if there is no compensation by calories from other foods.
[c]Patients with high blood cholesterol levels may need to use the Step II diet to achieve further reductions in LDL-cholesterol levels; in the Step II diet, saturated fats are reduced to less than 7 percent of total calories, and cholesterol levels to less than 200 mg/day. All of the other nutrients are the same as in Step I.
[d]Protein should be derived from plant sources and lean sources of animal protein.
[e]Complex carbohydrates from different vegetables, fruits, and whole grains are good sources of vitamins, minerals, and fiber. A diet rich in soluble fiber, including oat bran, legumes, barley, and most fruits and vegetables may be effective in reducing blood cholesterol levels. A diet high in all types of fiber may also aid in weight management by promoting satiety at lower levels of calorie and fat intake. Some authorities recommend 20 to 30 grams of dietary fiber daily, with an upper limit of 35 grams.
[f]During weight loss, attention should be given to maintaining an adequate intake of vitamins and minerals. Maintenance of the recommended calcium intake of 1,000 to 1,500 mg/day is especially important for women who may be at risk of osteoporosis.

SOURCE: "Table 4. Low-Calorie Step I Diet," in *The Practical Guide: Identification, Evaluation, and Treatment of Overweight and Obesity in Adults*, National Institutes of Health, National Heart, Lung, and Blood Institute, North American Association for the Study of Obesity, October 2000, http://www.nhlbi.nih.gov/guidelines/obesity/prctgd_b.pdf (accessed October 22, 2007)

Similarly, diets that involve stringent portion control effectively reduce calories to produce weight loss.

## Low-Calorie Diets

Traditional dietary therapy for weight loss generally seeks to create a deficit of five hundred to one thousand calories per day with the intent of promoting weight loss of between one to two pounds per week. Low-calorie diets for men usually range from twelve hundred to sixteen hundred calories per day; for women low-calorie diets contain between one thousand and twelve hundred calories per day. Table 5.5 is an example of the recommended percentages of nutrients in a low-calorie diet that aims to decrease risk factors for hypertension and high cholesterol as well as cause weight loss.

The most successful low-calorie diets take individual food preferences into account to custom-tailor the diet. Table 5.6 and Table 5.7 show examples of how traditional American cuisine may be used to create a low-calorie diet containing twelve hundred and sixteen hundred calories per day, respectively. Table 5.8 incorporates regional southern cuisine into a reduced-calorie diet. Table 5.9 illustrates how Asian-American cuisine may be adapted to twelve-hundred- and sixteen-hundred-calorie-per-day diets, and Table 5.10 shows how Mexican-American cuisine may be adapted for low-calorie diets. Table 5.11 is a sample of a reduced-calorie diet that vegetarians who eat milk and eggs but no meat or fish can use to lose weight. Food exchanges, such as those shown in Table 5.12, enable dieters to enjoy a variety of foods in their reduced-calorie meals, which can prevent boredom and the tendency to abandon the diet.

Research reveals that reducing fat in the diet is an effective way to reduce calories and that when low-calorie diets are combined with low-fat diets, better weight loss is achieved than through calorie reduction alone. Furthermore, even though very-low-calorie diets

that provide about five hundred calories per day have been demonstrated to produce greater initial weight loss than the low-calorie diets, the long-term weight loss is not different between the two regimens.

## Low-Carbohydrate Diets

During 2004 and 2005 several rigorous research studies reported that low-carbohydrate diets were as effective, or even more effective, in producing short-term weight loss than low-fat diets. The low-carbohydrate diets owed much of their success to adherence—dieters were better able to stick with their diets, and as a result achieved better results. Another hypothesis about the success of low-carbohydrate regimens is that dieters do not feel as hungry as they do on other diets because protein is the most satisfying of the three macronutrients: carbohydrates, fats, and proteins.

The scientific premise of low-carbohydrate diets is that consuming certain carbohydrates can cause surges in

**TABLE 5.6**

**Sample reduced calorie menus, traditional American cuisine—1,200 calories**

| | Calories | Fat (grams) | % Fat | Exchange for |
|---|---|---|---|---|
| **Breakfast** | | | | |
| • Whole wheat bread, 1 medium slice | 70 | 1.2 | 15 | (1 bread/starch) |
| • Jelly, regular, 2 tsp | 30 | 0 | 0 | (1/2 fruit) |
| • Cereal, shredded wheat, 1/2 cup | 104 | 1 | 4 | (1 bread/starch) |
| • Milk, 1%, 1 cup | 102 | 3 | 23 | (1 milk) |
| • Orange juice, 3/4 cup | 78 | 0 | 0 | (1 1/2 fruit) |
| • Coffee, regular, 1 cup | 5 | 0 | 0 | (free) |
| **Breakfast total** | **389** | **5.2** | **10** | |
| **Lunch** | | | | |
| • Roast beef sandwich: | | | | |
| Whole wheat bread, 2 medium slices | 139 | 2.4 | 15 | (2 bread/starch) |
| Lean roast beef, unseasoned, 2 oz | 60 | 1.5 | 23 | (2 lean protein) |
| Lettuce, 1 leaf | 1 | 0 | 0 | (1 vegetable) |
| Tomato, 3 medium slices | 10 | 0 | 0 | |
| Mayonnaise, low calorie, 1 tsp | 15 | 1.7 | 96 | (1/3 fat) |
| • Apple, 1 medium | 80 | 0 | 0 | (1 fruit) |
| • Water, 1 cup | 0 | 0 | 0 | (free) |
| **Lunch total** | **305** | **5.6** | **16** | |
| **Dinner** | | | | |
| • Salmon, 2 ounces edible | 103 | 5 | 44 | (2 lean protein) |
| • Vegetable oil, 1 1/2 tsp | 60 | 7 | 100 | (1 1/2 fat) |
| • Baked potato, 3/4 medium | 100 | 0 | 0 | (1 bread/starch) |
| • Margarine, 1 tsp | 34 | 4 | 100 | (1 fat) |
| • Green beans, seasoned, with margarine, 1/2 cup | 52 | 2 | 4 | (1 vegetable) (1/2 fat) |
| • Carrots, seasoned | 35 | 0 | 0 | (1 vegetable) |
| • White dinner roll, 1 small | 70 | 2 | 28 | (1 bread/starch) |
| • Iced tea, unsweetened, 1 cup | 0 | 0 | 0 | (free) |
| • Water, 2 cups | 0 | 0 | 0 | (free) |
| **Dinner total** | **454** | **20** | **39** | |
| **Snack** | | | | |
| • Popcorn, 2 1/2 cups | 69 | 0 | 0 | (1 bread/starch) |
| • Margarine, 3/4 tsp | 30 | 3 | 100 | (3/4 fat) |
| **Total** | **1,247** | **34–36** | **24–26** | |

| | | | |
|---|---|---|---|
| Calories | 1,247 | Saturated fat, % Kcals | 7 |
| Total carbohydrate, % Kcals | 58 | Cholesterol, mg | 96 |
| Total fat, % Kcals | 26 | Protein, % Kcals | 19 |
| *Sodium, mg | 1,043 | | |

Note: Calories have been rounded.
1,200: 100% RDA met for all nutrients except vitamin E 80%, vitamin $B_2$ 96%, vitamin $B_6$ 94%, calcium 68%, iron 63%, and zinc 73%.
*No salt added in recipe preparation or as seasoning. Consume at least 32 ounces of water.

SOURCE: "Appendix D. Traditional American Cuisine—1, 200 Calories," in *The Practical Guide: Identification, Evaluation, and Treatment of Overweight and Obesity in Adults*, National Institutes of Health, National Heart, Lung, and Blood Institute, North American Association for the Study of Obesity, October 2000, http://www.nhlbi.nih.gov/guidelines/obesity/prctgd_b.pdf (accessed October 22, 2007)

blood sugar and insulin that not only stimulate appetite and weight gain but also may the increase risk for diabetes and heart disease. At first, low-carbohydrate diets viewed all carbohydrates as equally harmful. Increasingly, however, low-carbohydrate diets distinguished between simple and complex carbohydrates, which contain simple or complex sugars.

Examples of single sugars from foods include fructose, which is found in fruits, and galactose, which is found in milk products. Double sugars include lactose in dairy products; maltose, which is found in certain vegetables and in beer; and sucrose (table sugar). Examples of complex carbohydrates, which are often referred to as starches, include breads, cereals, legumes, brown rice, and pastas. Simple carbohydrates occur naturally in

fruits, milk products, and vegetables; at the same time, these foods also contain vitamins and minerals. The simple carbohydrates most nutritionists call "empty calories" are the processed and refined sugars found in candy, table sugar, and sodas, as well as foods such as white flour, sugar, and polished white rice.

Besides distinguishing between simple and complex carbohydrates, low-carbohydrate regimens rely on a measure known as the glycemic index (GI), which ranks foods based on how rapidly their consumption raises blood glucose levels. The GI measures how much blood sugar increases over a period of two or three hours after a meal. Carbohydrate foods that break down quickly during digestion have the highest GI. The GI may be used to determine if a particular food will trigger the problematical

**TABLE 5.7**

**Sample reduced calorie menus, traditional American cuisine—1,600 calories**

| | Calories | Fat (grams) | % Fat | Exchange for |
|---|---|---|---|---|
| **Breakfast** | | | | |
| • Whole wheat bread, 1 medium slice | 70 | 1.2 | 15.4 | (1 bread/starch) |
| • Jelly, regular, 2 tsp | 30 | 0 | 0 | (1/2 fruit) |
| • Cereal, shredded wheat, 1 cup | 207 | 2 | 8 | (2 bread/starch) |
| • Milk, 1%, 1 cup | 102 | 3 | 23 | (1 milk) |
| • Orange juice, 3/4 cup | 18 | 0 | 0 | (1 1/2 fruit) |
| • Coffee, regular, 1 cup | 5 | 0 | 0 | (free) |
| • Milk, 1%, 1 oz | 10 | 0.3 | 27 | (1/8 milk) |
| **Breakfast total** | **502** | **6.5** | **10** | |
| **Lunch** | | | | |
| • Roast beef sandwich: | | | | |
| Whole wheat bread, 2 medium slices | 139 | 2.4 | 15 | (2 bread/starch) |
| Lean roast beef, unseasoned, 2 oz | 60 | 1.5 | 23 | (2 lean protein) |
| American cheese, low fat and low sodium, 1 slice, 3/4 oz | 46 | 1.8 | 36 | (1 lean protein) |
| Lettuce, 1 leaf | 1 | 1 | 0 | |
| Tomato, 3 medium slices | 10 | 0 | 0 | (1 vegetable) |
| Mayonnaise, low calorie, 2 tsp | 30 | 3.3 | 99 | (2/3 fat) |
| • Apple, 1 medium | 8 | 0 | 0 | (1 fruit) |
| • Water, 1 cup | 0 | 0 | 0 | (free) |
| **Lunch total** | **366** | **9** | **22** | |
| **Dinner** | | | | |
| • Salmon, 3 ounces edible | 155 | 7 | 40 | (3 lean protein) |
| • Vegetable oil, 1 1/2 tsp | 60 | 7 | 100 | (1 1/2 fat) |
| • Baked potato, 3/4 medium | 100 | 0 | 0 | (1 bread/starch) |
| • Margarine, 1 tsp | 34 | 4 | 100 | (1 fat) |
| • Green beans, seasoned, with margarine, 1/2 cup | 52 | 2 | 4 | (1 vegetable) (1/2 fat) |
| • Carrots, seasoned, with margarine, 1/2 cup | 52 | 2 | 4 | (1 vegetable) (1/2 fat) |
| • White dinner roll, 1 medium | 80 | 3 | 33 | (1 bread/starch) |
| • Ice milk, 1/2 cup | 92 | 3 | 28 | (1 bread/starch) (1/2 fat) |
| • Iced tea, unsweetened, 1 cup | 0 | 0 | 0 | (free) |
| • Water, 2 cups | 0 | 0 | 0 | (free) |
| **Dinner total** | **625** | **28** | **38** | |
| **Snack** | | | | |
| • Popcorn, 2 1/2 cups | 69 | 0 | 0 | (1 bread/starch) |
| • Margarine, 1/2 tsp | 58 | 6.5 | 100 | (1 1/2 fat) |
| **Total** | **1,613** | **50** | **28** | |

| | | | | |
|---|---|---|---|---|
| Calories | 1,613 | Saturated fat, % kcals | 8 | |
| Total carbohydrate, % kcals | 55 | Cholesterol, mg | 142 | |
| Total fat, % kcals | 29 | Protein, % kcals | 19 | |
| *Sodium, mg | 1,341 | | | |

Note: Calories have been rounded.
1,600: 100% RDA met for all nutrients except vitamin E 99%, iron 73%, and zinc 91%.
No salt added in recipe preparation or as seasoning. Consume at least 32 ounces of water.

SOURCE: "Appendix D. Traditional American Cuisine—1, 600 Calories," in *The Practical Guide: Identification, Evaluation, and Treatment of Overweight and Obesity in Adults*, National Institutes of Health, National Heart, Lung, and Blood Institute, North American Association for the Study of Obesity, October 2000, http://www.nhlbi.nih.gov/guidelines/obesity/prctgd_b.pdf (accessed October 22, 2007)

"carbohydrate–blood sugar–insulin cascade." High-GI foods are those that are rapidly digested and absorbed or transformed metabolically into glucose. These include refined starchy foods such as bread, cereal, pasta, and table sugar. In general, fiber-rich foods are low glycemic. Most vegetables, legumes, and fruits are low-GI foods.

The measurement of GI is a relatively recent practice. It began during the 1990s, following the discovery that specific carbohydrates such as potatoes and cornflakes raised blood sugar faster than others such as brown rice and oatmeal. Harvard University School of Public Health researchers used GI to calculate glycemic load—a measure that considers the food's GI and the amount of carbohydrate contained in a single serving. For example,

many whole fruits, vegetables, and grains have low glycemic loads, which when consumed prompt a moderate rise in blood glucose and insulin. When the same fruits, vegetables, and grains are squeezed or pulverized into juice or flour, their glycemic load increases—effectively rendering them with the same high glycemic load of sugar water.

After consuming a meal with a high glycemic load, blood sugar rises higher and faster than it does after eating a meal with a low glycemic load. In an effort to recover from the resulting peaks and plummets, the brain transmits a hunger signal long before the next meal is due. Wildly fluctuating blood sugar and insulin may result in overeating, which in turn causes overweight. For people who are overweight or physically inactive,

**TABLE 5.8**

**Sample reduced calorie menus, Southern cuisine**

| | 1,600 calories | 1,200 calories |
|---|---|---|
| **Breakfast** | | |
| • Oatmeal, prepared with 1% milk, low fat | 1/2 cup | 1/2 cup |
| • Milk, 1%, low fat | 1/2 cup | 1/2 cup |
| • English muffin | 1 medium | — |
| • Cream cheese, light, 18% fat | 1 T | — |
| • Orange juice | 3/4 cup | 1/2 cup |
| • Coffee | 1 cup | 1 cup |
| • Milk, 1%, low fat | 1 oz | 1 oz |
| **Lunch** | | |
| • Baked chicken, without skin | 2 oz | 2 oz |
| • Vegetable oil | 1 tsp | 1/2 tsp |
| • Salad: | | |
| Lettuce | 1/2 cup | 1/2 cup |
| Tomato | 1/2 cup | 1/2 cup |
| Cucumber | 1/2 cup | 1/2 cup |
| • Oil and vinegar dressing | 2 tsp | 1 tsp |
| • White rice | 1/2 cup | 1/4 cup |
| • Margarine, diet | 1/2 tsp | 1/2 tsp |
| • Baking powder biscuit, prepared with vegetable oil | 1 small | 1/2 small |
| • Margarine | 1 tsp | 1 tsp |
| • Water | 1 cup | 1 cup |
| **Dinner** | | |
| • Lean roast beef | 3 oz | 2 oz |
| • Onion | 1/4 cup | 1/4 cup |
| • Beef gravy, water-based | 1 T | 1 T |
| • Turnip greens | 1/2 cup | 1/2 cup |
| • Margarine, diet | 1/2 tsp | 1/2 tsp |
| • Sweet potato, baked | 1 small | 1 small |
| • Margarine, diet | 1/2 tsp | 1/4 tsp |
| • Ground cinnamon | 1 tsp | 1 tsp |
| • Brown sugar | 1 tsp | 1 tsp |
| • Corn bread prepared with margarine, diet | 1/2 medium slice | 1/2 medium slice |
| • Honeydew melon | 1/4 medium | 1/8 medium |
| • Iced tea, sweetened with sugar | 1 cup | 1 cup |
| **Snack** | | |
| • Saltine crackers, unsalted tops | 4 crackers | 4 crackers |
| • Mozzarella cheese, part skim, low sodium | 1 oz | 1 oz |
| Calories | 1,653 | 1,225 |
| Total carbohydrate, % kcals | 53 | 50 |
| Total fat, % kcals | 28 | 31 |
| *Sodium, mg | 1,231 | 867 |
| Saturated fat, % kcals | 8 | 9 |
| Cholesterol, mg | 172 | 142 |
| Protein, % kcals | 20 | 21 |

1,600: 100% RDA met for all nutrients except vitamin E 97%, magnesium 98%, iron 78%, and Zinc 90%.
1,200: 100% RDA met for all nutrients except vitamin E 82%, vitamin B₁ & B₂ 95%, vitamin B₃ 99%, vitamin B₆ 88%, magnesium 83%, iron 56%, and zinc 70%.
*No salt added in recipe preparation or as seasoning. Consume at least 32 ounces of water.

SOURCE: "Appendix D. Southern Cuisine—Reduced Calorie," in *The Practical Guide: Identification, Evaluation, and Treatment of Overweight and Obesity in Adults*, National Institutes of Health, National Heart, Lung, and Blood Institute, North American Association for the Study of Obesity, October 2000, http://www.nhlbi.nih.gov/guidelines/obesity/prctgd_b.pdf (accessed October 22, 2007)

**TABLE 5.9**

**Sample reduced calorie menus, Asian American cuisine**

| | 1,600 calories | 1,200 calories |
|---|---|---|
| **Breakfast** | | |
| • Banana | 1 small | 1 small |
| • Whole wheatbread | 2 slices | 1 slice |
| • Margarine | 1 tsp | 1 tsp |
| • Orange juice | 3/4 tsp | 3/4 tsp |
| • Milk 1%, low fat | 3/4 cup | 3/4 cup |
| **Lunch** | | |
| • Beef noodle soup, canned, low sodium | 1/2 cup | 1/2 cup |
| • Chinese noodle and beef salad: | | |
| Roast beef | 3 oz | 2 oz |
| Peanut oil | 1 1/2 tsp | 1 tsp |
| Soya sauce, low sodium | tsp | 1 tsp |
| Carrots | 1/2 cup | 1/2 cup |
| Zucchini | 1/2 cup | 1/2 cup |
| Onion | 1/4 cup | 1/4 cup |
| Chinese noodles, soft type | 1/4 cup | 1/4 cup |
| • Apple | 1 medium | 1 medium |
| • Tea, unsweetened | 1 cup | 1 cup |
| **Dinner** | | |
| • Pork stir-fry with vegetables: | | |
| Pork cutlet | 2 oz | 2 oz |
| Peanut oil | 1 tsp | 1 tsp |
| Soya sauce, low sodium | 1 tsp | 1 tsp |
| Broccoli | 1/2 cup | 1/2 cup |
| Carrots | 1 cup | 1 cup |
| Mushrooms | 1/4 cup | 1/2 cup |
| • Steamed white rice | 1 cup | 1/2 cup |
| • Tea, unsweetened | 1 cup | 1 cup |
| **Snack** | | |
| • Almond, cookies | 2 cookies | — |
| • Milk 1%, low fat | 1/2 cup | 1/2 cup |
| Calories | 1,609 | 1,220 |
| Total carbohydrate, % kcals | 56 | 55 |
| Total fat, % kcals | 27 | 27 |
| *Sodium, mg | 1,296 | 1,043 |
| Saturated fat, % kcals | 8 | 8 |
| Cholesterol, mg | 148 | 117 |
| Protein, % kcals | 20 | 21 |

1,600: 100% RDA net for all nutrients except zinc 95%, iron 87%, and calcium 93%
1,200: 100% RDA net for all nutrients except vitamin E 75%, calcium 84%, magnesium 98%, iron 66%, and zinc 77%
*No salt added in recipe preparation or as seasoning. Consume at least 32 ounces of water.

SOURCE: "Appendix D. Asian American Cuisine—Reduced Calorie," in *The Practical Guide: Identification, Evaluation, and Treatment of Overweight and Obesity in Adults*, National Institutes of Health, National Heart, Lung, and Blood Institute, North American Association for the Study of Obesity, October 2000, http://www.nhlbi.nih.gov/guidelines/obesity/prctgd_b.pdf (accessed October 22, 2007)

another potential danger of consuming foods with high glycemic loads is that they may already be insulin resistant, and overexertion of insulin-producing cells in the pancreas required to metabolize the high glycemic loads may ultimately exhaust their insulin-producing cells, leading to diabetes.

The proponents of low-carbohydrate diets observe that consuming foods with low glycemic loads stabilizes blood sugar and insulin to prevent the fluctuations that can cause overeating and may increase the risk for diabetes. They also assert that reliance on low-fat diets inadvertently led to diets that were high in simple carbohydrates and indirectly promoted the observed increase in overweight and diabetes in the United States.

**Low-Fat Diets**

Low-fat diets reduce caloric intake by reducing fat consumption. Fat has nine calories per gram, whereas

**TABLE 5.10**

**TABLE 5.11**

**Sample reduced calorie menus, Mexican American cuisine**

| | 1,600 calories | 1,200 calories |
|---|---|---|
| **Breakfast** | | |
| • Cantaloupe | 1 cup | 1/2 cup |
| • Farina, prepared with 1% low fat milk | 1/2 cup | 1/2 cup |
| • White bread | 1 slice | 1 slice |
| • Margarine | 1 tsp | 1 tsp |
| • Jelly | 1 tsp | 1 tsp |
| • Orange juice | 1 1/2 cup | 3/4 cup |
| • Milk, 1%, low fat | 1/2 cup | 1/2 cup |
| **Lunch** | | |
| • Beef enchilada: | | |
| Tortilla, corn | 2 tortillas | 2 tortillas |
| Lean roast beef | 2 1/2 oz | 2 oz |
| Vegetable oil | 2/3 tsp | 2/3 tsp |
| Onion | 1 T | 1 T |
| Tomato | 4 T | 4 T |
| Lettuce | 1/2 cup | 1/2 cup |
| Chili peppers | 2 tsp | 2 tsp |
| Refried beans, prepared with vegetable oil | 1/4 cup | 1/4 cup |
| • Carrots | 5 sticks | 5 sticks |
| • Celery | 6 sticks | 6 sticks |
| • Milk, 1%, low fat | 1/2 cup | — |
| • Water | — | 1 cup |
| **Dinner** | | |
| • Chicken taco: | | |
| Tortilla, corn | 1 tortilla | 1 tortilla |
| Chicken breast, without skin | 2 oz | 1 oz |
| Vegetable oil | 2/3 tsp | 2/3 tsp |
| Cheddar cheese, low fat and low sodium | 1 oz | 1/2 oz |
| Guacamole | 2 T | 2 T |
| Salsa | 1 T | 1 T |
| • Corn, seasoned with | 1/2 cup | 1/2 cup |
| margarine | 1/2 tsp | — |
| • Spanish rice without meat | 1/2 cup | 1/2 cup |
| • Banana | 1 large | 1/2 large |
| • Coffee | 1 cup | 1/2 cup |
| • Milk, 1% | 1 oz | 1 oz |

| Calories | 1,638 | Calories | 1,239 |
|---|---|---|---|
| Total carbohydrate, % kcals | 56 | Total carbohydrate, % kcals | 58 |
| Total fat, % kcals | 27 | Total fat, % kcals | 26 |
| *Sodium, mg | 1,616 | *Sodium, mg | 1,364 |
| Saturated fat, % kcals | 9 | Protein, % kcals | 8 |
| Cholesterol, mg | 153 | Cholesterol, mg | 91 |
| Protein, % kcals | 20 | Protein, % kcals | 19 |

1,600: 100% RDA met for all nutrients except vitamin in E 97% and Zinc 84%.
1,200: 100% RDNA met for all nutrients except vitamin E 71%, vitamin B₁ & B₃ 91%, vitamin B₂ & iron 90%, and calcium 92%.
*No salt in recipe preparation or as seasoning. Consume at least 32 ounces of water.

SOURCE: "Appendix D. Mexican American Cuisine—Reduced Calorie," in *The Practical Guide: Identification, Evaluation, and Treatment of Overweight and Obesity in Adults*, National Institutes of Health, National Heart, Lung, and Blood Institute, North American Association for the Study of Obesity, October 2000, http://www.nhlbi.nih.gov/guidelines/obesity/prctgd_b.pdf (accessed October 22, 2007).

**Sample reduced calorie menus, lacto-ovo vegetarian cuisine**

| | 1,600 calories | 1,200 calories |
|---|---|---|
| **Breakfast** | | |
| • Orange | 1 medium | 1 medium |
| • Pancakes, made with 1% lowfat milk and eggs whites | 3 4" circles | 2 4" circles |
| • Pancake syrup | 2 T | 1 T |
| • Margarine, diet | 1 1/2 tsp | 1 1/2 tsp |
| • Milk, 1%, lowfat | 1 cup | 1/2 cup |
| • Coffee | 1 cup | 1 cup |
| • Milk, 1%, lowfat | 1 oz | 1 oz |
| **Lunch** | | |
| • Vegetable soup, canned, low sodium | 1 cup | 1/2 cup |
| • Bagel | 1 medium | 1/2 medium |
| • Processed American cheese, lowfat | 3/4 oz | — |
| • Spinach salad: | | |
| Spinach | 1 cup | 1 cup |
| Mushrooms | 1/2 cup | 1/2 cup |
| • Salad dressing, regular calorie | 2 tsp | 2 tsp |
| • Apple | 1 medium | 1 medium |
| • Iced tea, unsweetened | 1 cup | 1 cup |
| **Dinner** | | |
| • Omelette: | | |
| Egg whites | 4 large eggs | 4 large eggs |
| Green pepper | 2 T | 2 T |
| Onion | 2 T | 2 T |
| Mozzarella cheese, made from part skim milk, low sodium | 1 oz | 1/2 oz |
| Vegetable oil | 1 T | 1/2 T |
| • Brown rice, seasoned with | 1/2 cup | 1/2 cup |
| margarine, diet | 1/2 tsp | 1/2 tsp |
| • Carrots, seasoned with | 1/2 cup | 1/2 cup |
| Margarine, diet | 1/2 tsp | 1/2 tsp |
| • Whole wheat bread | 1 slice | 1 slice |
| • Margarine, diet | 1 tsp | 1 tsp |
| • Fig bar cookie | 1 bar | 1 bar |
| • Tea | 1 cup | 1 cup |
| • Honey | 1 tsp | 1 tsp |
| • Milk, 1%, lowfat | 3/4 cup | 3/4 cup |

| Calories | 1,650 | Calories | 1,205 |
|---|---|---|---|
| Total carbohydrate, % kcals | 56 | Total carbohydrate, % kcals | 60 |
| Total fat, % kcals | 27 | Total fat, % kcals | 25 |
| *Sodium, mg | 1,829 | *Sodium, mg | 1,335 |
| Saturated fat, % kcals | 8 | Saturated fat, % kcals | 7 |
| Cholesterol, mg | 82 | Cholesterol, mg | 44 |
| Protein, % kcals | 19 | Protein, % kcals | 18 |

1,600: 100% RDA met for all nutrients except vitamin E 92%, vitamin B₃ 97%, vitamin B₆ 67%, iron 73%, and zinc 68%.
1,200: 100% RDA met for all nutrients except vitamin E 75%, vitamin B₁ 92%, vitamin B₃ 69%, vitamin B₆ 59%, iron 54%, and zinc 46%.
*No salt added in recipe preparation or as seasoning. Consume at least 32 ounces of water.

SOURCE: "Appendix D. Lacto-Ovo Vegetarian Cuisine—Reduced Calorie," in *The Practical Guide: Identification, Evaluation, and Treatment of Overweight and Obesity in Adults*, National Institutes of Health, National Heart, Lung, and Blood Institute, North American Association for the Study of Obesity, June 1998, http://www.nhlbi.nih.gov/guidelines/obesity/practgde.htm (accessed October 22, 2007)

protein and carbohydrates have four calories per gram. These diets rely on the high-fiber content of complex carbohydrates to satisfy dieters. High-fiber foods also slow the absorption of carbohydrates, so they do not provoke a rapid rise in blood sugar and insulin.

Table 5.13 shows some of the food substitutions that may be made to reduce the dietary fat content. Besides making substitutions, many fat-free or low-fat food products are available—from fat-free frozen desserts to reduced-fat peanut butter. However, dieters are often cautioned that fat-free or reduced-fat foods are not calorie-free and that their consumption will not result in weight loss when more of the reduced-fat foods are consumed than would be eaten of the full-fat versions. For example, eating twice as many baked tortilla chips would actually result in higher caloric intake than a single serving of regular tortilla chips. (See Table 5.14.)

**TABLE 5.12**

## Food exchange list

Within each group, these foods can be exchanged for each other. You can use this list to give yourself more choices.

**Vegetables** contain 25 calories and 5 grams of carbohydrate. One serving equals:
- 1/2 cup   Cooked vegetables (carrots, broccoli, zucchini, cabbage, etc.)
- 1 cup   Raw vegetables or salad greens
- 1/2 cup   Vegetable juice

**If you're hungry, eat more fresh or steamed vegetables.**

**Fat free and very low fat milk** contains 90 calories and 12 grams of carbohydrate per serving. One serving equals:
- 8 oz   Milk, fat free or 1% fat
- 1/4 cup   Yogurt, plain nonfat or low fat
- 1 cup   Yogurt, artificially sweetened

**Very lean protein** choices have 35 calories and 1 gram of fat per serving. One serving equals:
- 1 oz   Turkey breast or chicken breast, skin removed
- 1 oz   Fish fillet (flounder, sole, scrod, cod, haddock, halibut)
- 1 oz   Canned tuna in water
- 1 oz   Shellfish (clams, lobster, scallop, shrimp)
- 3/4 cup   Cottage cheese, nonfat or lowfat
- 2 each   Egg whites
- 1/4 cup   Egg substitute
- 1 oz   Fat free cheese
- 1/2 cup   Beans—cooked (black beans, kidney, chickpeas, or lentils): count as 1 starch/bread and 1 very lean protein

**Medium fat proteins** have 75 calories and 5 grams of fat per serving. One serving equals:
- 1 oz   Beef (any prime cut), corned beef, ground beef**
- 1 oz   Pork chop
- 1 each   Whole egg (medium)**
- 1 oz   Mozzarella cheese
- 1/4 cup   Ricotta cheese
- 4 oz   Tofu (note that this is a heart-healthy choice)

**\*\*Choose these very infrequently.**

**Fats** contain 45 calories and 5 grams of fat per serving. One serving equals:
- 1 tsp   Oil (vegetable, corn, canola, olive, etc.)
- 1 tsp   Butter
- 1 tsp   Stick margarine
- 1 tsp   Mayonnaise
- 1 T   Reduced fat margarine or mayonnaise
- 1 T   Salad dressing
- 1 T   Cream cheese
- 2 T   Lite cream cheese
- 1/8   Avocado
- 8 large   Black olives
- 10 large   Stuffed green olives
- 1 slice   Bacon

**Fruits** contain 15 grams of carbohydrates and 60 calories. One serving equals:
- 1 small   Apple, banana, orange, nectarine
- 1 medium   Fresh peach
- 1   Kiwi
- 1/2   Grapefruit
- 1/2   Mango
- 1 cup   Fresh berries (strawberries, raspberries, or blueberries)
- 1 cup   Fresh melon cubes
- 1/8   Honeydew melon
- 4 oz   Unsweetened juice
- 4 tsp   Jelly or jam

**Lean protein** choices have 55 calories and 2 to 3 grams of fat per serving. One serving equals:
- 1 oz   Chicken—dark meat, skin removed
- 1 oz   Turkey—dark meat, skin removed
- 1 oz   Salmon, swordfish, herring, catfish, trout
- 1 oz   Lean beef (flank steak, London broil, tenderloin, roast beef)*
- 1 oz   Veal, roast, or lean chop*
- 1 oz   Lamb, roast, or lean chop*
- 1 oz   Pork, tenderloin, or fresh ham*
- 1 oz   Lowfat luncheon meats (with 3 grams or less of fat per ounce)
- 1/4 cup   4.5% cottage cheese
- 2 medium   Sardines

**\*Limit to 1 to 2 times per week.**

**Starches** contain 15 grams of carbohydrate and 80 calories per serving. One serving equals:
- 1 slice   Bread (white, pumpernickel, whole wheat, rye)
- 2 slice   Reduced calorie or "lite" bread
- 1/4 (1 oz)   Bagel (varies)
- 1/2   English muffin
- 1/2   Hamburger bun
- 3/4 cup   Cold cereal
- 1/3 cup   Rice, brown or white—cooked
- 1/3 cup   Barley or couscous—cooked
- 1/3 cup   Legumes (dried beans, peas, or lentils)—cooked
- 1/2 cup   Pasta—cooked
- 1/2 cup   Bulgur—cooked
- 1/2 cup   Corn, sweet potato, or green peas
- 3 oz   Baked sweet or white potato
- 3/4 oz   Pretzels
- 3 cups   Popcorn, hot-air popped or microwave (80-percent light)

SOURCE: "Appendix E. Food Exchange List," in *The Practical Guide: Identification, Evaluation, and Treatment of Overweight and Obesity in Adults*, National Institutes of Health, National Heart, Lung, and Blood Institute, North American Association for the Study of Obesity, October 2000, http://www.nhlbi.nih.gov/guidelines/obesity/prctgd_b.pdf (accessed October 22, 2007)

### Low-Fat versus Low-Carbohydrate Diets

In the absence of rigorous scientific research and studies demonstrating the long-term safety and effectiveness of low-carbohydrate and low-fat diets, many investigators and health professionals hesitate to proclaim one diet's superiority over all others. There is consensus that even though some diets may produce greater initial weight loss, most perform similarly over time.

In "Efficacy and Safety of Low-Carbohydrate Diets: A Systematic Review" (*Journal of the American Medical Association*, vol. 289, no. 14, April 9, 2003), Dena M. Bravata et al. report the results of their analysis of data about diet-induced changes in weight, serum lipids, fasting serum glucose and fasting serum insulin levels, and blood pressure among adults using low-carbohydrate diets. The investigators undertook the research in response to concerns about low carbohydrates expressed by the American Dietetic Association and the American Heart Association. Both organizations had warned that low-carbohydrate diets may lead to abnormal metabolic functioning that in turn may prompt serious medical consequences, particularly for participants with cardiovascular disease, Type 2 diabetes mellitus, hyperlipidemia (an excess of fats called lipids, chiefly cholesterol and triglycerides, in the blood), or hypertension. Specifically, it has

**TABLE 5.13**

**Low calorie, lower fat food alternatives**

| Instead of... | | Replace with... |
|---|---|---|
| • Evaporated whole milk | | • Evaporated fat free (skim) or reduced fat (2%) milk |
| • Whole milk | | • Low fat (1%), reduced fat (2%), or fat free (skim) milk |
| • Ice cream | | • Sorbet, sherbet, lowfat or fat free frozen yogurt, or ice milk (check label for calorie content) |
| • Whipping cream | | • Imitation whipped cream (made with fat free [skim] milk) or lowfat vanilla yogurt |
| • Sour cream | | • Plain lowfat yogurt |
| • Cream cheese | **Dairy** | • Neufchatel or "light" cream cheese or fat free cream cheese |
| • Cheese (cheddar, Swiss, jack) | **Products** | • Reduced calorie cheese, low calorie processed cheese, etc. |
| | | • Fat free cheese |
| • American cheese | | • Fat free American cheese or other types of fat free cheeses |
| • Regular (4%) cottage cheese | | • Lowfat (1%) or reduced fat (2%) cottage cheese |
| • Whole milk mozzarella cheese | | • Part skim low-moisture mozzarella cheese |
| • Whole milk ricotta cheese | | • Part skim milk ricotta cheese |
| • Coffee cream (half and half) or nondairy creamer (liquid, power) | | • Low fat (1%) or reduced fat (2%) milk or nonfat dry milk power |
| • Ramen noodles | | • Rice or noodles (spaghetti, macaroni, etc.) |
| • Pasta with white sauce (alfredo) | **Cereals, grains** | • Pasta with red sauce (marinara) |
| • Pasta with cheese sauce | **and pasta** | • Pasta with vegetables (primavera) |
| • Granola | | • Bran flakes, crispy rice, etc. |
| | | • Cooked grits or oatmeal |
| | | • Whole grains (e.g., couscous, barley, bulgur, etc.) |
| | | • Reduced fat granola |
| • Cold cuts or lunch meats (bologna, salami, liverwurst, etc.) | | • Lowfat cold cuts (95% to 97% fat free lunch meats, lowfat pressed meats) |
| • Hot dogs (regular) | | • Lower fat hot dogs |
| • Bacon or sausage | | • Canadian bacon or lean ham |
| • Regular ground beef | | • Extra lean ground beef such as ground round or ground turkey (read labels) |
| • Chicken or turkey with skin, duck, or goose | | • Chicken or turkey without skin (white meat) |
| • Oil-packed tuna | | • Water-packed tuna (rinse to reduce sodium content) |
| • Beef (chuck, rib, brisket) | **Meat, fish,** | • Beef (round, loin) (trimmed of external fat) (choose select grades) |
| • Pork (spareribs, untrimmed loin) | **and poultry** | • Pork tenderloin or trimmed, lean smoked ham |
| • Frozen breaded fish or fried fish (homemade or commercial) | | • Fish or shellfish, unbreaded (fresh, frozen, canned in water) |
| • Whole eggs | | • Egg whites or egg substitutes |
| • Frozen TV dinners (containing more than 13 gram of fat per serving) | | • Frozen TV dinners (containing less than 13 grams of fat per serving and lower in sodium) |
| • Chorizo sausage | | • Turkey sausage, drained well (read label) |
| | | • Vegetarian sausage (made with tofu) |
| • Croissants, brioches, etc. | | • Hard French rolls or soft "brown 'n serve" rolls |
| • Donuts, sweet rolls, muffins, scones, or pastries | | • English muffins, bagels, reduced fat or fat free muffins or scones |
| • Party crackers | | • Lowfat crackers (choose lower in sodium) |
| • Saltine or soda crackers (choose lower in sodium) | **Baked goods** | |
| • Cake (pound, chocolate, yellow) | | • Cake (angel food, white, gingerbread) |
| • Cookies | | • Reduced fat or fat free cookies (graham crackers, ginger snaps, fig bars) (compare calorie level) |
| • Nuts | **Snacks and** | • Popcorn (air-popped or light microwave), fruits, vegetables |
| • Ice cream, e.g., cones or bars | **sweets** | • Frozen yogurt, frozen fruit, or chocolate pudding bars |
| • Custards or puddings (made with whole milk) | | • Puddings (made with skim milk) |
| • Regular margarine or butter | | • Light-spread margarines, diet margarine, or whipped butter, tub or squeeze bottle |
| • Regular mayonnaise | | • Light or diet mayonnaise or mustard |
| • Regular salad dressings | **Fats, oils, and** | • Reduced calorie or fat free salad dressings, lemon juice, or plain, herb-flavored, or wine vinegar |
| | **salad dressings** | |
| • Butter or margarine on toast or bread | | • Jelly, jam, or honey on bread or toast |
| • Oils, shortening, or lard | | • Nonstick cooking spray for stir-frying or sautéing |
| | | • As a substitute for oil or butter, use applesauce or prune puree in baked goods |
| • Canned cream soups | | • Canned broth-based soups |
| • Canned beans and franks | **Miscellaneous** | • Canned baked beans in tomato sauce |
| • Gravy (home made with fat and/or milk) | | • Gravy mixes made with water or homemade with the fat skimmed off and fat free milk included |
| • Fudge sauce | | • Chocolate syrup |
| • Avocado on sandwiches | | • Cucumber slices or lettuce leaves |
| • Guacamole dip or refried beans with lard | | • Salsa |

SOURCE: "Appendix C. Instead of . . . Replace with . . . ," in *The Practical Guide: Identification, Evaluation, and Treatment of Overweight and Obesity in Adults*, National Institutes of Health, National Heart, Lung, and Blood Institute, North American Association for the Study of Obesity, October 2000, http://www.nhlbi.nih.gov/guidelines/obesity/prctgd_b.pdf (accessed October 22, 2007)

been cautioned that low-carbohydrate diets cause the accumulation of ketones, which may result in abnormal metabolism of insulin, impaired liver and kidney function, and salt and water depletion that may cause postural hypotension (sudden drop in blood pressure when rising from sitting) as well as fatigue, constipation, and kidney stones. It has also been posited that excessive consumption of animal proteins and fats may promote hyperlipidemia and that higher dietary protein loads may impair kidney function.

**TABLE 5.14**

**Calories in fat free or reduced fat and regular food**

| Fat free or reduced fat | Calories | Regular | Calories |
|---|---|---|---|
| Reduced fat peanut butter, 2 T | 187 | Regular peanut butter, 2 T | 191 |
| Cookies | | Cookies | |
|    Reduced fat chocolate chip cookies, 3 cookies (30 g) | 118 |    Regular chocolate chip cookies, 3 cookies (30 g) | 142 |
|    Fat free fig cookies, 2 cookies (30 g) | 102 |    Regular fig cookies, 2 cookies (30 g) | 111 |
| Ice cream | | Ice cream | |
|    Nonfat vanilla frozen yogurt (1% fat), 1/2 cup | 100 |    Regular whole milk vanilla frozen yogurt (3–4% fat), 1/2 cup | 104 |
|    Light vanilla ice cream (7% fat), 1/2 cup | 111 |    Regular vanilla ice cream (11% fat), 1/2 cup | 133 |
|    Fat free caramel topping, 2 T | 103 |    Caramel topping, homemade with butter, 2 T | 103 |
| Low fat granola cereal, approx. 1/2 cup (55 g) | 213 | Regular granola cereal, approx 1/2 cup (55 g) | 257 |
| Low fat blueberry muffin, 1 small (2 1/2 inch) | 131 | Regular blueberry muffin, 1 small (2 1/2 inch) | 138 |
| Baked tortilla chips, 1 oz. | 113 | Regular tortilla chips, 1 oz. | 143 |
| Low fat cereal bar, 1 bar (1.3 oz.) | 130 | Regular cereal bar, 1 bar (1.3 oz.) | 140 |

SOURCE: "Fat Free or Reduced Fat [versus] Regular," in *The Practical Guide: Identification, Evaluation, and Treatment of Overweight and Obesity in Adults*, National Institutes of Health, National Heart, Lung, and Blood Institute, North American Association for the Study of Obesity, October 2000, http://www.nhlbi .nih.gov/guidelines/obesity/prctgd_b.pdf (accessed October 22, 2007)

Bravata et al. find that diets that restricted calorie intake and were longer in duration were associated with weight loss. They also observe that when lower-carbohydrate diets resulted in weight loss, it was likely because of the restriction of caloric intake and longer duration rather than changes in carbohydrate intake. The investigators note that at least in the short term, low-carbohydrate diets were not associated with the anticipated adverse effects on lipid levels, glucose levels, or blood pressure. Furthermore, their findings suggest that people without diabetes tolerated a lower-carbohydrate diet better than higher-carbohydrate alternatives and that this diet may be an effective means of achieving short-term weight loss without significant adverse effects on serum lipid levels, glycemic control, or blood pressure. They caution, however, that there is still inadequate evidence to recommend or condemn the use of low-carbohydrate diets among people with diabetes or for long-term use.

According to Peggy Peck, in "Four Popular Diets Equally Effective for Weight Loss" (*Medscape Today*, November 10, 2003, http://www.medscape.com/view article/464193), Michael L. Dansinger et al. compared the effectiveness of four popular diets: Atkins (low carbohydrates), the Zone (moderate carbohydrates), Ornish (low-fat vegetarian), and Weight Watchers (moderate fat). Study participants were asked to follow the diets they were given as best they could for two months, and they were given official diet cookbooks and assigned to small group classes for diet education. For the remaining ten months, the participants were told to follow their assigned diets "to whatever extent they wanted." Dansinger et al. report that nearly one-quarter (22%) of the participants had dropped out of each diet after just two months, and by twelve months half of the participants assigned to low-carbohydrate or low-fat vegetarian diets had dropped out, as had 35% of participants assigned to the moderate carbohydrates and moderate fat diets. For those participants who adhered, weight loss and reduction in cardiac risk scores as measured by reductions in LDL cholesterol and insulin levels were comparable for participants on the low-carbohydrate, moderate-carbohydrate, and moderate-fat plans. According to the study, the Ornish diet "does not increase HDL, while the other diets do achieve significant increases in HDL."

Dansinger et al. conclude that their research "demonstrated that all these diets work." They also reiterate the importance of tailoring the selection of a weight-loss diet to ensure adherence, asserting "that means that physicians can work with patients to select the diet that is best suited to the patient. For example, if you have a patient who likes meat, it is unlikely that he or she will comply with the Ornish diet."

In 2004 two published studies reaffirmed the safety and efficacy of low-carbohydrate diets. In the first study, "A Low-Carbohydrate, Ketogenic Diet versus a Low-Fat Diet to Treat Obesity and Hyperlipidemia: A Randomized, Controlled Trial" (*Annals of Internal Medicine*, vol. 140, no. 10, May 18, 2004), William Yancy Jr. et al. assigned 120 study participants to a low-carbohydrate, high-protein diet or a low-fat, low-cholesterol, low-calorie diet. The low-carbohydrate group was allowed unlimited calories, animal foods (meat, fowl, fish, and shellfish), and eggs, as well as four ounces of hard cheese, two cups of salad vegetables (lettuce, spinach, or celery), and one cup of low-carbohydrate vegetables (broccoli, cauliflower, or squash). The low-fat, low-cholesterol, low-calorie group consumed less than 30% of daily caloric intake from fat, less than 10% of calories from saturated fat, and less than three hundred milligrams of cholesterol daily. After six months, weight loss was greater in the low-carbohydrate diet group than in the low-fat diet group. Compared to the low-fat diet group, the low-carbohydrate diet group had greater decreases in serum triglyceride levels and greater increases in HDL cholesterol levels.

In the second study, "A Low-Carbohydrate, Keto-genic Diet versus a Low-Fat Diet to Treat Obesity and Hyperlipidemia: A Randomized, Controlled Trial" (*Annals of Internal Medicine*, vol. 140, no. 10, May 18, 2004), William Yancy Jr. et al. assigned 132 obese adults to either restrict carbohydrate intake to less than thirty grams per day (low-carbohydrate diet) or to restrict calo-ric intake by five hundred calories per day with less than 30% of calories from fat (conventional diet). After one year, weight loss was greater in the low-carbohydrate diet group, and Yancy et al. found that the low-carbohydrate diet group fared better in terms of a greater decrease in triglyceride levels.

In another study, Y. Wady Aude et al. confirm in "The National Cholesterol Education Program Diet vs. a Diet Lower in Carbohydrates and Higher in Protein and Monounsaturated Fat" (*Archives of Internal Medicine*, vol. 164, no. 19, October 25, 2004) that modified low-carbohydrate diets produced greater weight loss than the U.S. National Cholesterol Education Program diet, which replaces saturated fat with carbohydrates.

However, the article "Study: Low-Fat Tops Low-Carb for Keeping Pounds Off" (*USA Today*, November 16, 2004) notes that Suzanne Phelan of Brown Medical School asserted at a meeting of the North American Asso-ciation for the Study of Obesity that low-fat diets produce better long-term weight loss than low-carbohydrate diets. She and her colleagues studied twenty-seven hundred people who entered the National Weight Control Registry, which records successful efforts to lose at least thirty pounds and maintain the loss for at least one year. All the subjects reported eating about fourteen hundred calo-ries per day, but the portion derived from fat rose from 24% in 1995 to more than 29% in 2003, whereas the portion from carbohydrates fell, from 56% to 49%. The number who were on low-carbohydrate diets (less than ninety grams per day) rose from 6% to 17% during this same period.

Even though the type of diet—low fat or low carbo-hydrate—made no difference in how people lost weight initially, those who increased their fat intake over a year regained the most weight. The researchers noted that the subjects ate fewer carbohydrates, because the amount of protein in their diets remained the same. As a result, the researchers concluded that the minority of successful dieters use low-carbohydrate regimes.

Even though there is no single winner in the diet wars, research dispels some fears about the safety and effective-ness of the low-carbohydrate diet. Low-carbohydrate diets appear to be safe and effective in the short term, but long-term outcomes are still unclear. Some results suggest that higher protein and fat intakes lead to lower total caloric intake by producing earlier satiety, but these diets have not been shown to alter fundamental eating behav-iors, nor have they demonstrated, as many of their propo-nents argue, the ability to modify caloric balance such that weight loss persists when more calories are consumed than expended.

Finally, in "Comparison of the Atkins, Ornish, Weight Watchers, and Zone Diets for Weight Loss and Heart Disease Risk Reduction" (*Journal of the American Medical Association*, vol. 293, no. 1, January 5, 2005), Michael L. Dansinger et al. assert that adherence to a diet for one year, rather than the specific type of diet, is the single most important determinant of weight loss and reduction of risk of cardiovascular disease. The research-ers find that the amount of weight lost was associated with the level of dietary adherence but not with diet type. Dansinger et al. conclude that "one way to improve dietary adherence rates in clinical practice may be to use a broad spectrum of diet options, to better match individual patient food preferences, lifestyles, and cardi-ovascular risk profiles.... Our findings challenge the concept that one type of diet is best for everybody and that alternative diets can be disregarded. Likewise, our findings do not support the notion that very low carbohy-drate diets are better than standard diets, despite recent evidence to the contrary."

Gabrielle M. Turner-McGrievy, Neal D. Barnard, and Anthony R. Scialli indicate in "A Two-Year Randomized Weight Loss Trial Comparing a Vegan Diet to a More Moderate Low-Fat Diet" (*Obesity*, vol. 15, 2007), a weight-loss maintenance study that compared vegan diets to the National Cholesterol Education Pro-gram (NCEP) diet, a low-calorie, low-fat diet that is high in carbohydrates, that a vegan diet was associated with significantly greater weight loss than the NCEP diet after the one- and two-year follow-ups.

# CHAPTER 6
# PHYSICAL ACTIVITY, DRUGS, SURGERY, AND OTHER TREATMENTS FOR OVERWEIGHT AND OBESITY

*Lack of activity destroys the good condition of every human being, while movement and methodical physical exercise save it and preserve it.*

—Plato

One credible hypothesis about the source of the epidemic of overweight and obesity in the United States is the progressive decrease in physical activity expended in daily life—for work, transportation, and household chores. Some researchers contend that the average caloric intake of Americans has not substantially increased; instead, by reducing daily physical activity, the caloric imbalance between calories consumed and expended has shifted to favor weight gain. Even though no data conclusively prove this hypothesis, evidence does support it.

Among the recent studies that support the premise that Americans' sedentary lifestyle has precipitated the obesity epidemic is research that examined the diets of an Amish community in Ontario, Canada. In "Physical Activity in an Old Order Amish Community" (*Medicine and Science in Sports and Exercise*, vol. 36, no. 1, January 2004), David R. Bassett, Patrick L. Schneider, and Gertrude E. Huntington describe the "Amish paradox"— that despite a diet that is high in fat, calories, and refined sugar, the Amish community had a scant 4% obesity rate, compared to 31% in the general U.S. population. The researchers chose this particular Amish population because it has rejected technological advances such as automobiles and electricity, and its physically demanding lifestyle is comparable to the way Americans lived 150 years ago. (Other Amish communities that have assumed occupations less physically active than farming have obesity rates that are similar to those found in the general U.S. population.) Bassett, Schneider, and Huntington analyzed the daily routines of about one hundred Amish people and found that men averaged about eighteen thousand steps per day and women about fourteen thousand, compared to the recommended ten thousand steps per day

that most Americans struggle to achieve. The Amish men performed about ten hours per week of vigorous exercise and women spent about three-and-a-half hours engaged in heavy lifting, shoveling, digging, shoeing horses, or tossing straw bales. Men devoted an additional forty-three hours per week and women an average of thirty-nine hours to moderate physical activities such as gardening, performing farm-related chores, or doing laundry.

## PHYSICAL ACTIVITY

In sharp contrast to the Amish farmers, many Americans are not physically active. The Centers for Disease Control and Prevention (CDC) defines in "How Active Do Adults Need to Be to Gain Some Benefit?" (May 22, 2007, http://www.cdc.gov/nccdphp/dnpa/physical/recommendations/adults.htm) the minimum recommended physical activity level for adults as "moderate-intensity physical activity for 30 minutes or more on 5 or more days of the week, or vigorous-intensity physical activity for 20 minutes or more on 3 or more days of the week." Regardless, Table 6.1 shows that the percent of men and women that are physically inactive during leisure time increases with age and income. In 2004 more than half (52.7%) of adults aged sixty-five and older said they were physically inactive during leisure time, compared to about one-third (35.4%) of adults aged eighteen to forty-four. Women were more physically inactive than men of the same age across all age groups. However, the proportion of the U.S. population that reported no leisure-time physical activity has decreased from 31% in 1989 to 25% in 2005. (See Figure 6.1.)

The 2007 National Health Interview Survey data reveal that among adults aged eighteen to sixty-four who engage in regular leisure-time physical activity, the gender gap is closing, with comparable percentages of men and women reporting regular physical activity. (See Figure 6.2.) Figure 6.3 shows that non-Hispanic white

TABLE 6.1

**Physical activity among adults age 18 and older, by selected characteristics, 1998–2004**

[Data are based on household interviews of a sample of the civilian noninstitutionalized population]

| Characteristic | Inactive[a] | | | Some leisure-time activity[a] | | | Regular leisure-time activity[a] | | |
|---|---|---|---|---|---|---|---|---|---|
| | 1998 | 2003 | 2004 | 1998 | 2003 | 2004 | 1998 | 2003 | 2004 |
| | Percent of adults | | | | | | | | |
| Total, age-adjusted[b, c] | 40.5 | 37.6 | 39.5 | 30.0 | 29.5 | 30.4 | 29.5 | 32.8 | 30.2 |
| Total, crude[c] | 40.2 | 37.6 | 39.4 | 30.0 | 29.6 | 30.4 | 29.8 | 32.8 | 30.2 |
| **Age** | | | | | | | | | |
| 18–44 years | 35.2 | 32.9 | 35.4 | 31.4 | 30.3 | 31.5 | 33.5 | 36.8 | 33.1 |
| 18–24 years | 32.8 | 29.6 | 33.1 | 30.1 | 28.2 | 30.2 | 37.1 | 42.3 | 36.6 |
| 25–44 years | 35.9 | 34.0 | 36.1 | 31.8 | 31.0 | 31.9 | 32.4 | 34.9 | 31.9 |
| 45–64 years | 41.2 | 38.2 | 39.2 | 30.6 | 30.5 | 31.0 | 28.2 | 31.3 | 29.7 |
| 45–54 years | 38.9 | 36.5 | 37.5 | 31.4 | 30.8 | 32.8 | 29.8 | 32.8 | 29.7 |
| 55–64 years | 44.9 | 40.8 | 41.8 | 29.3 | 30.1 | 28.5 | 25.8 | 29.2 | 29.8 |
| 65 years and over | 55.4 | 51.4 | 52.7 | 24.7 | 25.3 | 25.6 | 19.9 | 23.3 | 21.7 |
| 65–74 years | 49.1 | 45.8 | 44.3 | 26.5 | 25.8 | 29.0 | 24.4 | 28.4 | 26.7 |
| 75 years and over | 63.3 | 57.5 | 62.1 | 22.4 | 24.8 | 21.8 | 14.3 | 17.7 | 16.1 |
| **Sex[b]** | | | | | | | | | |
| Male | 37.8 | 35.4 | 38.1 | 28.7 | 29.2 | 30.5 | 33.5 | 35.4 | 31.4 |
| Female | 42.9 | 39.5 | 40.6 | 31.1 | 29.9 | 30.3 | 26.0 | 30.6 | 29.1 |
| **Sex and age** | | | | | | | | | |
| Male: | | | | | | | | | |
| 18–44 years | 32.0 | 30.9 | 34.0 | 30.7 | 29.5 | 31.6 | 37.2 | 39.6 | 34.4 |
| 45–54 years | 37.7 | 36.4 | 38.3 | 29.6 | 30.5 | 33.4 | 32.6 | 33.2 | 28.3 |
| 55–64 years | 44.5 | 39.5 | 40.7 | 26.9 | 29.7 | 26.9 | 28.6 | 30.8 | 32.4 |
| 65–74 years | 45.3 | 43.0 | 41.7 | 23.6 | 24.9 | 29.1 | 31.1 | 32.1 | 29.2 |
| 75 years and over | 57.4 | 48.1 | 56.8 | 21.6 | 28.9 | 23.1 | 20.9 | 23.0 | 20.1 |
| Female: | | | | | | | | | |
| 18–44 years | 38.2 | 34.9 | 36.7 | 32.0 | 31.1 | 31.4 | 29.8 | 34.0 | 31.9 |
| 45–54 years | 39.9 | 36.5 | 36.7 | 33.0 | 31.1 | 32.3 | 27.1 | 32.4 | 31.0 |
| 55–64 years | 45.2 | 41.9 | 42.7 | 31.5 | 30.4 | 29.9 | 23.3 | 27.6 | 27.4 |
| 65–74 years | 52.2 | 48.0 | 46.5 | 28.7 | 26.6 | 28.9 | 19.0 | 25.4 | 24.6 |
| 75 years and over | 67.0 | 63.7 | 65.6 | 22.9 | 22.0 | 21.0 | 10.1 | 14.3 | 13.5 |
| **Race[b, d]** | | | | | | | | | |
| White only | 38.8 | 36.3 | 38.0 | 30.5 | 29.8 | 30.7 | 30.7 | 33.9 | 31.3 |
| Black or African American only | 52.2 | 48.5 | 50.5 | 25.2 | 26.1 | 26.1 | 22.6 | 25.5 | 23.3 |
| American Indian or Alaska Native only | 49.2 | 54.7 | 44.4 | 19.0 | 20.0 | 33.7 | 31.8 | 25.2 | 21.9 |
| Asian only | 39.4 | 35.9 | 39.1 | 35.2 | 31.1 | 33.4 | 25.4 | 33.1 | 27.5 |
| Native Hawaiian or other Pacific Islander only | — | * | * | — | * | * | — | * | * |
| 2 or more races | — | 33.3 | 28.8 | — | 34.1 | 38.7 | — | 32.6 | 32.4 |
| **Hispanic origin and race[b, d]** | | | | | | | | | |
| Hispanic or Latino | 55.5 | 51.9 | 52.8 | 23.4 | 23.6 | 24.8 | 21.1 | 24.4 | 22.3 |
| Mexican | 56.7 | 52.0 | 52.4 | 23.9 | 23.7 | 25.5 | 19.4 | 24.3 | 22.1 |
| Not Hispanic or Latino | 38.8 | 35.5 | 37.5 | 30.7 | 30.3 | 31.1 | 30.5 | 34.2 | 31.4 |
| White only | 36.7 | 33.4 | 35.3 | 31.3 | 30.9 | 31.6 | 32.0 | 35.8 | 33.1 |
| Black or African American only | 52.2 | 48.5 | 50.7 | 25.1 | 26.0 | 26.0 | 22.6 | 25.5 | 23.2 |
| **Education[e, f]** | | | | | | | | | |
| No high school diploma or GED | 64.8 | 61.2 | 63.8 | 19.4 | 20.6 | 21.5 | 15.8 | 18.1 | 14.7 |
| High school diploma or GED | 47.6 | 45.5 | 48.6 | 28.7 | 27.5 | 28.5 | 23.7 | 27.0 | 22.8 |
| Some college or more | 30.2 | 28.1 | 28.9 | 34.3 | 33.8 | 34.3 | 35.5 | 38.2 | 36.9 |
| **Percent of poverty level[b, g]** | | | | | | | | | |
| Below 100% | 59.4 | 55.1 | 56.7 | 20.5 | 22.0 | 22.8 | 20.1 | 22.9 | 20.4 |
| 100%–less than 200% | 52.2 | 50.5 | 52.4 | 26.2 | 24.8 | 26.4 | 21.6 | 24.7 | 21.2 |
| 200% or more | 34.7 | 31.4 | 33.4 | 32.4 | 32.0 | 32.6 | 33.0 | 36.7 | 34.0 |

adults (35%) were more likely than Hispanic adults (22%) and non-Hispanic African-American adults (20%) to participate in regular leisure-time physical activity.

## Physical Activity and Weight Loss

Increasing physical activity and exercise is an important element of regimens intended to produce weight loss, even though the addition of exercise to a diet program generally does not produce substantially greater weight loss—most weight lost is attributable to decreased caloric intake. By favorably affecting blood lipids, increased and sustained physical activity does offer many direct and indirect health benefits, including reducing risks for cardiovascular heart disease and Type 2 diabetes beyond the

**TABLE 6.1**

**Physical activity among adults age 18 and older, by selected characteristics, 1998–2004** [CONTINUED]

[Data are based on household interviews of a sample of the civilian noninstitutionalized population]

| Characteristic | Inactive[a] | | | Some leisure-time activity[a] | | | Regular leisure-time activity[a] | | |
|---|---|---|---|---|---|---|---|---|---|
| | 1998 | 2003 | 2004 | 1998 | 2003 | 2004 | 1998 | 2003 | 2004 |
| | | | | | Percent of adults | | | | |
| **Hispanic origin and race and percent of poverty level**[b, d, g] | | | | | | | | | |
| Hispanic or Latino: | | | | | | | | | |
| Below 100% | 68.6 | 64.2 | 63.7 | 18.0 | 19.1 | 20.9 | 13.4 | 16.7 | 15.5 |
| 100%–less than 200% | 60.8 | 58.8 | 59.2 | 21.2 | 21.0 | 23.8 | 18.0 | 20.3 | 17.0 |
| 200% or more | 45.6 | 41.8 | 43.1 | 27.6 | 27.3 | 27.7 | 26.8 | 30.9 | 29.3 |
| Not Hispanic or Latino: | | | | | | | | | |
| White only: | | | | | | | | | |
| Below 100% | 53.7 | 49.3 | 51.8 | 22.5 | 22.6 | 23.4 | 23.8 | 28.0 | 24.8 |
| 100%–less than 200% | 49.0 | 45.7 | 48.4 | 27.6 | 27.4 | 27.3 | 23.4 | 27.0 | 24.3 |
| 200% or more | 32.7 | 29.2 | 31.1 | 32.9 | 32.4 | 33.4 | 34.4 | 38.3 | 35.5 |
| Black or African American only: | | | | | | | | | |
| Below 100% | 64.3 | 61.3 | 61.3 | 17.4 | 20.9 | 21.6 | 18.3 | 17.8 | 17.1 |
| 100%–less than 200% | 55.6 | 55.3 | 59.5 | 24.4 | 22.9 | 22.9 | 19.9 | 21.8 | 17.6 |
| 200% or more | 46.0 | 40.7 | 42.4 | 28.7 | 29.1 | 29.2 | 25.3 | 30.2 | 28.4 |
| **Geographic region**[b] | | | | | | | | | |
| Northeast | 39.4 | 34.4 | 36.1 | 31.3 | 29.2 | 30.6 | 29.4 | 36.4 | 33.3 |
| Midwest | 37.3 | 34.7 | 34.7 | 31.7 | 32.2 | 34.1 | 31.0 | 33.1 | 31.2 |
| South | 46.9 | 42.6 | 46.7 | 27.1 | 27.7 | 27.1 | 26.0 | 29.7 | 26.2 |
| West | 33.9 | 34.9 | 35.3 | 31.6 | 29.9 | 31.6 | 34.6 | 35.2 | 33.2 |
| **Location of residence**[b] | | | | | | | | | |
| Within MSA[h] | 39.3 | 36.4 | 38.2 | 30.6 | 29.9 | 30.6 | 30.0 | 33.7 | 31.1 |
| Outside MSA[h] | 44.7 | 42.4 | 44.6 | 27.5 | 28.1 | 29.2 | 27.8 | 29.5 | 26.2 |

*Estimates are considered unreliable.
—Data not available.
[a]All questions related to leisure-time physical activity were phrased in terms of current behavior and lack a specific reference period. Respondents were asked about the frequency and duration of vigorous and light/moderate physical activity during leisure time. Adults classified as inactive reported no sessions of light/moderate or vigorous leisure-time activity of at least 10 minutes duration; adults classified with some leisure-time activity reported at least one session of light/moderate or vigorous physical activity of at least 10 minutes duration but did not meet the definition for regular leisure-time activity; adults classified with regular leisure-time activity reported 3 or more sessions per week of vigorous activity lasting at least 20 minutes or 5 or more sessions per week of light/moderate activity lasting at least 30 minutes in duration.
[b]Estimates are age adjusted to the year 2000 standard population using five age groups: 18–44 years, 45–54 years, 55–64 years, 65–74 years, and 75 years and over. Age-adjusted estimates in this table may differ from other age-adjusted estimates based on the same data and presented elsewhere if different age groups are used in the adjustment procedure.
[c]Includes all other races not shown separately and unknown education level.
[d]The race groups, white, black, American Indian or Alaska Native, Asian, Native Hawaiian or other Pacific Islander, and 2 or more races, include persons of Hispanic and non-Hispanic origin. Persons of Hispanic origin may be of any race. Starting with 1999 data, race-specific estimates are tabulated according to the 1997 Revisions to the Standards for the Classification of Federal Data on Race and Ethnicity and are not strictly comparable with estimates for earlier years. The five single race categories plus multiple race categories shown in the table conform to the 1997 Standards. Starting with 1999 data, race-specific estimates are for persons who reported only one racial group; the category 2 or more races includes persons who reported more than one racial group. Prior to 1999, data were tabulated according to the 1977 Standards with four racial groups and the Asian only category included Native Hawaiian or other Pacific Islander. Estimates for single race categories prior to 1999 included persons who reported one race or, if they reported more than one race, identified one race as best representing their race. Starting with 2003 data, race responses of other race and unspecified multiple race were treated as missing, and then race was imputed if these were the only race responses. Almost all persons with a race response of other race were of Hispanic origin.
[e]Estimates are for persons 25 years of age and over and are age adjusted to the year 2000 standard population using five age groups: 25–44 years, 45–54 years, 55–64 years, 65–74 years, and 75 years and over.
[f]GED stands for general educational development high school equivalency diploma.
[g]Percent of poverty level is based on family income and family size and composition using U.S. Census Bureau poverty thresholds. Missing family income data were imputed for 31%–36% of adults 18 years of age and over in 1998–2004.
[h]MSA is metropolitan statistical area.

SOURCE: "Table 72. Leisure-Time Physical Activity among Adults 18 Years of Age and Over, by Selected Characteristics: United States, Selected Years 1998–2004," in *Health, United States, 2006, with Chartbook on Trends in the Health of Americans*, Centers for Disease Control and Prevention, National Center for Health Statistics, 2006, http://www.cdc.gov/nchs/data/hus/hus06.pdf#summary (accessed October 9, 2007)

risk reduction possible through diet alone. Physical activity lowers low-density lipoprotein (LDL) cholesterol and triglycerides, increases high-density lipoprotein (HDL) cholesterol, reduces abdominal fat as measured by waist circumference, and may protect against a decrease in muscle mass during weight loss.

Like those who have been inactive or sedentary, overweight people are advised to initiate physical activity slowly and gradually. Walking and swimming at a slow pace are ideal activities because they are enjoyable, easy to schedule, and less likely to produce injuries than many competitive sports. Table 6.2 is an example of a walking program that progressively increases physical activity. Furthermore, because amounts of activity and the resulting health benefits are functions of the duration, intensity, and frequency, the same amounts of activity may be obtained in longer sessions of moderately intense activity

FIGURE 6.1

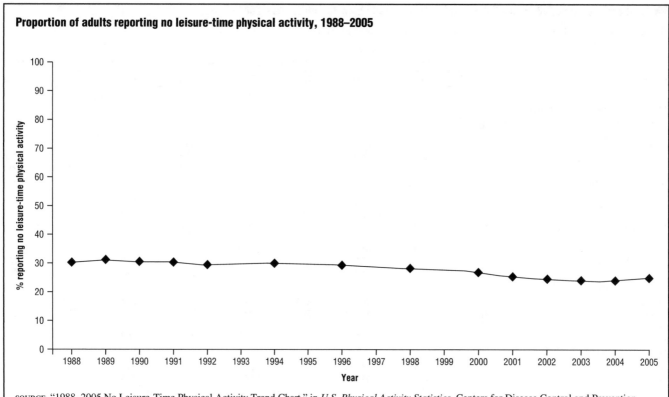

Proportion of adults reporting no leisure-time physical activity, 1988–2005

SOURCE: "1988–2005 No Leisure-Time Physical Activity Trend Chart," in *U.S. Physical Activity Statistics*, Centers for Disease Control and Prevention, Division of Nutrition, Physical Activity, and Obesity, National Center for Chronic Disease Prevention and Health Promotion, May 22, 2007, http://www.cdc.gov/nccdphp/dnpa/physical/stats/leisure_time.htm (accessed October 24, 2007)

such as brisk walking than in shorter sessions of more strenuous activities such as running. Table 6.3 shows how a moderate amount of activity—physical activity that uses about 150 calories of energy per day for a total of about 1,000 calories per week—can be obtained in a variety of ways. Table 6.3 also indicates how performing common household chores, and even self-care activities such as using a wheelchair, may be used to fulfill requirements for moderate amounts of physical activity. Changing routines to include walking up stairs rather than taking an elevator or parking farther than usual from work, school, or shopping are ways to increase physical activity incrementally. Even reducing sedentary time, such as hours spent in front of the television or computer, can serve to increase energy expenditure.

Table 6.3 also shows the relationship between the intensity and duration of physical activities by comparing the amount of time a 154-pound adult must spend performing each activity to expend 150 calories. It is interesting to note that just five additional minutes of walking at a moderate pace expends the same number of calories as walking at a brisk pace.

In "Effect of Exercise Duration and Intensity on Weight Loss in Overweight, Sedentary Women" (*Journal of the American Medical Association*, vol. 290, no. 10, September 10, 2003), John M. Jakicic et al. of the University of Pittsburgh confirm the weight-loss benefits of even moderate exercise. The study divided 201 women aged twenty-one to forty-five into four groups. Two groups of women expended one thousand calories per week walking at a moderate pace for forty minutes per day. The other two groups expended two thousand calories per week; one group walked at a moderate pace for sixty minutes per day and the other at a vigorous pace for forty-five minutes per day. All the study participants reduced their calorie consumption to between twelve hundred and fifteen hundred calories per day. Jakicic et al. find no differences based on different exercise durations and intensities; one group of women lost almost as much weight—about thirteen to twenty pounds over twelve months—from walking at a moderate pace as another group did from walking at a brisk pace.

In another study, Cris A. Slentz et al. find in "Effects of the Amount of Exercise on Body Weight, Body Composition, and Measures of Central Obesity: STRRIDE—A Randomized Controlled Study" (*Archives of Internal Medicine*, vol. 164, no. 1, January 12, 2004) a close relationship between exercise and weight loss—increasing amounts of exercise yielded greater benefits. The study randomly assigned 182 sedentary, overweight adults aged forty to sixty-five to one of four groups: a control group with no exercise; supervised low-dose/moderate-intensity

FIGURE 6.2

FIGURE 6.3

**Percentage of adults age 18 and older who engaged in regular physical activity, by age group and gender, January–March 2007**

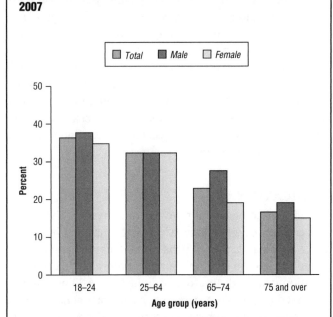

Notes: This measure reflects the definition used for the physical activity Leading Health Indicator (*Healthy People 2010*, (14)). Regular leisure-time physical activity is defined as engaging in light-moderate leisure-time physical activity for greater than or equal to 30 minutes at a frequency greater than or equal to five times per week or engaging in vigorous leisure-time physical activity for greater than or equal to 20 minutes at a frequency greater than or equal to three times per week. In Early Releases before September 2005 (based on the 2004 National Health Interview Survey (NHIS)), regular physical activity was calculated slightly differently than that of *Healthy People 2010*. The earlier Early Release estimates excluded from the analysis persons with unknown duration of light-moderate or vigorous leisure-time physical activity who were known to have not met the frequency recommendations for light-moderate or vigorous leisure-time physical activity (i.e., partial unknowns). With the current release, persons who were known to have not met the frequency recommendations are classified as "not regular," regardless of duration. The analyses excluded 183 persons (3.1%) with unknown physical activity participation.

SOURCE: P. Barnes, K. M. Heyman, and J. S. Schiller, "Figure 7.2. Percentage of Adults Aged 18 Years and Over Who Engaged in Regular Leisure-Time Physical Activity, by Age Group and Sex: United States, January–March 2007," in *Early Release of Selected Estimates Based on Data fom the January–March 2007 National Health Interview Survey*, Centers for Disease Control and Prevention, National Center for Health Statistics, September 26, 2007, http://www.cdc.gov/nchs/data/nhis/earlyrelease/earlyrelease200709.pdf (accessed October 24, 2007)

**Percentage of adults age 18 and older who engaged in regular physical activity, by race/ethnicity, January–March 2007**

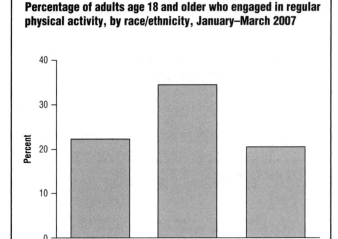

Notes: This measure reflects the definition used for the physical activity Leading Health Indicator (*Healthy People 2010*, (14)). Regular leisure-time physical activity is defined as engaging in light-moderate leisure-time physical activity for greater than or equal to 30 minutes at a frequency greater than or equal to five times per week or engaging in vigorous leisure-time physical activity for greater than or equal to 20 minutes at a frequency greater than or equal to three times per week. In Early Releases before September 2005 (based on the 2004 National Health Interview Survey (NHIS)), regular physical activity was calculated slightly differently than that of *Healthy People 2010*. The earlier Early Release estimates excluded from the analysis persons with unknown duration of light-moderate or vigorous leisure-time physical activity who were known to have not met the frequency recommendations for light-moderate or vigorous leisure-time physical activity (i.e., partial unknowns). With the current release, persons who were known to have not met the frequency recommendations are classified as "not regular," regardless of duration. The analyses excluded 183 persons (3.1%) with unknown physical activity participation. Estimates are age-sex adjusted using the projected 2000 U.S. population as the standard population and using five age groups: 18–24 years, 25–34 years, 35–44 years, 45–64 years, and 65 years and over.

SOURCE: P. Barnes, K. M. Heyman, and J. S. Schiller, "Figure 7.3. Age-Sex-Adjusted Percentage of Adults Aged 18 Years and Over Who Engaged in Regular Leisure-Time Physical Activity, by Race/Ethnicity: United States, January–March 2007," in *Early Release of Selected Estimates Based on Data fom the January–March 2007 National Health Interview Survey*, Centers for Disease Control and Prevention, National Center for Health Statistics, September 26, 2007, http://www.cdc.gov/nchs/data/nhis/earlyrelease/earlyrelease200709.pdf (accessed October 24, 2007)

exercise equivalent to walking twelve miles per week; low-dose/vigorous-intensity exercise equivalent to jogging twelve miles per week; or high-dose/vigorous-intensity exercise equivalent to jogging twenty miles per week. The subjects were advised to maintain their weight and not to change their diets. Slentz et al. followed the subjects for eight months and then measured weight, body fat, waist circumference, and lean muscle mass.

Weight change was a 3.5% loss in the high-dose/vigorous-intensity group and about a 1% loss in the two low-dose exercise groups, compared to a 1.1% gain in the control group. Increases in lean body mass were 1.4% in the two vigorous-intensity groups and 0.7% in the low-intensity group. Body fat mass increased by 0.5%

in the control group and decreased by 2% in the low-dose/moderate-intensity group, by 2.6% in the low-dose/vigorous-intensity group, and by 4.9% in the high-dose/vigorous-intensity group. Waist circumference increased by 0.8% in the control group and decreased by 1.6% in the low-dose/moderate-intensity group, by 1.4% in the low-dose/vigorous-intensity group, and by 3.4% in the high-dose/vigorous-intensity group. The three exercise groups had also significantly decreased waist and hip circumference measurements compared to the control group.

**RESEARCHERS RECONSIDER THE ROLE OF EXERCISE IN WEIGHT LOSS AND MAINTENANCE.** In "Long-Term Weight Losses Associated with Prescription of Higher Physical Activity Goals: Are Higher Levels of Physical Activity Protective against Weight Regain?" (*American Journal of*

*Clinical Nutrition*, vol. 85, no.4, April 2007), Deborah F. Tate et al. confirm that regular exercise and high levels of physical activity help maintain weight loss over time. However, does strenuous exercise really cause weight loss?

In an effort to answer this question, Neil A. King et al. asked thirty-five overweight people to exercise vigorously enough to burn five hundred calories per day for twelve weeks and reported their findings in "Individual Variabil-ity Following 12 Weeks of Supervised Exercise: Identification and Characterization of Compensation for Exercise-Induced Weight Loss" (*International Journal of Obesity*, September 11, 2007). Even though many of the subjects lost weight during the study, five gained weight—and there was not much variability between dietary changes made by subjects who lost as much as thirty pounds, those who lost just a few pounds, and those who gained. King et al. opine that their results demonstrate that there is considerable variability in the body's compensatory responses to exercise. In other words, moderate exercise may cause some people to lose weight, whereas others find that their weight is unchanged or even increases.

## MEDICATION

Pharmacotherapy for weight loss involves the use of prescription drugs as one of several strategies including diet, physical activity, behavioral therapy, counseling, and participation in group-support programs that in combination can work to effect weight loss. Adding weight-loss medications to a comprehensive treatment program consisting of diet, physical activity, and counseling can increase weight loss by five to twenty pounds during the first six months of treatment. The decision to add prescription drugs to a treatment program usually considers the individual's body mass index (BMI), other medical problems, or coexisting risk factors. Table 6.4 shows the therapies appropriate for people with differing BMIs and takes into account the presence of comorbidities (the coexistence of two or more diseases) such as diabetes, severe obstructive sleep apnea, or heart disease.

Most drugs used for weight loss are anorexiants (appetite suppressants), which act on neurotransmitters (chemical substances that convey impulses from one nerve cell to another) in the brain. Anorexiant drugs vary depending on which neurotransmitters they affect: some

**TABLE 6.2**

**A sample walking program**

| | Warm up | Exercising | Cool down | Total time |
|---|---|---|---|---|
| Week 1 | | | | |
| Session A | Walk 5 min. | Then walk briskly 5 min. | Then walk more slowly 5 min. | 15 min. |
| Session B | Repeat above pattern | | | |
| Session C | Repeat above pattern | | | |

*Continue with at least three exercise sessions during each week of the program.*

| | | | | |
|---|---|---|---|---|
| Week 2 | Walk 5 min. | Walk briskly 7 min. | Walk 5 min. | 17 min. |
| Week 3 | Walk 5 min. | Walk briskly 9 min. | Walk 5 min. | 19 min. |
| Week 4 | Walk 5 min. | Walk briskly 11 min. | Walk 5 min. | 21 min. |
| Week 5 | Walk 5 min. | Walk briskly 13 min. | Walk 5 min. | 23 min. |
| Week 6 | Walk 5 min. | Walk briskly 15 min. | Walk 5 min. | 25 min. |
| Week 7 | Walk 5 min. | Walk briskly 18 min. | Walk 5 min. | 28 min. |
| Week 8 | Walk 5 min. | Walk briskly 20 min. | Walk 5 min. | 30 min. |
| Week 9 | Walk 5 min. | Walk briskly 23 min. | Walk 5 min. | 33 min. |
| Week 10 | Walk 5 min. | Walk briskly 26 min. | Walk 5 min. | 36 min. |
| Week 11 | Walk 5 min. | Walk briskly 28 min. | Walk 5 min. | 38 min. |
| Week 12 | Walk 5 min. | Walk briskly 30 min. | Walk 5 min. | 40 min. |

Week 13 on: Gradually increase your brisk walking time to 30 to 60 minutes, three or four times a week. Remember that your goal is to get the benefits you are seeking and enjoy your activity.

SOURCE: "A Sample Walking Program," in *The Practical Guide: Identification, Evaluation, and Treatment of Overweight and Obesity in Adults*, National Institutes of Health, National Heart, Lung, and Blood Institute, North American Association for the Study of Obesity, October 2000, http://www.nhlbi.nih.gov/guidelines/obesity/prctgd_b.pdf (accessed October 24, 2007)

**TABLE 6.3**

**Examples of moderate amounts of physical activity**

| Common chores | Sporting activities | |
|---|---|---|
| Washing and waxing a car for 45–60 minutes | Playing volleyball for 45–60 minutes | Less vigorous more time* |
| Washing windows or floors for 45–60 minutes | Playing touch football for 45 minutes | |
| Gardening for 30–45 minutes | Walking 1 3/4 miles in 35 minutes (20 min/mile) | |
| Wheeling self in wheelchair for 30–40 minutes | Basketball (shooting baskets) for 30 minutes | |
| Pushing a stroller 1 1/2 miles in 30 minutes | Bicycling 5 miles in 30 minutes | |
| Raking leaves for 30 minutes | Dancing fast (social) for 30 minutes | |
| Walking 2 miles in 30 minutes (15 min/mile) | Water aerobics for 30 minutes | |
| Shoveling snow for 15 minutes | Swimming laps for 20 minutes | |
| Stairwalking for 15 minutes | Basketball (playing a game) for 15–20 minutes | |
| | Jumping rope for 15 minutes | More vigorous, less time |
| | Running 1 1/2 miles in 15 minutes | |

Note: A moderate amount of physical activity is roughly equivalent to physical activity that uses approximately 150 calories of energy per day, or 1,000 calories per week.
*Some activities can be performed at various intensities; the suggested durations correspond to expected intensity of effort.

SOURCE: "Appendix H. Examples of Moderate Amounts of Physical Activity," in *The Practical Guide: Identification, Evaluation, and Treatment of Overweight and Obesity in Adults*, National Institutes of Health, National Heart, Lung, and Blood Institute, North American Association for the Study of Obesity, October 2000, http://www.nhlbi.nih.gov/guidelines/obesity/prctgd_b.pdf (accessed October 24, 2007)

**TABLE 6.4**

**A guide to selecting weight loss treatment by body mass index (BMI)**

| Treatment | BMI category | | | | |
|---|---|---|---|---|---|
| | 25–26.9 | 27–29.9 | 30–34.9 | 35–39.9 | ≥40 |
| Diet, physical activity, and behavior therapy | With comorbidities | With comorbidities | + | + | + |
| Pharmacotherapy | | With comorbidities | + | + | + |
| Surgery | | | | With comorbidities | |

- Prevention of weight gain with lifestyle therapy is indicated in any patient with a BMI ≥25 kg/m², even without comorbidities, while weight loss is not necessarily recommended for those with a BMI of 25–29.9 kg/m² or a high waist circumference, unless they have two or more comorbidities.
- Combined therapy with a low-calorie diet (LCD), increased physical activity, and behavior therapy provide the most successful intervention for weight loss and weight maintenance.
- Consider pharmacotherapy only if a patient has not lost 1 pound per week after 6 months of combined lifestyle therapy.

The + represents the use of indicated treatment regardless of comorbidities.

SOURCE: "Table 3. A Guide to Selecting Treatment," in *The Practical Guide: Identification, Evaluation, and Treatment of Overweight and Obesity in Adults*, National Institutes of Health, National Heart, Lung, and Blood Institute, North American Association for the Study of Obesity, October 2000, http://www .nhlbi.nih.gov/guidelines/obesity/prctgd_b.pdf (accessed October 24, 2007)

affect catecholamines such as dopamine and norepinephrine; others affect serotonin; and a third class of drugs acts on more than one neurotransmitter. The drugs act by increasing the secretion of dopamine, norepinephrine, or serotonin, by inhibiting reuptake of neurotransmitters, or by a combination of both mechanisms. For example, sibutramine inhibits the reuptake of norepinephrine and serotonin.

Another class of weight-loss drugs blocks the absorption of fat. Orlistat, which was approved by the U.S. Food and Drug Administration (FDA) in 1999, decreases fat absorption in the digestive tract by about one-third. Because it also inhibits absorption of water and vitamins, some users suffer from cramping and diarrhea.

Rimonabant, another weight-loss drug, acts on the endocannabinoid system to block the "munchie receptor," which is believed to stimulate appetite among people who smoke marijuana. Because it blocks cravings, rimonabant has also been used to aid smoking cessation. Even though it is approved for use in Europe, at the close of 2007 the FDA had not granted marketing approval for the drug, largely due to reports of adverse side effects reported with its use, which were addressed in *Rimonabant Briefing Document* (June 13, 2007, http://www .fda.gov/ohrms/dockets/ac/07/briefing/2007-4306b1-fda-backgrounder.pdf).

The determination of which type of drug to prescribe is based on individual patient characteristics—sibutramine works best for people who are preoccupied with food and feel constantly hungry, orlistat may be effective for those who are unwilling to reduce fat from their diet, and rimonabant may help reduce food cravings. Even though drug therapy has not demonstrated remarkable effectiveness, only modestly enhancing weight loss over diet alone, consumer demand for weight-loss drugs is high. In February 2007 the FDA approved the over-the-counter (nonprescription) sale of orlistat.

Several weight-loss drugs that appeared effective and were popular among consumers have been withdrawn from the U.S. market due to the number and severity of adverse side effects associated with their use. During the 1990s a combination of two drugs—phentermine and fenfluramine, commonly known as "phen-fen"—was prescribed for long-term use (more than three months); however, rare but unacceptable side effects, including serious damage to the heart valves, prompted the withdrawal of fenfluramine and a similar drug, dexfenfluramine, in September 1997. Phentermine is still approved for short-term use.

In "Effective Obesity Treatments" (*American Psychologist*, vol. 62, no. 3, April 2007), Lynda H. Powell, James E. Calvin III, and James E. Calvin Jr. evaluate the results of published studies of the efficacy (effectiveness) of obesity treatments and conclude that "drug interventions result in modest weight loss with minimal risks but disproportionate clinical benefit. Combinations of lifestyle, drug, and, where appropriate, surgical interventions may be the most efficacious approach to achieving sustained weight loss for the widest diversity of patients."

**Research Focuses on New Weight-Loss Drugs**

By the close of 2007, only two weight-loss drugs, orlistat and sibutramine, were FDA-approved for long-term use, and evidence indicates that many users experience "rebound" weight gain when the use of either of these drugs is discontinued. During 2007, however, more than a dozen new drugs were in various stages of development.

Researchers have identified that the hormone ghrelin may be involved in establishing hunger and satiety (the feeling of fullness or satisfaction after eating) set points. When the stomach is empty, it releases ghrelin, which in turn triggers hunger signals in the brain. Blood levels of ghrelin peak before meals and decrease after eating. Because ghrelin appears to increase appetite and slow metabolism, an excess of it may sabotage long-term weight-loss efforts. Small studies show that ghrelin levels are higher in obese patients who have recently lost weight, compared to obese patients at a steady weight. In recent years, pharmaceutical companies have been seeking to create drugs that safely and effectively block ghrelin's effects. According to the article "Nastech Submits Investigational New Drug Application for PYY3-36 Nasal Spray to Treat Obesity" (*Medical News Today*, June 9, 2006), an analogous approach seeks to boost levels of a peptide known as PYY that produces the opposite effects of ghrelin. After eating, the stomach and digestive tract release PYY, conveying the satiety signal to the brain.

Mike Nagle reports in "Genaera Begins Human Trials of Obesity Drug" (DrugResearcher.com, August 5, 2007) that a new appetite-suppressing drug called trodusquemine was undergoing clinical trials in 2007. Trodusquemine is the first highly selective inhibitor of the protein tyrosine phosphatase 1B (PTP1B). In preclinical testing on obese mice, trodusquemine suppressed appetite, caused weight loss, and normalized fasting blood glucose and cholesterol levels.

Gabriele E. Sonnenberg, Glenn Matfin, and Rickey R. Reinhardt indicate in "Drug Treatments for Obesity: Where Are We Heading and How Do We Get There?" (*British Journal of Diabetes and Vascular Disease*, vol. 7, no, 3, 2007) that more effective drug therapy will likely target multiple systems, such as the gastrointestinal (digestive) system as well as the endocrine and neurological pathways. There is also enthusiasm for the development of a drug that increases the body's metabolic rate because it might enable people to forgo severely restricted diets and still realize weight loss.

## Nonprescription Weight-Loss Aids

The withdrawal of fenfluramine from the market prompted many consumers to seek alternative weight-loss aids, including herbal preparations that were marketed as dietary supplements and available over the counter. Some preparations combined ephedra, caffeine, and other ingredients. Ephedra (also known by its traditional Chinese medicine name, *ma huang*) is a naturally occurring substance that comes from botanicals. Products containing ephedra and ephedrine have been promoted to accelerate weight loss, increase energy, and improve athletic performance. The principal active ingredient in ephedrine is an amphetamine-like compound that stimulates the nervous system and heart. Because ephedrine has some anorectic and thermogenic properties, it may induce weight loss in some people, and some studies show that when ephedrine is combined with caffeine, the combination may lead to even more weight loss.

During 2003 the FDA and the National Institutes of Health (NIH) investigated reports of adverse effects linked to ephedra use. In *Ephedra and Ephedrine for Weight Loss and Athletic Performance Enhancement: Clinical Efficacy and Side Effects* (February 2003, http://www.ahrq.gov/downloads/pub/evidence/pdf/ephedra/ephedra.pdf), Paul Shekelle et al. of the RAND Corporation conclude that there is only limited evidence of health benefits resulting from ephedra use. These benefits do not outweigh the serious risks posed by its association with heart palpitations, psychiatric and upper gastrointestinal effects, tremors, and insomnia, especially in formulations in which it was combined with caffeine or taken with other stimulants. Shekelle et al. reviewed sixteen thousand adverse events and identified one seizure, two deaths, four heart attacks, five psychiatric cases, and nine strokes in which ephedra appeared to be the causative agent.

In another study, "The Relative Safety of Ephedra Compared with Other Herbal Products" (*Annals of Internal Medicine*, vol. 138, no. 6, March 18, 2003), Stephen Bent et al. compare the risk for adverse events attributable to ephedra and other herbal products. The researchers find that even though ephedra products comprised 0.8% of all dietary supplement sales, they accounted for 64% of adverse events associated with dietary supplements. Bent et al. conclude that "the risk for an adverse reaction after the use of ephedra is substantially greater than with other herbal products."

According to the press release "FTC Charges Direct Marketers of Ephedra Weight Loss Products with Making Deceptive Efficacy and Safety Claims" (July 1, 2003, http://www.ftc.gov/opa/2003/07/ephedra.shtm), in July 2003 the Federal Trade Commission (FTC) charged marketers of weight-loss products that contain ephedra with making deceptive efficacy and safety claims. The FTC deemed as examples of false advertising claims that ephedra causes rapid, substantial, and permanent weight loss without diet or exercise, and that "clinical studies" or "medical research" proved these claims. The FTC also challenged claims that the ephedra weight-loss products are "100% safe," "perfectly safe," or have "no side effects."

The press release "FDA Announces Plans to Prohibit Sales of Dietary Supplements Containing Ephedra" (http://www.hhs.gov/news/press/2003pres/20031230.html) notes that on December 30, 2003, the U.S. Department of Health and Human Services and the FDA notified manufacturers of dietary supplements containing ephedra that the sale of these dietary supplements would be banned sixty days following publication of the year-end notice. That same day, the FDA issued an alert to consumers advising them to stop using ephedra products immediately.

In early 2004 dieters flocked to health food stores and Internet sites selling dietary supplements and bought entire inventories of supplements containing ephedra in anticipation of the ban of its sale. Many of the supplements' fans asserted that the ban was prompted by the publicity surrounding the ephedra-related death of the Baltimore Orioles pitcher Steve Bechler on February 17, 2003. Bechler was twenty-three years old when he collapsed from heatstroke at the Orioles' spring training camp in Florida. Two weeks later, the FDA ordered warning labels be placed on products containing ephedra and set in motion plans to ban its sale.

Many health professionals and consumer watchdog agencies applauded the FDA action. However, they also observed that the FDA first proposed warning labels and a dosage curb for ephedra in 1997, but the supplement industry effectively blocked the move. The December 2003 action was a historic occasion—it was the first time the FDA completed the steps necessary to ban the sale of a dietary supplement.

**WILL HOODIA BE THE NEXT BEST THING FOR DIETERS?** With ephedra withdrawn from the market, dieters waited for the introduction of another nonprescription drug to replace it. In 2003 reports of an appetite suppressant derived from hoodia, a bitter-tasting cactus that grows in the South African Kalahari desert, generated interest. The San Bushmen of the Kalahari, one of the world's oldest and most primitive tribes, have been eating hoodia for thousands of years to stave off hunger during long hunting trips.

The plant contains the molecule p57, which is thought to act on the hypothalamus to mimic the sense of satiety that normally results only from eating food. The first clinical trials of hoodia were considered successful when subjects given hoodia consumed an average of one thousand calories fewer per day than those given a placebo. By 2007 dozens of Web sites offered pills, powders, and liquids containing various amounts of hoodia.

## SURGERY

Weight-loss surgery is considered a treatment option only for people for whom all other treatment methods have failed and who suffer from clinically severe obesity—BMI of 40 or greater or BMI of 35 or greater in the presence of comorbidities. (Clinically severe obesity was formerly known as morbid obesity, indicating its potential to cause disease.) Two types of surgical procedures have been demonstrated effective in producing weight loss maintained for five years: restrictive techniques, which restrict gastric volume, and malabsorptive procedures, which not only limit food intake but also alter digestion. An example of the first type is banded gastroplasty, in which an inflatable band that can be adjusted to different diameters is placed around the stomach. The Roux-en-Y gastric bypass is an example of the second type. (See Figure 6.4.) On average, patients maintain a weight loss of 25% to 40% of their preoperative body weight after these procedures.

**FIGURE 6.4**

**Surgical weight loss procedures**

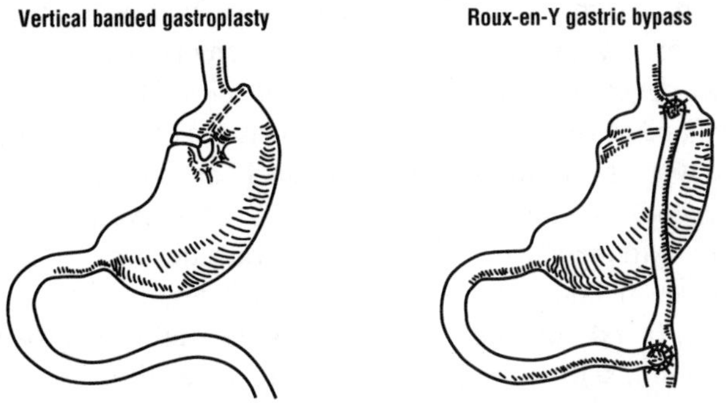

Vertical banded gastroplasty

Roux-en-Y gastric bypass

SOURCE: "Figure 5. Surgical Procedures in Current Use," in *The Practical Guide: Identification, Evaluation, and Treatment of Overweight and Obesity in Adults*, National Institutes of Health, National Heart, Lung, and Blood Institute, North American Association for the Study of Obesity, October 2000, http://www.nhlbi.nih.gov/guidelines/obesity/prctgd_b.pdf (accessed October 24, 2007)

The surgery not only improves patients' quality of life by causing weight loss and the resolution of many weight-related conditions such as sleep apnea, joint pain, and diabetes, but also reduces their risk of death. Lars Sjöström et al. find in "Effects of Bariatric Surgery on Mortality in Swedish Obese Subjects" (*New England Journal of Medicine*, vol.357, no, 8, August 23, 2007), a study following 2,010 bariatric surgery patients and a control group consisting of 2,037 obese subjects who did not have surgery, that bariatric surgery was associated with a reduction in overall mortality.

Because the surgical procedures are not without risk, physicians generally recommend surgery only when the risks of obesity far outweigh the risks associated with the surgery. The National, Heart, Lung, and Blood Institute explains that surgical complications vary depending on the weight and overall health of the surgical patient. According to Daniel Leslie, Todd A. Kellogg, and Sayeed Ikramuddin, in "Bariatric Surgery Primer for the Internist: Keys to the Surgical Consultation" (*Medical Clinics of North America*, vol. 91, no. 3, May 2007), young people without comorbidities and a BMI equal to or less than 50 have the lowest reported mortality rates—less than 1%. Not unexpectedly, those with a BMI equal to or greater than 60 with comorbidities such as diabetes or high blood pressure have higher mortality rates.

People who undergo weight-loss surgeries require lifelong medical monitoring. After surgery, they are no longer able to eat in the way to which they were accustomed. Those who have undergone gastric bypass experience "dumping syndrome" with symptoms such as sweating, palpitations, lightheadedness, and nausea when they ingest significant amounts of calorie-dense food, and most become conditioned not to eat such foods. Patients who have had gastric restriction surgery are unable to eat more than a limited amount of food at a single sitting without vomiting, so they must eat several small meals per day to maintain adequate nutrition. Those who do not adhere to a prescribed regimen of vitamins and minerals may develop vitamin and iron deficiencies. There are also postoperative and long-term complications of surgery such as wound infections, problems such as hernias at the incision site, and gallstones. Generally, however, patients fare extremely well, experiencing dramatic improvement and even complete resolution of diabetes, hypertension (high blood pressure), and infertility, as well as improved mobility, self-esteem, and overall quality of life.

The article "Gastric Bypass Surgery May Cause Post-op Nutrient Deficiencies" (Reuters Health Information. October 15, 2007) notes that besides the possibility of nutritional deficiencies, research reveals that bypass surgery may cause a condition known as "small intestinal bacterial overgrowth," which interferes with nutrient absorption. Deficiencies in vitamin D, calcium, and zinc may in turn increase the risk of developing other serious conditions such as hypothyroidism (deficiency of the thyroid hormone, which is produced by the thyroid gland) and osteoporosis.

In "Death Rates and Causes of Death after Bariatric Surgery for Pennsylvania Residents, 1995 to 2004" (*Archives of Surgery*, vol. 142, no. 10, October 2007), a study of more than sixteen thousand patients who had undergone bariatric surgery, Bennet I. Omalu et al. report an excessive number of patient deaths attributable to coronary artery disease and suicide. Heart disease was the leading cause of death and was responsible for 20% of deaths that occurred thirty days or more after the surgery. This rate is nearly three times higher than in the general population. Another 7% of deaths were attributable to suicide or drug overdose, and the researchers speculate that continued obesity and/or weight regain might have contributed to both the heart disease and suicide deaths.

### Number of Surgeons and Surgeries Soars

Nanci Hellmich states in "Study: Gastric Bypass Reduces Death Risk in the Morbidly Obese" (*USA Today*, August 22, 2007) that 23,100 bariatric procedures were performed in 1997; by 2007 the number was estimated to be more than 205,000. For thousands of patients, weight-loss surgery has eliminated debilitating diseases and improved their quality of life. With the number of candidates for bariatric surgery increasing, the number of procedures is expected to continue to grow, even in view of data that reveal that the risks may be greater than previously believed.

In "Assessing the Value of Weight Loss among Primary Care Patients" (*Journal of General Internal Medicine*, vol. 19, no. 12, December 2004), Christina C. Wee et al. find that even the risk of death does not dissuade many patients from undergoing bariatric surgery. In an effort to quantify the value people place on modest weight loss, the researchers interviewed 365 patients at a large hospital-based primary care practice, one-third of whom were obese. The subjects were asked to imagine a treatment that would guarantee them effortless weight loss of varying amounts of weight. For each amount, they were asked if they would be willing to accept a risk of death to achieve it. If so, how much of a risk of death?

Willingness to risk death or trade years of life to lose weight significantly increased with higher BMI, and the more weight the subjects imagined they could lose, the greater the risk they would take to achieve it. Eighteen percent of overweight and 33% of obese people said they would risk death for even a modest 10% weight loss, compared to just 4% of normal-weight subjects willing to risk death to lose 10% of their weight.

Many of the overweight and obese participants in the survey also said they would give up some of their remaining years of life if they could live those years weighing slightly less. Thirty-one percent of obese patients and 8.3% of overweight patients said they would trade up to 5% of their remaining life to be 10% thinner. Wee et al. conclude that many people, especially those who are obese, value modest weight loss and suggest that physicians emphasize the benefits of modest weight loss when counseling their patients.

## COUNSELING AND BEHAVIORAL THERAPY

Weight-loss counseling and behavioral therapy aim to assist people to develop the skills needed to identify and modify eating and activity behaviors, and change thinking patterns that undermine weight-control efforts. Behavioral strategies include self-monitoring of weight, food intake, and physical activity; identifying and controlling stimuli that provoke overeating; problem identification and problem solving; and using family and social support systems to reinforce weight-control efforts. Counseling and behavioral therapy are often perceived as necessary components of comprehensive weight-loss treatment, but are also viewed as labor intensive because educating and supporting people seeking to lose weight is time consuming. The effort also requires the active participation of everyone who may be involved in treatment—the affected individuals, their families, physicians, nurses, nutritionists, dieticians, exercise instructors, and mental health professionals. In view of the considerable resources that must be allocated to deliver counseling and behavioral therapy, it is important to know if these approaches effectively promote weight loss.

Kathleen M. McTigue et al. considered the evidence supporting the efficacy of counseling and behavioral therapy as well as other treatment methods and reported their findings in "Screening and Interventions for Obesity in Adults: Summary of the Evidence for the U.S. Preventive Services Task Force" (*Annals of Internal Medicine*, vol. 139, no. 11, December 2, 2003). The investigators report that counseling to promote change in diet, exercise, or both, and behavioral therapy to help patients acquire the skills, motivations, and support to change diet and exercise patterns enabled obese patients to achieve modest but clinically significant and sustained (one to two years) weight loss. Furthermore, they observe that because control groups also frequently received some form of counseling, education, or support, they might have underestimated the effectiveness of counseling. Not unexpectedly, more intensive programs, with more frequent contact, were generally more successful, as were those incorporating behavioral therapy.

Interestingly, McTigue et al. find that treating patients on an individual basis rather than on a group basis did not appear to affect outcomes. This finding offers credence to the theory that the benefits of mutual aid and peer support provided by group programs may be as powerful as the personalized, one-to-one attention afforded patients in individual counseling sessions. If this is true, then group programs might be a laborsaving, cost-effective alternative to individual weight-loss counseling.

McTigue et al. conclude that "all obesity therapies carry promise and burden, which must be balanced in clinical decision-making. Counseling approaches appear the least harmful and produce modest, clinically important weight loss, but entail cost in time and resources. Pharmacotherapy promotes modest additional weight loss, but long-term drug use may be needed to sustain this benefit with unknown long-term adverse events and appreciable cost. Only surgical options consistently result in large amounts of long-term weight reduction; however, they carry a low risk for severe complications and are expensive. Body size, health status, and prior weight-loss history may all influence obesity treatment."

## Comparing Weight-Loss Using a Self-Help Program and a Commercial Program

Stanley Heshka et al. report the results of their research to determine the efficacy of commercial weight-loss programs in "Weight Loss with Self-Help Compared with a Structured Commercial Program" (*Journal of the American Medical Association*, vol. 289, no. 14, April 9, 2003). Their study randomly assigned one group of obese men and women to a self-help program consisting of two twenty-minute counseling sessions with a nutritionist and provision of self-help resources such as public library materials, Web sites, and telephone numbers of health organizations that offered free weight-control information. The other group was assigned to attend Weight Watchers, a commercial weight-loss program consisting of a food plan, an exercise plan consistent with NIH-recommended physical activity guidelines, regular weight monitoring, printed educational materials, and a behavior modification plan, delivered at weekly meetings.

Subjects were evaluated regularly during the course of the two-year study—at 12, 26, 52, 78, and 104 weeks. The primary outcome measure used to evaluate the effectiveness of the programs was change in body weight; however, BMI, waist circumference, and body fat as quantified by bioimpedance analysis (electrical resistance) were also recorded. Other secondary measures were blood pressure, total cholesterol, HDL cholesterol, triglycerides, insulin, and quality of life measured using the "Medical Outcomes Study Short-Form 36 Health Survey and Impact of Weight on Quality of Life Questionnaire."

After one year of participation in the study, subjects in the commercial program had greater weight loss than those in the self-help group. Similarly, waist circumference and BMI decreased more in the commercial group than in the self-help group. Blood pressure and serum insulin showed greater improvement in the commercial group, compared to the self-help group at year one, but only insulin was significantly different at year two. Total cholesterol and the HDL/total cholesterol ratio improved in both groups. The commercial group maintained a weight loss of 9.5 to 11 pounds at the end of the first year and was 5.9 to 6.6 pounds lower than its initial weight at the end of the second year. Subjects who attended 78% or more of the commercial group sessions maintained a mean weight loss of almost 11 pounds at the end of the two-year study. Heshka et al. conclude that even though the structured commercial weight-loss program provided only modest weight loss, it was more effective than brief counseling and self-help for overweight and obese adults over a two-year period.

## Weight-Loss Counseling to Change Behavior

The NIH designed a practical protocol, known as an algorithm, for obtaining and organizing information necessary for effective weight-loss counseling. The algorithm is based on the "five As":

- Assessing obesity risk
- Asking about readiness to lose weight
- Advising about a weight-control program
- Assisting to establish appropriate intervention
- Arranging for follow-up

The National Heart, Lung, and Blood Institute recommends in *Practical Guide to the Identification, Evaluation,* and *Treatment of Overweight and Obesity in Adults* (October 2000, http://www.nhlbi.nih.gov/guide lines/obesity/prctgd_c.pdf) that health-care professionals consider a variety of psychosocial, environmental, and health-related issues when performing a "behavioral assessment" of an individual for whom weight loss is indicated. These issues include:

- Whether the individual is seeking to lose weight on his or her own or in response to pressure from family members, an employer, or a physician. This is an important consideration because people who feel coerced into seeking weight-loss treatment are not as likely to achieve success as those who seek it on their own initiative.

- Identifying the source of the individual's desire to lose weight to better understand his or her motivation and goals. Because many people have suffered from overweight or obesity for years before seeking treatment, pinpointing the stimulus to lose weight can assist the health-care professional to motivate and support the individual's weight-loss efforts.

- Assessing the individual's stress level to determine if external stressors such as family-, financial-, or work-related problems might prevent the individual from concentrating on weight loss. It is also important to determine if the individual is suffering from depression or other mental health problems because it is usually advisable to treat mood disorders or other mental health problems before embarking on a weight-loss program.

- Evaluating the individual for the presence of an eating disorder such as binge eating that may coexist with overweight or obesity. People suffering from eating disorders are more likely to require psychological treatment and nutritional counseling to ensure the success of weight-loss programs than those who do not have eating disorders.

- Determining the individual's understanding of the lifestyle and other changes required for weight loss. The success of treatment hinges on the individual's ability to successfully make the required changes, so it is vital to develop a treatment plan that includes realistic activities such as gradually increasing physical activity that the individual agrees are attainable.

- Setting and agreeing on realistic weight-loss goals and objectives. If an obese individual has unrealistic expectations about the amount of weight that will be lost, then he or she may become discouraged and abandon efforts to lose weight. Health professionals should temper unrealistic expectations by informing individuals about the considerable health and lifestyle benefits of even modest weight loss.

Successful weight loss is more likely to occur when health-care professionals—physicians, nurses, nutrition-ists, dieticians, and mental health professionals—actively involve people seeking to lose weight in a collaborative effort to establish short-term goals and attain them. "Shaping" is a behavioral technique in which a series of short-term objectives are identified that ultimately lead to a treatment goal, such as incrementally increasing physical activity from ten minutes per day to forty-five minutes per day over time. "Self-monitoring" is the practice of observing and recording behaviors such as caloric intake, food choices, amounts consumed, and emotional or other triggers to eat as well as physical activity performed and daily or weekly monitoring of body weight.

Finally, the National Heart, Lung, and Blood Institute reminds health professionals to acknowledge the challenges of accomplishing weight loss and encourages everyone involved in treatment to "focus on positive changes and adapt a problem-solving approach toward shortfalls.... Emphasize that weight control is a journey, not a destination, and that some missteps are inevitable opportunities to learn how to be more successful."

**Weight-Loss Counseling Online**

An expanding array of diet, counseling, and support group programs are available on the Internet; however, little research has compared them or determined their efficacy. In "A Randomized Trial Comparing Human E-Mail Counseling, Computer-Automated Tailored Counseling, and No Counseling in an Internet Weight Loss Program" (*Archives of Internal Medicine*, vol. 166, no. 15, 2006), Deborah F. Tate, Elizabeth H. Jackvony, and Rena R. Wing sought to determine whether computer-generated feedback, delivered via the Internet, would prove to be a viable alternative to human counseling via e-mail. They compared the effects of custom-tailored computer-automated interactions with an Internet program that provided weight-loss counseling from a human via e-mail.

All the subjects received one weight-loss group session, coupons for meal replacements, and access to an interactive Web site. The human e-mail counseling and computer-automated feedback groups also had access to an electronic diary and message board. The human e-mail counseling group received weekly e-mail feedback from a counselor, the computer-automated feedback group received automated, custom-tailored messages, and a control group received no counseling at all. Recommendations included calorie-restricted diets of between twelve hundred and fifteen hundred calories per day, daily exercise equivalent to walking for thirty minutes, and instructions about how to use meal replacement products. All participants were encouraged to self-monitor their diets and exercise using diaries and calorie books. Both groups accessed the same Web site, which featured weekly reporting and graphs of weight, weekly e-mail prompts to report weight, weekly weight-loss tips via e-mail, recipes, and a weight-loss e-buddy network system that enabled users to interact with other dieters with similar characteristics via e-mail.

The primary outcome measure used to compare the groups was change in body weight from baseline and at three and six months. Both the human and automated e-counseling groups had greater reductions in weight than the control group at each weigh-in. Tate, Jackvony, and Wing conclude that automated computer feedback was as effective as human e-mail counseling.

### Complementary and Alternative Therapies

Many complementary and alternative medicine practices such as yoga, Dahn—a holistic mind-body training method—and "mindful eating," which teaches greater awareness of bodily sensations such as hunger and satiety and helps people identify "emotional eating," have been used to promote weight loss. Acupuncture (the Chinese practice of inserting extremely thin, sterile needles into any of 360 specific points on the body) and hypnosis are, however, the only alternative medical practices that have been studied as potential treatments for obesity. Several studies report that acupuncture does not appear to have any benefit greater than a placebo.

Hypnosis is an altered state of consciousness. It is a state of heightened awareness and suggestibility and enables focused concentration that may be used to alter perceptions of hunger and satiety and to modify behavior. Hypnosis is considered a mainstream treatment for addictions and overeating. There are conflicting data about its effectiveness—some studies find that it adds little, if any, benefit beyond that of placebo. Others conclude that hypnosis may have some initial benefit for people seeking weight loss, but that it has little sustained effect.

In "Complementary Therapies for Reducing Body Weight: A Systematic Review" (*International Journal of Obesity*, vol. 29, no. 9, 2005), Max H. Pittler and Edzard Ernst review the published literature describing a variety of complementary and alternative medicine therapies for weight loss. They find that subjects receiving hypnotherapy lost more weight than subjects in a control group that did not receive hypnotherapy; that the addition of hypnotherapy to cognitive behavioral therapy led to a small reduction in body weight; and that patients in a small hypnotherapy group aimed at stress management lost significantly more weight than those in a control group.

## MIGHT WEIGHT LOSS BE HARMFUL?

Successful weight-loss treatments generally result in reduced blood pressure, reduced triglycerides, increased HDL cholesterol, and reduced total cholesterol and LDL cholesterol. Weight loss of as little as 5% to 10% of

initial weight produces measurable health benefits and may prevent illnesses among people at risk. These findings suggest that treatment should not exclusively focus on the medical consequences of obesity, but that obesity itself should be treated. The NIH recommends weight loss for people with a BMI greater than 30 and for those with a BMI greater than 25 with two or more obesity-related risk factors. The NIH guidelines recommend that for people with a BMI between 25 and 30 without other risk factors, the focus should be on prevention of further weight gain, rather than on weight loss.

In "Obesity: What Mental Health Professionals Need to Know" (*American Journal of Psychiatry*, vol. 157, no. 6, June 2000), Michael J. Devlin, Susan Z. Yanovski, and G. Terence Wilson report that critics cite the health and psychological risks of weight cycling (the repeated loss and regain of body weight) as even greater than the risks associated with obesity. They assert that multiple unsuccessful efforts to lose weight demoralize people and make future weight loss even more challenging, and that the dietary treatment of obesity may trigger or worsen binge eating among people who are obese. Devlin, Yanovski, and Wilson also offer several studies that find an association between weight cycling and increased morbidity and mortality as evidence of the dangers of dieting.

In "Screening and Interventions for Obesity in Adults: Summary of the Evidence for the U.S. Preventive Services Task Force" (*Annals of Internal Medicine*, vol. 139, no. 11, 2003), Kathleen M. McTigue et al. review the studies that reveal a link between weight cycling and mortality. The investigators find that some studies failed to distinguish between intentional and unintentional weight loss. In the research considering the relationship between weight cycling with intentional weight loss, some studies found unfavorable effects on coronary heart disease and its risk factors and others did not. McTigue et al. also find data suggesting that a weight-cycling risk increases inversely with BMI—the higher the BMI, the lower the risk of weight cycling. If these findings are correct, then people suffering from obesity as opposed to overweight are at less risk of morbidity and mortality attributable to weight cycling.

### Is It Better to Be Overweight?

Two studies indicate that there is less risk associated with overweight than previously thought. The first, Katherine M. Flegal et al.'s "Excess Deaths Associated with Underweight, Overweight, and Obesity" (*Journal of the American Medical Association*, vol. 293, no. 15, April 20, 2005), finds that increased risk of death from obesity was mostly among the extremely obese, a group constituting just 8% of Americans. The researchers also find that extreme thinness carried a slight increase in the risk of death. Flegal et al.'s study does not explain how or why being slightly overweight affords protection, but they speculate that it is because most people die when they are over seventy. Being mildly overweight in old age may be protective, because it gives rise to more muscle and bone.

The second study, "Secular Trends in Cardiovascular Disease Risk Factors According to Body Mass Index in U.S. Adults" (*Journal of the American Medical Association*, vol. 293, no.15, April 20, 2005) by Edward W. Gregg et al., examines forty-year trends in cardiovascular disease (CVD) risk factors by BMI groups among adults aged twenty to seventy-four years and finds that except for diabetes, CVD risk factors have declined considerably over the past forty years in all BMI groups. Even though obese people still have higher risk-factor levels than lean people, the levels of these risk factors are much lower than in previous decades. Gregg et al. observe that obese people in the twenty-first century have better CVD risk-factor profiles than their leaner counterparts did twenty to thirty years ago; however, they suggest that other factors, such as effective treatment to reduce cholesterol and blood pressure as well as the decreased prevalence of smoking, might explain the improved profiles of obese people.

# CHAPTER 7

# THE ECONOMICS OF OVERWEIGHT AND OBESITY

The economic impact of obesity is considerable. According to the American Obesity Association (AOA), in "Costs of Obesity" (May 2, 2005, http://obesity1 .tempdomainname.com/treatment/cost.shtml), the World Bank estimates the cost of obesity in the United States was 12% of the national health-care budget in the late 1990s. The increasing prevalence of overweight and obesity in the United States has resulted in a corresponding increase in direct and indirect health-care costs. Direct health-care costs are those incurred for preventive measures, diagnostic, and treatment services. Examples of direct health-care costs are physician office visits, hospital and nursing home charges, prescription drug costs, and special hospital beds to accommodate obese patients. Indirect costs are measured in terms of decreased earnings: lost wages and lower productivity resulting from the inability to work because of illness or disability, as well as the value of future earnings lost by premature mortality (death).

There are also personal costs of obesity: obese workers may earn less than their healthy-weight counterparts because of job discrimination. Many insurance companies, particularly in the life insurance sector, charge higher premiums with increasing degrees of overweight. When obesity compromises physical functioning and limits activities of daily living, affected individuals may require assistance from home health aides, durable medical equipment such as walkers or wheelchairs, or other costly adaptations to accommodate disability.

## THE HIGH COST OF OVERWEIGHT AND OBESITY

The National Center for Chronic Disease Prevention and Health Promotion (NCCDHP) calculates and compares in *Chronic Disease Prevention* (August 16, 2007, http://www.cdc.gov/nccdphp/press/index.htm) the economic burden of several chronic diseases including obesity. Table 7.1 shows that the direct health costs resulting from overweight and obesity—$75 billion—are comparable to those resulting from tobacco use. Because obesity has been linked to all the other chronic conditions described in Table 7.1 except tobacco use, it may be argued that some percentage of the costs attributed to arthritis, cancer, diabetes, heart disease, and stroke are also attributable to obesity. For example, Kathleen M. McTigue et al. estimate in "Screening and Interventions for Obesity in Adults: Summary of the Evidence for the U.S. Preventive Services Task Force" (*Annals of Internal Medicine*, vol.139, no. 11, December 2, 2003) that the direct costs of obesity are 5.7% of the total U.S. health care expenditures; however, the lifetime costs of cardiovascular disease increase by 20% with mild obesity, 50% with moderate obesity, and by almost 200% with clinically severe or extreme obesity. The NCCDHP also reports that hospital costs for treatment of overweight and obese children and teens more than tripled from 1983 to 2003.

According to Anne M. Wolf, JoAnn E. Manson, and Graham A. Colditz, in "The Economic Impact of Overweight, Obesity, and Weight Loss" (Robert H. Eckel, ed., *Obesity: Mechanisms and Clinical Management*, 2003), the estimated annual medical spending attributable to overweight and obesity was about $93 billion in 2002. Wolf, Manson, and Colditz estimate the total cost as $117 billion, with an additional $33 billion spent on weight-loss products and services. Estimates of the medical care costs, direct and indirect as well as total cost of overweight and obesity in the United States, vary depending on how the conditions are defined, whether overweight and obesity are considered together or separately, and which costs and obesity-related conditions are included in the estimates and projections. For example, Wolf, Manson, and Colditz's total cost is based on epidemiological studies that defined obesity and overweight as a body mass index (BMI) equal to or greater than 29.

**TABLE 7.1**

**Economic and health burden of chronic disease, selected years, 1979–2007**

| Disease/risk factors | Morbidity (illness) | Mortality (death) | Direct cost/indirect cost |
|---|---|---|---|
| Arthritis | Arthritis affects 1 in 5, or 46 million, US adults, making it one of the most common chronic conditions. Over 40%, or nearly 19 million, adults with arthritis are limited in their activities because of their arthritis. By 2030, nearly 67 million (25%) of U.S. adults will have doctor-diagnosed arthritis. In addition, adults with arthritis-attributable activity limitation are projected to increase from 16.9 million (7.9%) to 25 million (9.3% of the US adult population) by 2030. | From 1979–1998, the annual number of arthritis and other related rheumatic conditions (AORC) deaths rose from 5,537 to 9,367. In 1998, the crude death rate from AORC was 3.48 per 100,000 population. | The total costs attributable to arthritis and other rheumatic conditions (AORC) in the United States in 2003 was approximately $128 billion ($80.8 billion in medical care expenditures and $47 billion in earnings losses). This equaled 1.2% of the 2003 U.S. gross domestic product. |
| Cancer | About 1.3 million people in the U.S. are diagnosed with cancer each year. | Cancer is the second leading cause of death in the United States. In 2003, an estimated 556,000 people died of cancer. | NIH (National Institutes of Health) estimates that the overall costs for cancer in the year 2006 at 206 billion: of this amount, $78 billion for direct medical costs and more than $128 billion for indirect costs such as lost productivity. |
| Diabetes | More than 20.8 million Americans have diabetes, and about 6.2 million don't know that they have the disease. | Diabetes is the sixth leading cause of death. Over 200,000 people die each year of diabetes-related complications. | The estimated economic cost of diabetes in 2002 was $132 billion. Of this amount, $92 billion was due to direct medical costs and $40 billion to indirect costs such as lost workdays, restricted activity, and disability due to diabetes. |
| Heart disease and stroke | More than 79 million Americans currently live with a cardiovascular disease. | More than 1.4 million Americans die of cardiovascular diseases each year, which amounts to one death every 36 seconds. | The cost of cardiovascular disease and stroke in the United States in 2007 is projected to be $431.8 billion including direct and indirect costs. |
| Overweight/obesity | In 2003–2004 over 66 million adults, or 32% of the adult population, were obese. Over 125 million or 17.1% of children and adolescents 2–19 years of age are overweight. | The latest study from CDC (Centers for Disease Control and Prevention) scientists estimates that about 112,000 deaths are associated with obesity each year in the United States. | Direct health costs attributable to obesity have been estimated at $52 billion in 1995 and $75 billion in 2003. Among children and adolescents, annual hospital costs related to overweight and obesity more than tripled over the past two decades. |
| Tobacco | An estimated 45.1 million adults in the United States smoke cigarettes even though this single behavior will result in death or disability for half of all regular users. | Tobacco use is responsible for approximately 438,000 deaths each year. | The economic burden of tobacco use is enormous: more than $75 billion in medical expenditures and another $92 billion in indirect costs. |

SOURCE: "Quick Facts: Economic and Health Burden of Chronic Disease," in *Chronic Disease Prevention*, Centers for Disease Control and Prevention, National Center for Chronic Disease Prevention and Health Promotion, August 16, 2007, http://www.cdc.gov/nccdphp/press/index.htm (accessed October 29, 2007)

In "State-Level Estimates of Annual Medical Expenditures Attributable to Obesity" (*Obesity Research*, vol. 12, no. 1, January 2004), an analysis of medical spending attributable to obesity, Eric A. Finkelstein, Ian C. Fiebelkorn, and Guijing Wang estimate that in 2003 Medicare and Medicaid spent $75 billion treating obesity-related diseases. In this study the researchers calculated state-level estimates of total, Medicare, and Medicaid obesity-attributable medical expenditures.

According to the press release "Obesity Costs States Billions in Medical Expenses" (January 21, 2004, http://www.cdc.gov/od/oc/media/pressrel/r040121.htm) by the Centers for Disease Control and Prevention (CDC), the 1999–2000 National Health and Nutrition Examination Survey indicates that among Medicare recipients, obesity prevalence ranges from 12% in Hawaii to 30% in Washington, D.C. The percentage of annual medical expenditures in each state attributable to obesity ranges from 4% in Arizona to 6.7% in Alaska. Medicare expenditures connected to obesity range from 3.9% in Arizona to 9.8% in Delaware. For Medicaid recipients, the percentages are considerably higher due to the higher prevalence of obesity among Medicaid recipients—from 7.7% in Rhode Island (where 21% of Medicaid recipients are obese) to 15.7% in Indiana (where 44% of Medicaid recipients are obese).

State-level estimates range from totals of $87 million in Wyoming to $7.7 billion in California. Obesity-attributable Medicare estimates range from $15 million in Wyoming to $1.7 billion in California, and Medicaid estimates range from $23 million in Wyoming to $3.5 billion in New York. (It is important to remember that state-level spending is largely a function of population, so it is reasonable that a less populous state such as Wyoming will spend less state and federal dollars than a population-dense state such as California or New York.)

**Obesity Costs in New Mexico**

Eldo E. Frezza, Mitchell S. Wachtel, and Bradley T. Ewing developed an economic model intended to assess the impact of obesity on a state's economy. They evaluated the cost of obesity in terms of lost business output, employment, and income for the state of New Mexico and reported their findings in "The Impact of Morbid Obesity on the State Economy: An Initial Evaluation" (*Surgery for Obesity Related Diseases*, vol. 2, no. 5, September–October 2006). The investigators find that obesity cost the state more than seventy-three hundred jobs and its economic effect exceeded $1.3 billion—the impact on labor accounted for nearly $200 million and reduced state and local tax revenues totaled more than $48 million—accounting for 2.5% of New Mexico's gross state product.

**Obesity Increases Health Expenditures**

In "Differences in Disease Prevalence As a Source of the U.S.-European Health Care Spending Gap" (*Health Affairs*, vol. 26, no. 6, October 2, 2007), Kenneth E. Thorpe, David H. Howard, and Katya Galactionova examine spending in the United States and Europe for the ten most costly medical conditions. Their analysis reveals that nearly twice as many adults in the United States are obese compared to those in Europe—33% of Americans, compared to 17% of people in ten of the largest European countries—which results in higher numbers of Americans being afflicted with cancer, diabetes, and other chronic conditions. The treatment of obesity-related chronic diseases adds $100 billion to $150 billion to U.S. annual health expenditures.

**MEDICAL CARE AND HEALTH-RELATED COSTS**

Besides estimates of total direct and indirect costs of overweight and obesity, the National Institute of Diabetes and Digestive and Kidney Diseases (NIDDK), the U.S. government's lead agency responsible for biomedical research on nutrition and obesity, specifies in *Statistics Related to Overweight and Obesity* (June 2000, http://www.medhelp.org/NIHlib/GF-367.html) the portion that obesity-related diseases contribute to these costs. In 2000 heart disease related to overweight and obesity generated direct costs of $6.9 billion (17% of the total direct cost of heart disease, independent of stroke), and the total cost of Type 2 diabetes was $63.1 billion (direct cost, $32.4 billion; indirect cost, $30.7 billion). A significant contribution to increasing diabetes-related costs is hospitalization. Table 7.2 shows hospital discharges in 1990, 2000, and 2004 that were attributable to diabetes. Increases were registered among men aged sixty-five to seventy-four and among both men and women aged seventy-five years and over.

In contrast, the total costs of overweight and obesity in 2000 that were related to other types of diseases were: osteoarthritis, $17.2 billion (direct cost, $4.3 billion; indirect cost, $12.9 billion); hypertension (high blood pressure), $3.2 billion (17 percent of the total cost of

hypertension); colon cancer, $2.8 billion (direct cost, $1 billion; indirect cost, $1.8 billion); breast cancer, $2.3 billion (direct cost, $840 million; indirect cost, $1.5 billion); and endometrial cancer, $790 million (direct cost, $286 million; indirect cost, $504 million).

According to the Weight-control Information Network (WIN), in *Statistics Related to Overweight and Obesity* (June 2007, http://win.niddk.nih.gov/publications/PDFs/stat904z.pdf), the cost of lost productivity related to obesity among Americans aged seventeen to sixty-four is about $3.9 billion annually. This dollar figure translates into $239 million in days of restricted activity, $89.5 million in bed-days (days when people remained in bed rather than performing their activities of daily living), $62.7 million in physician office visits, and $39.3 million in lost workdays related to obesity.

In another study, Roland Sturm of the RAND Corporation compares in "The Effects of Obesity, Smoking, and Problem Drinking on Chronic Medical Problems and Health Care Costs" (*Health Affairs*, vol. 21, no. 2, March–April 2002) the effects of obesity, smoking, heavy alcohol consumption, and poverty on chronic health conditions and health expenditures. Sturm finds that obese individuals spent more on both health-care services and medication than daily smokers and heavy drinkers. For example, obese individuals spent about 36% more than the general population on health-care services, compared to a 21% increase for daily smokers and a 14% increase for heavy drinkers. Furthermore, obese people spent 77% more on medications. The only variable with a greater effect on health-care expenditures was aging—and aging trumped obesity only on expenditures for medications. Sturm concludes that obesity generates significantly higher health-care expenditures and affects more individuals than smoking, heavy drinking, or poverty.

Even though it is well documented that obese people incur higher health-care costs at a given point in time, until recently the effects of rising rates of obesity on spending growth had not been quantified. Kenneth E. Thorpe et al. find in "The Impact of Obesity on Rising Medical Spending" (*Health Affairs*, October 20, 2004) that health-care spending was about 36% higher for obese adults under sixty-five. Furthermore, they seek to estimate the share of spending growth attributable to three obesity-related comorbidities (the coexistence of two or more diseases): diabetes, hyperlipidemia, and heart disease including hypertension. Their analysis reveals that increases in the proportion of, and spending on, obese people relative to people of normal weight accounted for 27% of the increase in per capita spending between 1987 and 2001. This increase was attributable to spending for heart disease (41%), diabetes (38%), and hyperlipidemia (22%). Increases in obesity prevalence alone accounted for 12% of the growth in health-care spending. Thorpe

TABLE 7.2

**Rates of discharges and days of care in non-federal short-stay hospitals, by gender, age, and selected first-listed diagnoses, selected years, 1990–2004**

[Data are based on a sample of hospital records]

| Sex, age, and first-listed diagnosis | Discharges | | | Days of care | | |
|---|---|---|---|---|---|---|
| | 1990 | 2000 | 2004 | 1990 | 2000 | 2004 |
| **Both sexes** | Number per 1,000 population | | | | | |
| Total, age adjusted[a, b] | 125.2 | 113.3 | 118.4 | 818.9 | 557.7 | 568.7 |
| Total, crude[b] | 122.3 | 112.8 | 119.2 | 784.0 | 554.6 | 574.1 |
| **Male** | | | | | | |
| All ages[a, b] | 113.0 | 99.1 | 102.6 | 805.8 | 535.9 | 541.1 |
| Under 18 years[b] | 46.3 | 40.9 | 43.6 | 233.6 | 195.6 | 201.5 |
| Pneumonia | 3.7 | 2.6 | 2.7 | 16.7 | 8.5 | 8.9 |
| Asthma | 3.3 | 3.5 | 3.4 | 9.3 | 7.4 | 7.3 |
| Injuries and poisoning | 6.8 | 5.0 | 5.2 | 30.1 | 21.4 | 18.5 |
|   Fracture, all sites | 2.2 | 1.8 | 1.7 | 9.3 | 7.2 | 4.4 |
| **18–44 years[b]** | 57.9 | 45.0 | 46.5 | 351.7 | 217.5 | 225.6 |
| HIV infection | *0.3 | 0.6 | 0.4 | *3.0 | *5.4 | 3.9 |
| Alcohol and drug[c] | 3.7 | 4.0 | 3.2 | 33.1 | 19.1 | 14.3 |
| Serious mental illness[d] | 3.4 | *5.3 | 5.7 | 47.1 | *43.6 | 43.4 |
| Diseases of heart | 3.0 | 2.7 | 2.9 | 16.3 | 9.4 | 10.1 |
| Intervertebral disc disorders | 2.6 | 1.5 | 1.2 | 10.7 | 3.2 | 2.4 |
| Injuries and poisoning | 13.1 | 7.3 | 8.0 | 65.7 | 33.2 | 40.8 |
|   Fracture, all sites | 4.0 | 2.5 | 2.8 | 22.7 | 12.8 | 15.5 |
| **45–64 years[b]** | 140.3 | 112.7 | 118.4 | 943.4 | 570.4 | 612.4 |
| HIV infection | *0.1 | *0.5 | 0.4 | * | * | *4.1 |
| Malignant neoplasms | 10.6 | 6.2 | 6.0 | 99.1 | 42.1 | 38.6 |
|   Trachea, bronchus, lung | 2.7 | 0.9 | 0.8 | 19.1 | 5.2 | 5.4 |
| Diabetes | 2.9 | 3.7 | 3.1 | 21.2 | 22.5 | 16.6 |
| Alcohol and drug[c] | 3.5 | 3.5 | 4.4 | 29.7 | 15.8 | 19.6 |
| Serious mental illness[d] | 2.5 | *4.0 | 4.7 | 34.8 | *34.6 | 42.2 |
| Diseases of heart | 31.7 | 26.4 | 23.9 | 185.0 | 101.5 | 95.7 |
|   Ischemic heart disease | 22.6 | 17.7 | 14.7 | 128.2 | 63.8 | 54.3 |
|     Acute myocardial infarction | 7.4 | 5.9 | 4.6 | 55.8 | 27.8 | 23.7 |
|   Heart failure | 3.1 | 3.4 | 3.9 | 21.0 | 17.3 | 20.7 |
| Cerebrovascular diseases | 4.1 | 3.8 | 3.4 | 40.7 | 19.8 | 17.6 |
| Pneumonia | 3.4 | 3.4 | 3.7 | 27.1 | 20.3 | 19.6 |
| Injuries and poisoning | 11.6 | 8.8 | 10.9 | 82.6 | 49.8 | 68.9 |
|   Fracture, all sites | 3.3 | 2.5 | 3.0 | 24.2 | 16.2 | 19.1 |
| **65–74 years[b]** | 287.8 | 264.9 | 268.5 | 2,251.5 | 1,489.7 | 1,442.3 |
| Malignant neoplasms | 27.9 | 17.6 | 19.1 | 277.6 | 121.2 | 128.7 |
|   Large intestine and rectum | 3.0 | 3.0 | 3.1 | 34.2 | 27.3 | 29.1 |
|   Trachea, bronchus, lung | 6.4 | 2.8 | 3.0 | 55.7 | 19.2 | 19.4 |
|   Prostate | 5.1 | 3.7 | 2.9 | 33.1 | 14.0 | 9.9 |
| Diabetes | 4.4 | 4.7 | 5.5 | 39.8 | 29.0 | 29.9 |
| Serious mental illness[d] | 2.5 | *3.4 | 2.8 | 43.8 | 39.9 | 26.4 |
| Diseases of heart | 69.4 | 70.6 | 65.0 | 487.2 | 331.9 | 280.9 |
|   Ischemic heart disease | 42.0 | 39.7 | 35.0 | 285.2 | 171.2 | 141.8 |
|     Acute myocardial infarction | 14.0 | 12.5 | 10.8 | 122.4 | 66.5 | 56.7 |
|   Heart failure | 11.8 | 13.6 | 13.9 | 93.1 | 77.6 | 70.9 |
| Cerebrovascular diseases | 13.8 | 13.2 | 12.5 | 114.8 | 59.0 | 56.9 |
| Pneumonia | 11.2 | 12.7 | 12.2 | 106.9 | 81.5 | 75.0 |
| Hyperplasia of prostate | 14.4 | 5.4 | 4.0 | 65.0 | 15.0 | 11.3 |
| Osteoarthritis | 5.5 | 10.3 | 11.3 | 48.5 | 48.8 | 42.8 |
| Injuries and poisoning | 17.6 | 17.9 | 17.4 | 139.0 | 105.7 | 109.7 |
|   Fracture, all sites | 4.5 | 4.7 | 4.2 | 45.9 | 29.9 | 34.1 |
|     Fracture of neck of femur (hip) | 1.5 | *2.0 | 1.7 | *18.1 | *15.9 | 10.9 |

et al. conclude that future cost-containment efforts should address the increasing prevalence of obesity and the institution of effective approaches to weight loss for people who are obese.

### Hospital Costs of Childhood and Adolescent Obesity

Guijing Wang and William H. Dietz of the CDC examine trends in obesity-linked diseases among children and adolescents and their related economic costs. In

"Economic Burden of Obesity in Youths Aged 6 to 17 Years: 1979–1999" (*Pediatrics*, vol. 109, no. 5, May 2002), the researchers report the results of an analysis and comparison of data from the 1979–81 and 1997–99 National Hospital Discharge Surveys conducted by the National Center for Health Statistics. When Wang and Dietz adjust hospital costs to reflect 2001 dollars, they find that hospital costs linked to childhood obesity and three specific obesity-related illness—diabetes, sleep

TABLE 7.2

**Rates of discharges and days of care in non-federal short-stay hospitals, by gender, age, and selected first-listed diagnoses, selected years, 1990–2004** [CONTINUED]

[Data are based on a sample of hospital records]

| Sex, age, and first-listed diagnosis | Discharges | | | Days of care | | |
|---|---|---|---|---|---|---|
| | 1990 | 2000 | 2004 | 1990 | 2000 | 2004 |
| **Male—Con.** | | | Number per 1,000 population | | | |
| **75 years and over**[b] | 478.5 | 467.4 | 483.1 | 4,231.6 | 2,888.0 | 2,815.5 |
| Malignant neoplasms | 41.0 | 21.9 | 21.3 | 408.3 | 165.2 | 149.0 |
|   Large intestine and rectum | 5.4 | 4.2 | 3.9 | 80.7 | 44.1 | 34.0 |
|   Trachea, bronchus, lung | 5.4 | 3.0 | 3.9 | 53.4 | 18.3 | 27.0 |
|   Prostate | 9.7 | 3.2 | 2.4 | 65.6 | *19.4 | 8.7 |
| Diabetes | 4.6 | 6.5 | 6.9 | 51.2 | 43.2 | 37.2 |
| Serious mental illness[d] | *2.6 | 2.9 | 2.4 | *40.5 | *32.6 | 24.5 |
| Diseases of heart | 106.2 | 113.3 | 113.1 | 855.7 | 600.9 | 579.0 |
|   Ischemic heart disease | 49.1 | 53.0 | 45.1 | 398.1 | 276.1 | 237.9 |
|     Acute myocardial infarction | 23.1 | 23.0 | 21.6 | 227.5 | 136.5 | 152.0 |
|   Heart failure | 31.8 | 30.9 | 36.6 | 248.6 | 178.6 | 193.7 |
| Cerebrovascular diseases | 30.2 | 30.2 | 24.9 | 298.3 | 171.2 | 129.9 |
| Pneumonia | 38.1 | 36.7 | 38.8 | 391.3 | 228.6 | 232.5 |
| Hyperplasia of prostate | 17.9 | 6.8 | 6.0 | 109.2 | 21.6 | 17.6 |
| Osteoarthritis | 6.6 | 7.2 | 11.8 | 60.7 | 28.7 | 47.1 |
| Injuries and poisoning | 31.2 | 33.6 | 33.4 | 341.3 | 257.7 | 207.8 |
|   Fracture, all sites | 13.7 | 14.4 | 14.3 | 145.1 | *119.2 | 91.2 |
|     Fracture of neck of femur (hip) | 8.5 | 8.4 | 8.8 | 97.8 | 63.3 | 60.9 |
| **Female** | | | | | | |
| **All ages**[a, b] | 139.0 | 127.7 | 134.9 | 840.5 | 581.0 | 599.6 |
| **Under 18 years**[b] | 46.4 | 39.6 | 42.4 | 218.7 | 161.5 | 184.4 |
| Pneumonia | 2.9 | 2.4 | 2.4 | 13.7 | 9.5 | 7.9 |
| Asthma | 2.2 | 2.4 | 2.0 | 6.8 | 5.5 | 4.6 |
| Injuries and poisoning | 4.3 | 3.1 | 3.4 | 16.7 | *12.0 | 12.6 |
|   Fracture, all sites | 1.3 | 0.9 | 0.8 | 6.4 | 2.3 | 2.7 |
| **18–44 years**[b] | 146.8 | 124.8 | 136.2 | 582.0 | 401.1 | 445.5 |
| HIV infection | * | 0.3 | 0.2 | * | *2.1 | 1.6 |
| Delivery | 69.9 | 64.5 | 71.4 | 195.0 | 160.2 | 186.2 |
| Alcohol and drug[c] | 1.6 | *2.1 | 2.0 | 14.1 | *10.8 | *9.2 |
| Serious mental illness[d] | 3.7 | *5.4 | 5.9 | 54.3 | *41.1 | 43.1 |
| Diseases of heart | 1.3 | 1.7 | 1.7 | 7.2 | 6.3 | 8.3 |
| Intervertebral disc disorders | 1.5 | 1.0 | 1.1 | 7.3 | 2.4 | 2.8 |
| Injuries and poisoning | 6.7 | 4.3 | 5.3 | 36.6 | 18.1 | 22.4 |
|   Fracture, all sites | 1.6 | 1.0 | 1.1 | 10.7 | 4.5 | 4.8 |
| **45–64 years**[b] | 131.0 | 110.2 | 117.3 | 886.5 | 533.6 | 571.8 |
| HIV infection | * | * | *0.3 | * | * | * |
| Malignant neoplasms | 12.7 | 6.1 | 5.8 | 107.4 | 34.7 | 36.9 |
|   Trachea, bronchus, lung | 1.7 | 0.5 | 0.7 | 14.8 | 3.4 | 5.0 |
|   Breast | 2.8 | 1.3 | 0.8 | 12.1 | 2.6 | 2.8 |
| Diabetes | 2.9 | 2.9 | 2.8 | 25.8 | 15.0 | 12.4 |
| Alcohol and drug[c] | 1.0 | 1.5 | 1.6 | 8.0 | *7.1 | 7.8 |
| Serious mental illness[d] | 4.0 | 4.6 | 5.5 | 60.5 | 42.7 | 51.8 |
| Diseases of heart | 16.6 | 14.6 | 13.3 | 101.1 | 59.5 | 57.1 |
|   Ischemic heart disease | 9.9 | 7.8 | 6.6 | 57.4 | 29.5 | 25.6 |
|     Acute myocardial infarction | 2.8 | 2.0 | 2.0 | 21.6 | 10.0 | 11.6 |
|   Heart failure | 2.2 | 2.9 | 2.7 | 16.3 | 13.6 | 15.0 |
| Cerebrovascular diseases | 3.0 | 3.5 | 2.8 | 32.1 | 19.5 | 16.2 |
| Pneumonia | 3.3 | 3.6 | 3.3 | 26.1 | 20.7 | 18.8 |
| Injuries and poisoning | 9.4 | 7.7 | 9.2 | 63.3 | 41.2 | 50.8 |
|   Fracture, all sites | 3.1 | 2.7 | 2.4 | 25.0 | 13.3 | 13.8 |

apnea, and gallbladder disease—had more than tripled since 1981, from $35 million to $127 million per year.

Days spent in the hospital for obesity-related disease more than doubled and the average length of hospital stays increased by about a third, from 5.3 to 7 days. Wang and Dietz observe that this increase in average length of stay occurred during a time when U.S. hospital stays overall were shortening and assert that longer lengths of stay for children with obesity-related medical problems underscored the severity of these problems.

Wang and Dietz conclude that the increase in the percentage of discharges with obesity-related diseases was most likely a reflection of the medical consequences of the obesity epidemic. They state, "Although the numbers of percentage are small, the increases are substantial, especially for obesity (197% increase), sleep apnea (436%), and gallbladder disease (228%). These data may suggest that the increasing prevalence of obesity in children and adolescents has led to increased hospital stays related to obesity-associated diseases. The increasing proportion of

**TABLE 7.2**

**Rates of discharges and days of care in non-federal short-stay hospitals, by gender, age, and selected first-listed diagnoses, selected years, 1990–2004** [CONTINUED]

[Data are based on a sample of hospital records]

| Sex, age, and first-listed diagnosis | Discharges | | | Days of care | | |
|---|---|---|---|---|---|---|
| | 1990 | 2000 | 2004 | 1990 | 2000 | 2004 |
| Female—Con. | Number per 1,000 population | | | | | |
| 65–74 years[b] | 241.1 | 246.1 | 251.4 | 1,959.3 | 1,397.1 | 1,374.0 |
| Malignant neoplasms | 20.9 | 14.1 | 13.8 | 189.8 | 101.0 | 101.3 |
| Large intestine and rectum | 2.4 | 1.7 | 1.9 | 34.9 | 15.2 | 17.3 |
| Trachea, bronchus, lung | 2.6 | 2.4 | 2.2 | 26.9 | *17.5 | 16.5 |
| Breast | 3.9 | 2.8 | 1.5 | 17.6 | * | 2.9 |
| Diabetes | 5.8 | 4.6 | 5.0 | 46.8 | 26.1 | 27.9 |
| Serious mental illness[d] | 3.9 | 4.0 | 3.9 | 62.8 | 46.3 | 47.3 |
| Diseases of heart | 45.1 | 52.1 | 42.6 | 316.9 | 256.0 | 199.7 |
| Ischemic heart disease | 24.4 | 23.3 | 18.4 | 153.8 | 113.9 | 77.2 |
| Acute myocardial infarction | 7.5 | 8.0 | 5.8 | 58.1 | 52.8 | 32.2 |
| Heart failure | 9.5 | 12.8 | 10.4 | 84.0 | 69.1 | 59.7 |
| Cerebrovascular diseases | 11.3 | 12.3 | 9.1 | 96.0 | 59.4 | 44.4 |
| Pneumonia | 8.5 | 11.3 | 12.5 | 79.6 | 71.4 | 72.3 |
| Osteoarthritis | 7.8 | 10.0 | 15.1 | 74.5 | 47.2 | 60.9 |
| Injuries and poisoning | 17.8 | 18.3 | 19.6 | 166.2 | 109.9 | 114.6 |
| Fracture, all sites | 8.4 | 7.7 | 8.8 | 97.3 | 43.8 | 47.4 |
| Fracture of neck of femur (hip) | 3.6 | 3.2 | 3.4 | *59.6 | 21.1 | 23.6 |
| 75 years and over[b] | 409.6 | 458.8 | 462.4 | 3,887.1 | 2,830.8 | 2,653.9 |
| Malignant neoplasms | 22.1 | 17.6 | 15.6 | 257.3 | 125.7 | 103.6 |
| Large intestine and rectum | 4.6 | 3.4 | 2.8 | 69.8 | 28.4 | 28.8 |
| Trachea, bronchus, lung | 2.1 | 1.9 | 2.2 | 20.6 | 14.0 | 14.3 |
| Breast | 3.9 | 2.5 | 2.4 | 22.0 | *8.9 | 5.9 |
| Diabetes | 4.6 | 6.3 | 6.1 | 55.3 | 34.0 | 31.1 |
| Serious mental illness[d] | 4.2 | 4.7 | 3.5 | 78.4 | 49.2 | 37.4 |
| Diseases of heart | 84.6 | 99.1 | 94.6 | 672.8 | 523.4 | 485.2 |
| Ischemic heart disease | 33.7 | 35.5 | 31.0 | 253.2 | 185.5 | 155.6 |
| Acute myocardial infarction | 13.1 | 16.5 | 14.8 | 125.9 | 110.7 | 91.0 |
| Heart failure | 28.6 | 32.5 | 31.9 | 240.8 | 183.4 | 171.9 |
| Cerebrovascular diseases | 29.6 | 27.6 | 24.5 | 302.0 | 156.8 | 129.2 |
| Pneumonia | 23.5 | 30.1 | 28.0 | 255.8 | 206.5 | 179.1 |
| Osteoarthritis | 6.2 | 9.9 | 13.2 | 62.2 | 46.3 | 54.7 |
| Injuries and poisoning | 46.3 | 44.7 | 45.1 | 489.2 | 275.4 | 284.0 |
| Fracture, all sites | 31.5 | 30.0 | 28.3 | 352.7 | 190.0 | 168.8 |
| Fracture of neck of femur (hip) | 18.8 | 17.9 | 16.3 | 236.3 | 125.3 | 101.4 |

*Estimates are considered unreliable.

[a]Estimates are age adjusted to the year 2000 standard population using six age groups: under 18 years, 18–44 years, 45–54 years, 55–64 years, 65–74 years, and 75 years and over.

[b]Includes discharges with first-listed diagnoses not shown in table.

[c]Includes abuse, dependence, and withdrawal. These estimates are for non-federal short-stay hospitals only and do not include alcohol and drug discharges from other types of facilities or programs such as the Department of Veterans Affairs or day treatment programs.

[d]These estimates are for non-federal short-stay hospitals only and do not include serious mental illness discharges from other types of facilities or programs such as the Department of Veterans Affairs or long-term hospitals.

Notes: Excludes newborn infants. Diagnostic categories are based on the International Classification of Diseases, Ninth Revision, Clinical Modification (ICD–9-CM).

Rates are based on the civilian population as of July 1. Starting with *Health, United States, 2003*, rates for 2000 and beyond are based on the 2000 census. Rates for 1990–1999 use population estimates based on the 1990 census adjusted for net under enumeration using the 1990 National Population Adjustment Matrix from the U.S. Census Bureau. Rates for 1990–1999 are not strictly comparable with rates for 2000 and beyond because population estimates for 1990–1999 have not been revised to reflect the 2000 census.

SOURCE: "Table 97. Discharges and Days of Care in Non-Federal Short-Stay Hospitals, by Sex, Age, and Selected First-Listed Diagnoses: United States, Selected Years 1990–2004," in *Health, United States, 2006, with Chartbook on Trends in the Health of Americans*, Centers for Disease Control and Prevention, National Center for Health Statistics, 2006, http://www.cdc.gov/nchs/data/hus/hus06.pdf#summary (accessed October 9, 2007)

hospital discharges with obesity-associated diseases in the last 20 years may also reflect the impact of increasing severity of obesity."

In "Incremental Hospital Charges Associated with Obesity As a Secondary Diagnosis in Children" (*Obesity*, vol. 15, no 7, 2007), Susan J. Woolford et al. of the University of Michigan find that even when obesity was the secondary diagnosis resulting in a hospital stay, obese children's lengths of stay were longer and their hospital costs were significantly higher than those of healthy-weight children. For example, hospital charges were significantly higher for discharges with obesity as a secondary diagnosis versus those without: appendicitis ($14,134 versus $11,049), asthma ($7,766 versus $6,043), and pneumonia ($12,228 versus $9,688).

### Insurance Coverage for Obesity Treatment

Even though the Medicare and Medicaid programs spend billions on obesity-related illnesses, neither entitlement program covers treatment for obesity itself. Medicaid does not cover obesity treatment, and under Medicare, hospital and physician services for obesity are generally excluded. Historically, Medicare has covered treatment when obesity results from a disease such as hypothyroidism (deficiency of the thyroid hormone, which is produced by the thyroid gland) or Cushing's disease (a condition in which excess cortisol, a hormone released in response to stress, is secreted by the pituitary gland) and when weight loss is medically necessary to treat a disease such as diabetes, hypertension, or heart disease. It also provides coverage for surgical treatment of obesity when it is medically appropriate and the surgery is to correct an illness that caused the obesity or was aggravated by the obesity.

Until 2004 Medicare justified excluding coverage for obesity treatment by asserting that obesity is not a disease; however, in *CMS Manual System: Pub. 100-03 Medicare National Coverage Determinations* (October 1, 2004, http://www.cms.hhs.gov/transmittals/downloads/R23NCD.pdf) the Centers for Medicare and Medicaid Services (CMS), which administers Medicare, eliminated language from its policy (that "obesity itself cannot be considered an illness") that had been used to deny coverage for weight-loss treatment. The decision stopped short of designating obesity a disease and does not specifically grant coverage for weight-loss treatment; regardless, it enables individuals, physicians, and companies to apply to Medicare for reimbursement for a variety of weight-loss therapies. Because private insurance companies often use Medicare as a model for their coverage and benefits, some health-care industry observers believe the Medicare decision will pressure other payers to cover weight-loss treatments.

The Medicare Prescription Drug, Improvement, and Modernization Act of 2003 excludes drugs used for weight loss; however, the CMS states in "Medicare Program; Policy and Technical Changes to the Medicare Prescription Drug Benefit" (*Federal Register*, vol. 72, no. 101, May 25, 2007) that weight-loss drugs may be covered by Medicare when they are prescribed for a "medically accepted indication" such as clinically severe obesity. However, according to the AOA, in the fact sheet "Obesity, Medicaid, and Medicare" (May 2, 2005, http://obesity1.tempdomainname.com/subs/fastfacts/Obesity_Medicare.shtml), the plans administering the outpatient prescription drug benefit that became effective in 2006 will not include weight-loss drugs.

In view of the high prevalence of obesity among the populations covered by Medicaid—the poor and minorities—and the significant Medicaid expenditures for obesity-related illnesses, many health-care industry observers believe it is shortsighted that many states specifically exclude coverage of antiobesity products in their Medicaid programs. For example, Morgan Downey indicates in "Insurance Coverage for Obesity Treatments" (May 2, 2005, http://obesity1.tempdomainname.com/treatment/insurance2.shtml) that ten states—Illinois, Indiana, Nevada, New Hampshire, New York, Ohio, Oklahoma, South Carolina, South Dakota, and Wyoming—do not cover antiobesity pharmaceuticals through Medicaid. California, Delaware, Hawaii, Kentucky, Maine, Massachusetts, Mississippi, Montana, New Mexico, Oregon, Rhode Island, Vermont, and Virginia cover orlistat, sibutramine, and phentermine in their Medicaid programs; however, in some states coverage is limited to people with clinically severe obesity, Type 2 diabetes, or hyperlipidemia, an excess of fats called lipids, chiefly cholesterol and triglycerides, in the blood. Some healthcare analysts and advocacy groups including the AOA contend that it is difficult to reconcile this limited coverage of obesity in light of Medicaid coverage for inpatient and outpatient alcohol detoxification and rehabilitation; chemical dependency treatment and drug rehabilitation; and services for sexual impotence.

According to the AOA, many health insurance plans do not provide reimbursement for weight-loss treatment. Furthermore, few private insurance indemnity plans or managed-care organizations (e.g., health maintenance organizations and preferred-provider organizations) appear to cover the costs of obesity treatment independent of whether the service is a medically supervised weight-loss program, surgery, or a prescription drug. The AOA notes that most employer-funded health insurance plans do not pay for obesity treatment or services, including medications, diet supplements, weight-control programs, or bariatric surgeries.

The Pharmacy Benefit Management Institute, Inc. (PBMI), an independent organization that is not affiliated with any employee benefits program or pharmaceutical manufacturer, periodically surveys employers to determine the extent, cost, and coverage of their pharmacy benefits. The PBMI publishes survey data and trends in *Prescription Drug Benefit Cost and Plan Design Report* (2007, http://www.pbmi.com/2007onlinereport/pdfs/2007_Cost_and_Plan_Design_Report.pdf). The 2007 survey queried 340 companies that provide coverage to 6.2 million beneficiaries. The PBMI study finds that 82.7% of employers exclude weight-loss products from their coverage.

The reluctance to cover antiobesity drugs is driven by concern about cost, in that many payers may determine that the rising prevalence of obesity and its comorbidities require higher prescription drug costs than drug treatment of obesity itself. For example, the article "Study: Metabolic Syndrome Brings Big Costs" (Associated Press, May 6, 2005) reports that Medco Health Solutions, a national prescription benefit management company, found that Americans with metabolic syndrome account for $4 out of every $10 spent on prescription drugs for adults. (Metabolic syndrome is the name given to conditions that often occur together—obesity, diabetes, high blood pressure, and high triglycerides that can lead to cardiovascular disease.) Drug treatment of metabolic syndrome skyrocketed 36% between 2002 and 2004, and prescription costs for adults with metabolic syndrome averaged $4,116 in 2004, which was 4.2 times the average.

**OVERWEIGHT WORKERS MAY PAY MORE FOR HEALTH INSURANCE COVERAGE.** New federal regulations, which were reported in "Nondiscrimination and Wellness Programs in Health Coverage in the Group Market; Final Rules" (*Federal Register*, vol. 71, no. 239, December 13, 2006) and which took effect on July 1, 2007, for some groups and on January 1, 2008, for others, permit companies to charge overweight employees more for their health insurance than their healthy-weight peers.

According to Eve Tahmincioglu, in "No Fatties Here" (MSNBC.com, September 10, 2007), Stephen Glick, the administrator of the Chamber Insurance Trust, asserts that most small business owners are choosing to incentivize rather than penalize obese workers by offering them gifts and other rewards for successful weight-loss efforts rather than forcing them to pay higher premiums.

### Obese People Pay More for Health Care

David Arterburn, Matthew L. Maciejewski, and Joel Tsevat of the University of Cincinnati find in "Impact of Morbid Obesity on Medical Expenditures in Adults" (*International Journal of Obesity*, vol. 29, no. 3, 2005) that adults with clinically severe obesity (also known as morbid obesity, defined as one hundred pounds or more over ideal body weight or a BMI greater than 40) had health-care costs that were nearly twice those of their normal-weight peers. The researchers analyzed the records of 16,262 adults from the 2000 Medical Expenditure Panel Survey. Per capita health-care expenditures were calculated for BMI categories, based on self-reported height and weight, and adjusted for age, gender, race, income, education level, type of health insurance, marital status, and smoking status.

### FUNDING OBESITY RESEARCH

During the last four decades, considerable progress has been made in identifying the causes of obesity and developing treatments. Despite the enhanced understanding of the origins of obesity, increasing numbers of Americans continue to become overweight and obese. The AOA, along with myriad medical professional organizations and advocacy groups, contends that public funding for obesity research is woefully inadequate in view of the size and scope of this public health problem. Besides insufficient National Institutes of Health (NIH)

funding for obesity research, the AOA cites inequities in research grants awarded by the NIH—even though more grants have been awarded to obesity research than in past years, obesity still receives a disproportionately small share of grant funding.

Table 7.3 shows NIH funding for a variety of diseases and research areas for fiscal years 2003 through 2007 as well as an estimates for 2008. Funding for obesity research peaked in fiscal year 2006 and was anticipated to decline slightly in fiscal year 2008.

### WEIGHING THE PRICE BUSINESS PAYS

Obese employees incur substantially higher health-care costs than normal-weight employees. Obesity significantly increases health expenditures and absenteeism. In "The Costs of Obesity among Full-Time Employees" (*American Journal of Health Promotion*, vol. 20, no. 1, September–October, 2005), Eric A. Finkelstein, Ian C. Fiebelkorn, and Guijing Wang find that about 30% of the total costs result from increased absenteeism, and even though workers with clinically severe obesity represent just 3% of the employed population, they account for 21% of the costs due to obesity. The investigators report that overweight and obesity-related costs ranged from $175 per year for overweight male employees to $2,485 per year for obese female employees. The costs of obesity alone (excluding overweight) for a company with one thousand employees were estimated as a staggering $285,000 per year.

According to the U.S. Department of Health and Human Services, in *Prevention Makes Common "Cents"* (September 2003, http://aspe.hhs.gov/health/prevention/), U.S. companies pay $13 billion per year for medical-care costs to treat obesity-related diseases, lower productivity, and absenteeism. Health insurance costs ($8 billion) make the greatest contribution to the total, followed by paid sick leave ($2.4 billion), life insurance ($1.8 billion), and disability insurance ($1 billion). The National Business Group on Health, a consortium of large employers that researches and develops solutions to health-service delivery challenges, states in the fact sheet "Healthy Weight, Healthy Lifestyles: Primary Fact Sheet for the Institute on the Costs and Health Effects of Obesity" (February 1, 2006, http://www.businessgrouphealth.org/pdfs/obesity_factsheet.pdf) that higher health-care utilization rates, such as 45% more inpatient hospital days, produce higher health-care expenditures—36% higher for inpatient and outpatient care and 77% higher prescription drug spending. About 8% of private employer medical claims are attributable to overweight and obesity, and in 2004 obesity-related disabilities cost employers an average of $8,720 per claimant per year for wage indemnity.

TABLE 7.3

**Estimates of funding for various diseases, conditions, research areas, fiscal years 2003–08**

[Dollars in millions and rounded]

| Research/disease areas | FY 2003 actual | FY 2004 actual | FY 2005 actual | FY 2006 actual | FY 2007 estimate | FY 2008 estimate |
|---|---|---|---|---|---|---|
| Acute respiratory distress syndrome | $77 | $72 | $72 | $74 | $74 | $73 |
| Agent orange & dioxin | 18 | 20 | 20 | 17 | 17 | 17 |
| Aging | 2,211 | 2,343 | 2,415 | 2,431 | 2,423 | 2,414 |
| Alcoholism | 493 | 503 | 512 | 511 | 509 | 508 |
| Allergic rhinitis (hay fever) | 2 | 2 | 3 | 4 | 4 | 4 |
| ALS | 40 | 47 | 42 | 44 | 44 | 43 |
| Alzheimer's disease | 658 | 633 | 656 | 643 | 643 | 642 |
| American Indians/Alaska Natives | 108 | 134 | 140 | 155 | 153 | 152 |
| Anorexia | 10 | 12 | 14 | 15 | 15 | 14 |
| Anthrax | 219 | 249 | 183 | 150 | 117 | 111 |
| Antimicrobial resistance | 181 | 203 | 217 | 221 | 221 | 220 |
| Aphasia | N/A | 5 | 3 | 15 | 15 | 15 |
| Arctic | 33 | 25 | 22 | 17 | 17 | 17 |
| Arthritis | 380 | 374 | 368 | 355 | 354 | 353 |
| Assistive technology | 126 | 131 | 138 | 182 | 181 | 181 |
| Asthma | 248 | 272 | 289 | 283 | 284 | 285 |
| Ataxiate langiectasia | 10 | 9 | 10 | 9 | 9 | 9 |
| Atherosclerosis | 318 | 326 | 322 | 337 | 338 | 338 |
| Attention deficit disorder (ADD) | 103 | 104 | 107 | 116 | 115 | 115 |
| Autism | 93 | 100 | 102 | 108 | 108 | 108 |
| Autoimmune disease | 591 | 584 | 589 | 598 | 597 | 593 |
| Basic behavioral and social science | 938 | 1,052 | 1,065 | 1,062 | 1,056 | 1,054 |
| Batten disease | 8 | 8 | 9 | 8 | 8 | 7 |
| Behavioral and social science | 2,684 | 2,932 | 3,044 | 3,001 | 2,993 | 2,981 |
| Biodefense | 1,554 | 1,629 | 1,696 | 1,766 | 1,731 | 1,723 |
| Bioengineering | 1,006 | 1,216 | 1,318 | 1,546 | 1,553 | 1,551 |
| Biotechnology | 9,893 | 10,685 | 10,889 | 9,974 | 9,946 | 9,920 |
| Brain cancer | 164 | 187 | 157 | 178 | 178 | 177 |
| Brain disorders | 4,740 | 4,821 | 4,784 | 4,732 | 4,711 | 4,704 |
| Breast cancer | 693 | 708 | 700 | 718 | 717 | 716 |
| Burden of illness | 424 | 429 | 433 | 508 | 507 | 506 |
| Cancer | 5,432 | 5,547 | 5,639 | 5,575 | 5,556 | 5,534 |
| Cardiovascular | 2,286 | 2,360 | 2,333 | 2,349 | 2,343 | 2,341 |
| Cerebral palsy | 18 | 22 | 23 | 18 | 21 | 20 |
| Cervical cancer | 92 | 94 | 96 | 97 | 96 | 96 |
| Charcot-Marie tooth disease | N/A | N/A | N/A | 7 | 6 | 6 |
| Child abuse and neglect research | N/A | N/A | 40 | 38 | 37 | 36 |
| Childhood leukemia | 70 | 62 | 60 | 53 | 53 | 53 |
| Chronic fatigue syndrome | 6 | 5 | 5 | 5 | 5 | 4 |
| Chronic liver disease and cirrhosis | 348 | 362 | 410 | 408 | 407 | 406 |
| Chronic obstructive pulmonary disease | 54 | 55 | 63 | 67 | 67 | 66 |
| Climate change | N/A | 63 | 57 | 50 | 50 | 50 |
| Clinical research | 8,028 | 8,495 | 8,719 | 8,785 | 8,807 | 8,786 |
| Clinical trials | 2,723 | 2,877 | 2,863 | 2,767 | 2,764 | 2,756 |
| Colo-rectal cancer | 295 | 297 | 284 | 269 | 270 | 269 |
| Complementary and alternative medicine | 296 | 309 | 306 | 301 | 298 | 298 |
| Conditions affecting unborn children | 111 | 113 | 108 | 103 | 103 | 102 |
| Contraception/reproduction | 330 | 355 | 340 | 335 | 334 | 329 |
| Cooley's anemia | 55 | 47 | 42 | 42 | 42 | 42 |
| Cost effectiveness research | N/A | 126 | 134 | 143 | 143 | 143 |
| Crohn's disease | 50 | 53 | 59 | 64 | 64 | 64 |
| Cystic fibrosis | 117 | 128 | 89 | 85 | 85 | 85 |
| Dental/oral and craniofacial disease | 401 | 410 | 415 | 413 | 410 | 409 |
| Depression | 288 | 302 | 329 | 335 | 334 | 334 |
| Diabetes | 910 | 996 | 1,055 | 1,038 | 1,035 | 1,031 |
| Diagnostic radiology | 717 | 750 | 788 | 712 | 711 | 710 |
| Diethylstilbestrol (DES) | 8 | 8 | 9 | 8 | 8 | 8 |
| Digestive diseases | 1,137 | 1,237 | 1,237 | 1,252 | 1,250 | 1,245 |
| Digestive diseases—(gallbladder) | 7 | 7 | 7 | 7 | 7 | 7 |
| Digestive diseases—(peptic ulcer) | 17 | 18 | 18 | 17 | 16 | 16 |
| Down syndrome | 23 | 19 | 15 | 14 | 14 | 13 |

In "Obesity and Workers' Compensation" (*Archives of Internal Medicine*, vol. 167, no. 8, April 23, 2007), Truls Østbye, John M. Dement, and Katrina M. Krause of the Duke University Medical Center sought to determine the relationship between BMI and the number and types of workers' compensation claims, associated costs, and lost workdays. They find that obese employees filed more workers' compensation claims. Employees with BMIs greater than or equal to 40 had twice the rate of claims as workers at healthy weights. The number of lost workdays was nearly thirteen times higher among obese workers, medical claims costs were seven times higher, and indemnity claims costs

TABLE 7.3

**Estimates of funding for various diseases, conditions, research areas, fiscal years 2003–08** [CONTINUED]

[Dollars in millions and rounded]

| Research/disease areas | FY 2003 actual | FY 2004 actual | FY 2005 actual | FY 2006 actual | FY 2007 estimate | FY 2008 estimate |
|---|---|---|---|---|---|---|
| Spinal cord injury | 89 | 89 | 89 | 66 | 65 | 64 |
| Spinal muscular atrophy | 13 | 14 | 15 | 15 | 15 | 15 |
| Stem cell research | 517 | 553 | 609 | 643 | 641 | 639 |
| Stem cell research—human embryonic | 20 | 24 | 40 | 38 | 37 | 37 |
| Stem cell research—non-human embryonic | 113 | 89 | 97 | 110 | 110 | 109 |
| Stem cell research—human non-embryonic | 191 | 203 | 199 | 206 | 206 | 205 |
| Stem cell research—non-human non-embryonic | 192 | 236 | 273 | 289 | 288 | 287 |
| Stem cell research involving umbilical cord blood/placenta | 17 | 19 | 18 | 19 | 19 | 19 |
| Stem cell research involving umbilical cord blood/placenta—human | 16 | 16 | 15 | 16 | 16 | 16 |
| Stem cell research involving umbilical cord blood/placenta—non-human | 2 | 3 | 3 | 4 | 4 | 4 |
| Stroke | 330 | 313 | 342 | 342 | 339 | 336 |
| Substance abuse | 1,462 | 1,496 | 1,508 | 1,490 | 1,485 | 1,484 |
| Sudden infant death syndrome | 69 | 81 | 84 | 77 | 76 | 75 |
| Suicide | 31 | 33 | 34 | 32 | 32 | 31 |
| Teenage pregnancy | 32 | 30 | 26 | 21 | 21 | 20 |
| Temporomandibular muscle/joint disorder | 16 | 17 | 20 | 17 | 17 | 17 |
| Tobacco | 531 | 536 | 531 | 515 | 514 | 513 |
| Topical microbicides | 58 | 66 | 66 | 88 | 88 | 99 |
| Tourette syndrome | 17 | 16 | 13 | 13 | 13 | 13 |
| Transmissible spongiform encephalopathy (TSE) | 31 | 33 | 37 | 35 | 35 | 35 |
| Transplantation | 504 | 530 | 545 | 551 | 550 | 547 |
| Tuberculosis | 122 | 137 | 158 | 150 | 150 | 149 |
| Tuberculosis vaccine | 13 | 18 | 26 | 22 | 22 | 21 |
| Tuberous sclerosis | 8 | 10 | 9 | 9 | 9 | 9 |
| Urologic diseases | 551 | 595 | 576 | 536 | 532 | 532 |
| Uterine cancer | 34 | 35 | 39 | 28 | 27 | 26 |
| Vaccine related | 1,066 | 1,610 | 1,450 | 1,449 | 1,486 | 1,507 |
| Vaccine related (AIDS) | 405 | 452 | 511 | 566 | 564 | 571 |
| Vector-borne diseases | 296 | 419 | 447 | 464 | 462 | 457 |
| Violence against women | 21 | 20 | 22 | 20 | 18 | 18 |
| Violence research | N/A | N/A | 121 | 113 | 110 | 109 |
| West Nile virus | 37 | 43 | 43 | 85 | 42 | 63 |
| Women's health | 3,497 | 3,478 | 3,551 | 3,498 | 3,489 | 3,496 |
| Youth violence | N/A | N/A | 69 | 67 | 66 | 65 |

*Includes research on HIV/AIDS, its associated opportunistic infections, malignancies, & clinical manifestations as well as basic science that also benefits a wide spectrum of non-AIDS disease research.
N/A=Data not available.

SOURCE: "Estimates of Funding for Various Diseases, Conditions, Research Areas," U.S. Department of Health and Human Services, National Institutes of Health, February 5, 2007, http://www.nih.gov/news/fundingresearchareas.pdf (accessed October 29, 2007).

million on obesity-attributable disability insurance during the late 1990s. Many industry observers believe that the price businesses pay for obesity-related disability is destined to rise as sharply as the prevalence of obesity has increased in the United States.

Darius N. Lakdawalla, Jayanta Bhattacharya, and Dana P. Goldman assert in "Are the Young Becoming More Disabled?" (*Health Affairs*, vol. 23, no. 1, 2004) that obesity is a key cause of the more than 50% increase in disability rates over the last two decades, particularly among younger Americans. After analyzing data from the National Health Interview Survey, an annual nationwide government survey of about thirty-six thousand households, the researchers identify disability trends among people aged eighteen to sixty-nine between 1984 and 2000 and find significant growth in reported disability rates among those under fifty years but not among the elderly.

Lakdawalla, Bhattacharya, and Goldman report that "obesity accounts for about half the increased disability among those ages eighteen to twenty-nine." For those thirty to thirty-nine years old, the number reporting disabilities increased from 118 per 10,000 people in 1984 to 182 per 10,000 people in 1996. Among people forty to forty-nine years old, the number rose from 212 per 10,000 to 278 per 10,000 during this same period. Among people aged fifty to fifty-nine, disability rose only among those who were obese. The number of disability cases resulting from musculoskeletal problems and diabetes grew more rapidly than those from other problems during the length of the study, and the proportion that was diabetes-related doubled. Lakdawalla, Bhattacharya, and Goldman caution that this increase in the disability rate could translate into higher health-care costs in the future. Because people with disabilities generally use more medical services, should this trend persist, it could generate additional costs to the nation's already enormous health-care bill.

Soham Al Snih et al. of the University of Texas Medical Branch looked at the relationship between obesity, disability, and mortality by following the health of 12,725 adults aged sixty-five or older. In "The Effect of Obesity on Disability vs. Mortality in Older Americans" (*Archives of Internal Medicine*, vol. 167, no. 8, April 23, 2007), the researchers report that over the course of eleven years, 3,570 subjects became disabled and 2,019 died. Subjects with a low BMI (less than 18.5, which is considered underweight) and obese subjects (BMI greater than 30) were significantly more likely to experience disability and death. Snih et al. conclude that "disability-free life expectancy is greatest among subjects with a BMI of 25 to less than 30."

## THE HIGH COST OF LOSING WEIGHT

The AOA estimates in "Consumer Protection" (May 2, 2005, http://obesity1.tempdomainname.com/subs/fast facts/Obesity_Consumer_Protect.shtml) that at any given moment approximately 40% of women and 25% of men are trying to lose weight, and that forty-five million Americans diet each year. Americans spend about $30 billion per year to lose or prevent weight gain. The market research firm Marketdata forecasts in *U.S. Weight Loss and Diet Control Market* (2005) that a substantial annual growth in the U.S. weight-loss industry will produce a $61 billion industry in 2008.

Marketdata reports that in 2004 Americans consumed more diet soft drinks and that their share of the total soft drink market reached a near historical high. Diet soft drinks dominated in terms of sales, generating more than $15 billion in 2004, and health clubs ranked second. The most rapid growth occurred in do-it-yourself, over-the-counter (nonprescription) diet aids, which are less costly alternatives to medically supervised weight-loss and commercial programs.

With no new prescription weight-loss drugs on the horizon and the growing popularity of the African herb hoodia, Marketdata forecasts a 16% growth in this segment. Citing the success of several heavily advertised products, Marketdata predicts growth of 11.5% per year, to $703 million in 2008.

Along with commercial weight-loss centers, medically supervised weight-loss programs, and prescription diet drugs, products such as diet books, audio and video programs, Web-based diet and nutrition services, low-calorie and low-carbohydrate food products, meal replacements, and over-the-counter appetite suppressants compete for consumer dollars. As the low-carbohydrate diet craze subsides, more dieters are returning to structured commercial programs such as Weight Watchers,

LA Weight Loss, Jenny Craig, and other chains. Marketdata estimates that revenue from weight-loss centers will grow 11% to $2 billion. An estimated 7.1 million American dieters use such programs. Small local or regional chains of ten to fifty centers are growing as well.

According to Marketdata, the most affluent dieters, primarily in big cities, are purchasing home delivery of diet foods. Companies including Zone Chefs, NutriSystem, Jenny Direct (Jenny Craig), Seed Live Cuisine, Sunfare, and Nutropia are catering to this market. The cost averages $10 to $40 per day for home-delivered diet food, and dieters can spend as much as $1,200 per month. Weight-loss camps are also expected to grow in popularity and enrollment as the childhood obesity rate climbs.

Marketdata estimates that in 2004, 20,500 registered dietitians offered some form of weight-loss counseling, either in private practices or as consultants or employees of health clubs, hospitals, and other health-related facilities. A typical customized, six-month plan costs an average of $802. Nutritionists, who are not licensed dieticians and whose training varies widely from people without degrees to highly trained professionals with graduate degrees in nutritional science, also provide weight-loss counseling. Their services average $643 for a six-month contract.

Another study, *The U.S. Market for Weight Loss Products and Trends* (2005) by Marigny Research Group, describes the total U.S. market for weight-loss products and forecasts emerging trends in the U.S. market for weight-loss foods and beverages, where the low-fat and low-carbohydrate diets and foods may have peaked in popularity, spurring consumers to explore low-sugar and low glycemic index products. Like the Marketdata report, this study confirms the preeminence of the low-calorie food and beverage market and predicts increasing use and greater acceptance of artificial sweeteners in noncarbonated beverages, including refreshment, sports, and energy drinks, and meal replacement bars.

In *Weight Loss Market: Products, Services, Foods, and Beverages* (2003), Jack Baen also anticipates continued growth of weight-loss centers. Because the popularity of low-carbohydrate diets peaked in 2004, many analysts believe that conventional, "sensible" diets such as Weight Watchers' tried-and-true formula of portion control, healthy diet, and exercise will continue to attract people seeking to lose weight, thereby reenergizing corporate finances. According to Eric Wahlgren, in "The Skinny on Weight Watchers" (*BusinessWeek*, November 17, 2003), Kathleen Heaney, an analyst with the Maxim Group in New York, opines that consumers "typically end up at Weight Watchers after several other diet attempts have failed" and asserts that if anything, Weight Watchers' potential market in the United States has been drastically underestimated—its potential is about one hundred million clients.

The article "Rating the Diets from A to Zone" (*Consumer Reports*, vol. 70, no. 6, June 2005), a review of popular diets, gives Weight Watchers high marks in terms of safety, efficacy, and flexibility—its program allows people who prefer not to cook to use its branded controlled-calorie meals. The article also recommends the low-fat Ornish diet for vegetarians and the Slim-Fast diet for people who are not inclined to cook because Slim-Fast drinks and bars replace part of breakfast and lunch and dieters need to prepare just one meal per day.

## Medical and Behavioral Treatments

Even though the greatest proportion of outlays for weight loss are for food products and commercial weight-loss programs, McTigue et al. observe that medical and behavioral treatment options for obesity involve considerable cost. The researchers state that "intensive counseling programs require significant time and staffing commitment. Based on average U.S. wholesale price, a 1-year supply of orlistat (120 mg 3 times daily) is $1,445.40 and of sibutramine (15 mg daily) is $1,464.78." It is important to note that consumers generally purchase prescription drugs at retail rather than at wholesale prices, so their costs are considerably higher than those reported by McTigue et al.

According to the WIN, in "Gastrointestinal Surgery for Severe Obesity" (December 2004, http://win.niddk. nih.gov/publications/gastric.htm), weight-loss surgery costs from about $20,000 to $35,000, and the availability of medical insurance coverage for these surgical procedures varies by state and health insurance provider. William E. Encinosa, Didem M. Bernard, and Claudia A. Steiner find in "National Trends in the Costs of Bariatric Surgery" (*Bariatrics Today*, vol. 3, 2005) that even though the number of bariatric surgeries has increased over 1,000%, from 16,000 procedures performed in 1992 to more than 180,000 performed in 2006, coverage policies remain uneven among insurers. National hospital costs for bariatric surgeries increased by more than ten times, from $173 million in 1998 to $1.7 billion in 2003. Surgical costs reflect both the fees associated with the invasive procedure and the long-term follow-up that patients who have undergone the surgery require.

## Long-Term Savings

Even though surgical treatment of obesity is a relatively recent phenomenon, research reveals that its costs are offset by a reduction in future utilization of health-care services and a resultant reduction in health-care costs. Jane Salodof MacNeil reports in "Slimmed-down Health Plan Members Lower Medical Costs" (*Medscape Today*, November 17, 2004) that Gregory A. Nichols

et al. of the Kaiser Permanente Northwest Center for Health Research in Portland, Oregon, compared medical costs for two groups of patients who participated in the Kaiser weight-loss program "Freedom from Diets" from 1996 to 2000: 458 patients who lost more than 5% of weight and 457 patients who failed to lose weight. The investigators also created a control group of 2,290 patients who did not participate in the program and did not lose weight. Nichols et al. found that the regional health plan saved nearly $850 overall in per-person medical costs the year after an overweight member lost 5% or more body weight in a voluntary program. The researchers calculated that the health plan would save $2,500 over five years and noted that the savings would be real, even in view of the observation that most patients would regain the lost weight. According to MacNeil, another Kaiser Permanente study described drug savings as the key cost efficiency during the first two years after bariatric surgery. Pharmaceutical costs decreased by $510 per person among surgical patients, but costs increased in candidates who did not have the surgery.

Other research confirms cost savings, but in "Obesity, Weight Management, and Health Care Costs: A Primer" (*Disease Management*, vol. 10, no. 3, 2007), a review of the available evidence, Keith H. Bachman cautions that "the cost-effectiveness of obesity-related interventions is highly dependent on the risk status of the treated population, as well as the length, cost, and effectiveness of the intervention. Bariatric surgery offers high initial costs and uncertain long-term cost savings. From the perspective of a payor, obesity management services are as cost-effective as other commonly offered health services, though not likely to offer cost savings."

## CATERING TO AN EXPANDING MARKET

*On one hand, we have to make the world safe for a fatter population, but the more we adjust our world to accept our weight, the harder it is to motivate us to do the healthier thing and lose the weight. If we tacitly readjust our world, in some sense we are responding to reality. At the same time, there is no doubt that making those adjustments makes it easier to live bigger.*

—Arthur Caplan of the University of Pennsylvania School of Medicine, "Plus-Size People, Plus-Size Stuff" (Associated Press, November 10, 2003)

Along with increased costs, many businesses have discovered that they must literally expand their products and services to meet the needs of overweight and obese consumers. The article "Plus-Size People, Plus-Size Stuff" (Associated Press, November 10, 2003) describes a wide array of products—from scales that weigh people as heavy as one thousand pounds and steering wheels for drivers who do not fit behind standard wheels to seat-belt extenders and supersized towels—designed to meet the needs of obese Americans.

Service industries have also responded. In "That Tough First Step" (*Los Angeles Times*, January 26, 2004), Jeannine Stein reports that gyms are reaching out to attract and meet the special needs of people who are overweight and want to exercise. Some provide personal trainers who assist overweight clients to use equipment safely, design realistic exercise regimens, and maintain motivation. Other gyms affiliate with medical centers and health professionals to offer nutritional counseling, support groups, and exercises suitable for people who are overweight, including aquatic exercise programs in pools. Health clubs, gyms, and fitness programs understand not only the health benefits they can offer overweight clients but also the financial benefits they can realize by tapping into this market of people who have previously stayed away from gyms.

According to Deborah Yao, in "Stores Target Plus-Size Market" (Associated Press, April 24, 2006), the market research firm NPD Group reports that from March 2005 to February 2006 sales of plus-sized women's apparel rose by nearly 7% to $19 billion. In "Demand for Plus-Size Girls' Apparel Expanding" (*Los Angeles Times*, October 9, 2007), Marshal Cohen, an analyst with the NPD Group, opines that the children's plus-sized market could grow to 18% of the total children's apparel market of more than $35 billion. The article "Expanding Plus-Size and Big-and-Tall Clothing Market Estimated to Reach $107 Billion by 2012" (PRNewswire, June 26, 2007) notes that another market research firm, Packaged Facts, predicts that the plus-sized clothing market will grow by 41% from 2006 to 2012, with commensurate growth in sales, from $47.1 billion in 2006 to almost $65 billion in 2012.

Hot Topic, a California-based company that specializes in clothing for teenagers and young women, launched in 2001 a chain of six stores called Torrid that offer fashion-forward plus-sized clothing for young women. By 2007 Hot Topic noted in the press release "Hot Topic, Inc. Reports 1st Quarter Loss of $0.02 Per Share; Provides Guidance for the Second Quarter of 2007" (May 23, 2007, http://investorrelations.hottopic.com/) that 131 Torrid stores offered an array of clothing and lingerie for young women who wear larger sizes.

In "The Widening of America, or How Size 4 Became a Size 0" (*New York Times*, January 20, 2004), Jane E. Brody asserts that Americans' increasing girth has prompted size inflation throughout the fashion and apparel industry. Brody reports that the apparel industry has accommodated expanding Americans by increasing sizes such that a women's size 4 in 2004 would previously have been a size 8, and a size 8 would formerly have been a size 12. Men's clothing has also expanded with pants that were formerly "regular" now designated as "slim cut" and easy fit, loose fit, and baggy styles to accommodate excess weight.

Similarly, the article "Obesity Products Help Americans Live Large" (Associated Press, April 17, 2006) states that an expanding girth has created market demand for products to contain it—such as "husky" baby seats, extra sturdy chairs, larger doorways to accommodate wider wheelchairs, and sponges on sticks to help larger bathers clean hard-to-reach areas.

Demands for larger, sturdier hospital beds and stretchers to accommodate extremely heavy patients, special imaging equipment such as computed tomography scans and magnetic resonance imaging to accommodate obese patients, bigger blood pressure cuffs, recliners constructed to hold 350 pounds, automobiles that comfortably seat obese drivers and passengers, and devices that enable people who cannot bend over to put on their socks and shoes have prompted the design and manufacture of these and other specialty products. Even morticians have observed and responded to the obesity epidemic. In "On the Final Journey, One Size Doesn't Fit All These Days" (*New York Times*, September 28, 2003), Warren St. John reports that when the founders of Goliath Casket Company in Lynn, Indiana, opened their business in the late 1980s, they sold just one triple-wide casket—the largest model they made—per year. During 2003 the company shipped about five of the oversized coffins, which measure forty-four inches across, compared to the twenty-four-inch standard model, per month. David A. Hazelett, the president of Astral Industries, another coffin builder in Indiana, acknowledges the issue and adds that the problem affects every aspect of the funeral industry. Hazelett explains that "the standard-size casket is meant to go in the standard-size vault, and the standard-size vault is meant to go into the standard-size cemetery plot." St. John reports that hearse manufacturers have increased the width of their vehicles' rear doors, cemeteries have increased their standard burial plot size to accommodate wider vaults, and mausoleums have constructed larger crypts to accommodate oversized coffins.

According to Daniel Connolly, in "Obesity Creates Need for Oversized Caskets" (*Birmingham Post-Herald*, April 20, 2005), Mike Hauser, the marketing director of Ridout funeral homes and cemeteries, explains that newer parts of the company's cemeteries are being laid out with wider spaces for graves to accommodate larger bodies. In the case of an extremely large casket and vault, families that purchased a family plot can allow the grave to take up two spaces rather than one. Connolly notes that the Batesville Casket Company in Indiana, one of the nation's largest casket makers, introduced thirteen new oversized models in 2004; by 2005 it offered a total of fifty-three oversized models. Connolly states that the Goliath Casket Company has also continued to increase the size of its offerings. Sales at

Goliath Casket doubled in 2004, and the company sold about eight hundred oversized caskets. The company makes forty-four-inch, forty-eight-inch, and fifty-two-inch-wide caskets, which are constructed with extra supports intended for body weights between 650 and 1,200 pounds. The fifty-two-inch-wide casket is slightly wider than a standard pickup bed size.

Naturally, these oversized accommodations carry additional costs, and as a result some families opt for cremation. For the most severely obese, cremation may not, however, be an option. St. John notes that Jack Springer, the executive director of the Cremation Association of North America, explains that most crematoria are not equipped to handle bodies weighing more than five hundred pounds.

# CHAPTER 8

# POLITICAL, LEGAL, AND SOCIAL ISSUES
# OF OVERWEIGHT AND OBESITY

*Obesity is the terror within. Unless we do something about it, the magnitude of the dilemma will dwarf 9-11 or any other terrorist attempt. Where will our soldiers and sailors and airmen come from? Where will our policemen and firemen come from if the youngsters today are on a trajectory that says they will be obese, laden with cardiovascular disease, increased cancers and a host of other diseases when they reach adult-hood?.*

—Richard Carmona, the former U.S. surgeon general, lecture at the University of South Carolina (March 2006)

## THE GLOBAL POLITICS OF OBESITY

At the international level, the World Health Organization (WHO) has developed an aggressive plan to combat an escalating global epidemic of overweight and obesity—"globesity"—throughout the world. The WHO guidelines on diet and exercise, *The Expert Consultation on Diet, Nutrition, and the Prevention of Chronic Disease* (2003, http://whqlibdoc.who.int/trs/WHO_TRS_916.pdf), advocate actions such as lowering the intake of sugar, salt, and saturated fats. They also recommend sharply limiting the marketing of food to children and using tax and pricing policies to influence food consumption. The WHO asserts that these measures are necessary to reverse rising rates of the obesity-related illnesses—heart disease, diabetes, and cancer—which are forecast to account for nearly three-quarters of deaths worldwide by 2020.

The WHO plan was developed by an international team of experts using the latest scientific evidence available and has been commended by public health officials throughout the world. It is not, however, favored by some food manufacturers because among its proposals are restrictions on advertising unhealthy foods to children and the imposition of taxes and farm subsidy changes aimed at increasing prices of sugary and high-fat foods. For example, the International Sugar Research Organization strenuously objects to the recommendation that sugar

amount to no more than 10% of food and drink calories consumed per day, calling instead for a 25% cap. Table 8.1 shows that the total U.S. consumption of caloric sweeteners peaked in 2002. The use of caloric sweeteners declined slightly in 2003 but increased again in 2004 and 2005; preliminary data for 2006 show that the use of caloric sweeteners remains relatively unchanged.

On January 2, 2004, the United States (http://www.commercialalert.org/bushadmincomment.pdf) expressed its opposition to the WHO plan and demanded significant changes to the initiative. William R. Steiger (1969–), the director of the Office of Global Health Affairs and special assistant to the secretary for international affairs at the U.S. Department of Health and Human Services (HHS), questioned the validity of some of the dietary recommendations. In a twenty-eight-page critique of the WHO plan, Steiger wrote, "There is also an unsubstantiated focus on 'good' and 'bad' foods, and a conclusion that specific foods are linked to non-communicable diseases and obesity." Steiger put forth the U.S. position that all foods can be part of a healthy and balanced diet and called for "greater personal responsibility in battling obesity." According to the WHO spokesperson David Porter, Steiger was the only member of the international scientific community to contest the proposed population nutrient intake goals.

U.S. opposition to the WHO proposal has been criticized as a clear effort to appease U.S. food and sugar suppliers. Some WHO scientists and consumer advocacy groups say the U.S. objections—specifically those about the recommendations to limit sugar consumption and reconsider food advertising aimed at young children—aim to protect industries that have recently been under attack rather than to improve public health. However, the food industry itself has publicly pledged to support the WHO plan. The Grocery Manufacturers of America, the world's largest

**TABLE 8.1**

**Total estimated deliveries of caloric sweeteners for domestic food and beverage use, by calendar year, 1966–2006**

| Calendar year | Sugar[a] | | Corn sweeteners | | | | Honey | Other edible syrups | Total caloric sweeteners[b] |
|---|---|---|---|---|---|---|---|---|---|
| | Raw value | Refined basis | High fructose corn syrup | Glucose syrup | Dextrose | Total | | | |
| | | | | | 1,000 short tons, dry basis | | | | |
| 1966 | 10,235 | 9,565 | 0 | 952 | 415 | 1,367 | 98 | 69 | 11,099 |
| 1967 | 10,474 | 9,789 | 3 | 984 | 428 | 1,415 | 89 | 50 | 11,342 |
| 1968 | 10,656 | 9,959 | 15 | 1,031 | 444 | 1,489 | 90 | 70 | 11,608 |
| 1969 | 10,950 | 10,234 | 33 | 1,061 | 459 | 1,553 | 101 | 61 | 11,949 |
| 1970 | 11,163 | 10,433 | 56 | 1,102 | 471 | 1,629 | 103 | 51 | 12,216 |
| 1971 | 11,345 | 10,603 | 86 | 1,163 | 482 | 1,731 | 93 | 52 | 12,478 |
| 1972 | 11,487 | 10,736 | 121 | 1,257 | 485 | 1,863 | 105 | 52 | 12,756 |
| 1973 | 11,429 | 10,681 | 218 | 1,384 | 489 | 2,092 | 95 | 53 | 12,922 |
| 1974 | 10,945 | 10,229 | 295 | 1,480 | 486 | 2,262 | 75 | 43 | 12,609 |
| 1975 | 10,302 | 9,628 | 527 | 1,515 | 473 | 2,515 | 108 | 43 | 12,294 |
| 1976 | 10,893 | 10,180 | 782 | 1,514 | 452 | 2,748 | 100 | 44 | 13,072 |
| 1977 | 11,099 | 10,373 | 1,057 | 1,517 | 429 | 3,003 | 100 | 44 | 13,519 |
| 1978 | 10,889 | 10,177 | 1,198 | 1,551 | 410 | 3,159 | 120 | 45 | 13,501 |
| 1979 | 10,756 | 10,052 | 1,660 | 1,519 | 399 | 3,578 | 117 | 44 | 13,791 |
| 1980 | 10,189 | 9,522 | 2,158 | 1,472 | 393 | 4,024 | 94 | 50 | 13,690 |
| 1981 | 9,769 | 9,130 | 2,626 | 1,486 | 390 | 4,501 | 96 | 46 | 13,773 |
| 1982 | 9,153 | 8,554 | 3,090 | 1,479 | 392 | 4,961 | 104 | 46 | 13,665 |
| 1983 | 8,812 | 8,236 | 3,655 | 1,523 | 398 | 5,577 | 116 | 47 | 13,975 |
| 1984 | 8,428 | 7,877 | 4,399 | 1,552 | 408 | 6,359 | 108 | 47 | 14,391 |
| 1985 | 8,003 | 7,479 | 5,386 | 1,607 | 418 | 7,411 | 104 | 48 | 15,043 |
| 1986 | 7,731 | 7,225 | 5,498 | 1,632 | 430 | 7,561 | 121 | 50 | 14,957 |
| 1987 | 8,103 | 7,573 | 5,792 | 1,679 | 441 | 7,912 | 104 | 55 | 15,644 |
| 1988 | 8,136 | 7,604 | 5,998 | 1,747 | 452 | 8,197 | 100 | 54 | 15,955 |
| 1989 | 8,304 | 7,761 | 5,960 | 1,587 | 438 | 7,985 | 95 | 53 | 15,894 |
| 1990 | 8,615 | 8,051 | 6,202 | 1,700 | 455 | 8,358 | 103 | 53 | 16,565 |
| 1991 | 8,622 | 8,058 | 6,376 | 1,776 | 463 | 8,615 | 116 | 53 | 16,842 |
| 1992 | 8,826 | 8,249 | 6,652 | 1,943 | 461 | 9,056 | 126 | 53 | 17,483 |
| 1993 | 8,886 | 8,305 | 7,086 | 2,050 | 481 | 9,617 | 135 | 56 | 18,112 |
| 1994 | 9,072 | 8,478 | 7,398 | 2,093 | 502 | 9,993 | 126 | 54 | 18,651 |
| 1995 | 9,258 | 8,652 | 7,676 | 2,176 | 528 | 10,380 | 120 | 57 | 19,209 |
| 1996 | 9,400 | 8,785 | 7,788 | 2,216 | 537 | 10,541 | 131 | 57 | 19,514 |
| 1997 | 9,481 | 8,861 | 8,240 | 2,364 | 511 | 11,116 | 129 | 58 | 20,163 |
| 1998 | 9,594 | 8,966 | 8,552 | 2,358 | 502 | 11,411 | 130 | 59 | 20,566 |
| 1999 | 9,912 | 9,264 | 8,897 | 2,281 | 488 | 11,666 | 147 | 60 | 21,138 |
| 2000 | 9,901 | 9,253 | 8,845 | 2,230 | 476 | 11,551 | 157 | 61 | 21,022 |
| 2001 | 9,839 | 9,195 | 8,920 | 2,205 | 469 | 11,595 | 134 | 61 | 20,986 |
| 2002 | 9,746 | 9,109 | 9,045 | 2,224 | 473 | 11,741 | 153 | 62 | 21,065 |
| 2003 | 9,479 | 8,859 | 8,849 | 2,209 | 449 | 11,507 | 146 | 63 | 20,575 |
| 2004 | 9,678 | 9,045 | 8,779 | 2,292 | 487 | 11,558 | 130 | 64 | 20,797 |
| 2005 | 10,001 | 9,346 | 8,756 | 2,261 | 478 | 11,494 | 155 | 66 | 21,062 |
| 2006[c] | 9,986 | 9,332 | 8,783 | 2,053 | 465 | 11,302 | 167 | 66 | 20,867 |

NA=not available.

Note: Per capita deliveries of sweeteners by U.S. processors and refiners and direct-consumption imports to food manufacturers, retailers, and other end users represent the per capita supply of caloric sweeteners. The data exclude deliveries to manufacturers of alcoholic beverages. Actual human intake of caloric sweeteners is lower because of uneaten food, spoilage, and other losses.

[a]Based on U.S. sugar deliveries for domestic food and beverage use.
[b]Total includes sugar, refined basis.
[c]Preliminary.

SOURCE: "Table 49. U.S. Total Estimated Deliveries of Caloric Sweeteners for Domestic Food and Beverage Use, by Calendar Year," in *Sugar and Sweeteners: Data Tables*, United States Department of Agriculture, Economic Research Service, March 15, 2007, http://www.ers.usda.gov/Briefing/Sugar/Data.htm (accessed October 29, 2007)

association of food and drink companies, which includes PepsiCo Inc. and Hershey Foods Corp., said it was committed to working with the WHO to combat obesity.

In "The Sweet and Lowdown on Sugar" (*New York Times*, January 23, 2004), Kelly Brownell and Marion Nestle compare the food industry's self-serving attempts to delay action on the WHO proposal to efforts made by the tobacco industry to defend the harmlessness of cigarettes. They assert that "by making its position on the W.H.O. indistinguishable from that of the food industry, the Bush administration undermines the efforts of more forward-thinking food companies and threatens public health. Its action underscores the need for government to create a wall between itself and the food industry when establishing nutrition and public health policy. Recommendations to cut back on sugars may not please food companies, but it's time to stop trading calories for dollars."

The WHO global strategy did not become official until it was endorsed by member states at the United Nations (UN) summit in May 2004. The plan is not binding, but it is considered a guiding document for public health efforts on the issue worldwide. Even though the draft gained broad international support, in January 2004 the WHO agreed to U.S. demands for additional time to comment on the final resolution. Nutritionists, public health agencies, and medical professional associations responded with shock and dismay that the United States had succeeded in stalling the global obesity-control plan. Despite U.S. efforts to delay its adoption, at the Fifty-seventh World Health Assembly, the WHO Global Strategy on Diet, Physical Activity, and Health was endorsed by resolution WHA57.17 (http://www.who.int/dietphysicalactivity/strategy/eb11344/strategy_english_web.pdf) in May 2004.

The strategy provides member states with a range of policy options to address two of the major risks responsible for the heavy and growing burden of chronic diseases attributable to unhealthy diet and physical inactivity. It explains how healthier diet, nutrition, and physical activity can help prevent and control these diseases. The document describes roles of WHO member states, UN agencies, civil society, educators, and the private sector to help reduce the occurrence of obesity. It recommends obesity-prevention measures, including effective food and agriculture policies, fiscal policies, surveillance systems, consumer education, and nutrition labeling. The strategy urges limiting the intake of sugars, fats, and salt in foods, and increasing the consumption of fruits, vegetables, legumes, whole grains, and nuts. It also emphasizes the need for countries to develop national strategies with a long-term, sustainable perspective to make the healthy choices the preferred alternatives at both the individual and community levels.

### Is Sugar the New Tobacco?

The WHO named sugar as the principal culprit in the current epidemic of obesity and obesity-related diseases, diabetes, and cardiovascular heart disease. The WHO approach to food is not, however, comparable to its strategy to combat tobacco use. The food strategy aims to provide member states and other interested stakeholders with a range of recommendations and policy options to promote healthier diets and more physical activity. It will be up to member states to decide how these should be further developed and implemented at the national level. Because the strategy was endorsed at the World Health Assembly, member states are responsible for determining which specific policy options are appropriate to their circumstances. The WHO will then provide technical support for the implementation of programs, as requested by member states.

### AMERICANS CRAVE SUGAR

Eric Margolis notes in "Sugar and Politics" (January 27, 2004, http://www.ericmargolis.com/archives/2004/01/index.php) that even though the United States encompasses just 5% of the world population, it accounts for 33% of total global sugar consumption, more than ten million tons annually. Table 8.2 shows monthly estimates of U.S. sugar supply and use during fiscal year 2008. Sugar is the most subsidized U.S. crop. At a rate of nearly $500 per acre annually, U.S. sugar producers receive $1.4 billion in federal subsidies each year. U.S. sugar prices are artificially inflated because of import restrictions that protect producers from foreign competition. Americans pay as much as four times more for domestic sugar than they would if foreign competitors were permitted to market sugar in the United States. Critics of these subsidies observe that the sugar industry makes generous contributions to members of Congress of both parties.

Sugar—sucrose, dextrose, fructose, corn syrup, or maltodextrin—is a key ingredient of many processed food products. Table 8.3 lists the names of added sugars that may be principal ingredients of processed foods. The Center for Science in the Public Interest (CSPI) reports in "Added-Sugars Consumption" (2007, http://www.cspinet.org/reports/sugar/addedsugar.html) that Americans' sugar consumption has been steadily increasing since the mid-1980s. The average American consumes at least sixty-four pounds of sugar per year, and the average teenage boy at least 109 pounds. American adults get 16% of their calories from added sugars, and children aged six to eleven get 18% of their calories from added sugars. Adolescents aged twelve to nineteen get 20% of their calories from added sugars. The CSPI also observes that people with diets high in added sugars consume lower levels of fiber, fewer vitamins, and less folate, magnesium, and calcium, among other nutrients. By displacing vital nutrients and foods in the diet, added sugars may increase the risk of osteoporosis, cancer, high blood pressure, heart disease, and other health problems.

Even though the health food industry has been warning the public about the perils of the overconsumption of refined sugars for more than thirty years, mainstream nutritionists and public health professionals have joined the ranks of those calling for reduced sugar consumption. Along with ending sugar subsidies, they want to sharply limit the advertising of sugary products to children, ban the sale of soft drinks in schools, and conduct widespread community public health education programs to inform Americans about the health risks of consuming excessive amounts of refined sugars.

### THE U.S. WAR ON OBESITY GAINS MOMENTUM

Besides generating international debate, the issue of obesity is receiving considerable attention from lawmakers, public health officials, and politicians throughout the United States. Some legislators and policy makers have chastised the administration of President George W. Bush (1946–) for allegedly yielding to the food industry and trying to dilute

TABLE 8.2

**Monthly estimates of fiscal 2008 sugar supply and use, May 2007–October 2007**

| | May 2007 | June 2007 | July 2007 | Aug. 2007 | Sept. 2007 | Oct. 2007 |
|---|---|---|---|---|---|---|
| Beginning stocks* | 1,715 | 1,692 | 1,615 | 1,627 | 1,772 | 1,749 |
| Total production | 8,255 | 8,255 | 8,293 | 8,292 | 8,342 | 8,446 |
| Beet sugar | 4,520 | 4,520 | 4,619 | 4,621 | 4,657 | 4,764 |
| Cane sugar | 3,735 | 3,735 | 3,674 | 3,671 | 3,684 | 3,683 |
| Florida | 1,870 | 1,870 | 1,774 | 1,774 | 1,774 | 1,774 |
| Louisiana | 1,430 | 1,430 | 1,430 | 1,430 | 1,430 | 1,430 |
| Texas | 206 | 206 | 198 | 198 | 198 | 198 |
| Hawaii | 229 | 229 | 271 | 268 | 282 | 280 |
| Puerto Rico | 0 | 0 | 0 | 0 | 0 | 0 |
| Total imports | 1,789 | 1,889 | 1,889 | 1,889 | 2,109 | 2,124 |
| Tariff-rate quota imports | 1,284 | 1,284 | 1,284 | 1,284 | 1,354 | 1,369 |
| Other program imports | 425 | 425 | 425 | 425 | 425 | 425 |
| Non-program imports | 80 | 180 | 180 | 180 | 330 | 330 |
| Total supply | 11,759 | 11,837 | 11,797 | 11,808 | 12,223 | 12,319 |
| Exports | 250 | 250 | 250 | 250 | 250 | 250 |
| Adjustments | 0 | 0 | 0 | 0 | 0 | 0 |
| Total deliveries | 10,170 | 10,170 | 10,170 | 10,170 | 10,170 | 10,170 |
| Domestic food and beverage | 10,000 | 10,000 | 10,000 | 10,000 | 10,000 | 10,000 |
| Other use | 170 | 170 | 170 | 170 | 170 | 170 |
| Total use | 10,420 | 10,420 | 10,420 | 10,420 | 10,420 | 10,420 |
| Ending stocks | 1,339 | 1,417 | 1,377 | 1,388 | 1,803 | 1,899 |
| Stocks/use ratio | 12.85 | 13.60 | 13.22 | 13.32 | 17.30 | 18.22 |

*As of May 2004, includes all stocks held by processors, millers, and refiners, including stocks held for others.

SOURCE: "Table 26. Monthly Estimates of Fiscal 2008 U.S. Sugar Supply and Use," in *Sugar and Sweeteners: Data Tables*, United States Department of Agriculture, Economic Research Service, October 12, 2007 http://www.ers.usda.gov/Briefing/Sugar/Data.htm (accessed October 29, 2007)

**TABLE 8.3**

**Names for added sugars that appear on food labels**

A food is likely to be high in sugars if one of these names appears first or second in the ingredient list or if several names are listed.

| | |
|---|---|
| Brown sugar | Invert sugar |
| Corn sweetener | Lactose |
| Corn syrup | Malt syrup |
| Dextrose | Maltose |
| Fructose | Molasses |
| Fruit juice concentrate | Raw sugar |
| Glucose | Sucrose |
| High-fructose corn syrup | Syrup |
| Honey | Table sugar |

SOURCE: "Box 21. Names for Added Sugars that Appear on Food Labels," in *Nutrition and Your Health: Dietary Guidelines for Americans*, 5th ed., U.S. Department of Health and Human Services and U.S. Department of Agriculture, 2000, http://www.health.gov/dietaryguidelines/dga2000/document/choose.htm (accessed October 29, 2007)

the WHO antiobesity plan. Among the many legislative initiatives being considered are proposals to mandate nutrition information on restaurant menus, improving school lunch programs, and the imposition of taxes on high-calorie, low-nutrition food items.

Skirmishes in the war on obesity do not center on whether there is a problem, but on how best to address it. Participants on one side characterize the food industry, advertisers, and the media as complicit—coercing consumers with seductive advertising and sugary, high-calorie treats. Their opponents believe that consumers should exercise personal responsibility and make their own choices about food and exercise.

In "The Ironic Politics of Obesity" (*Science*, vol. 299, no. 5608, February 7, 2003), Marion Nestle asserts that the war on obesity is unlikely to be won because healthful eating is not in the best interest of U.S. industry, and government agencies are beset by conflicts of interest. Nestle condemns the lack of government leadership, observing that the U.S. Department of Agriculture (USDA) offers confusing and conflicting advice to consumers. To fulfill its mission to promote U.S. agricultural products, the USDA simultaneously exhorts consumers to eat more, while issuing advice about diet, which for many overweight Americans means "eat less." This conflict of interest has produced vague federal dietary guidelines that advise Americans to "aim for a healthy weight [and] choose beverages and foods to moderate your intake of sugars." Nestle calls for "small taxes on junk foods and soft drinks (to raise funds for anti-obesity campaigns); restrictions on food marketing to children, especially in schools and on television; calorie labels on fast foods; and changes in farm subsidies to promote the consumption of fruits and vegetables."

In contrast, the WHO strategy does not stipulate any specific tax or subsidy. However, it observes that several countries have adopted fiscal measures to promote the availability of and access to various foods, and to increase or decrease consumption of certain types of food. The strategy notes that public policies can influence prices through measures such as tax policies and subsidies. The strategy acknowledges that decisions on policy options are the responsibility of individual member states, depending on their particular circumstances.

The Public Health Advocacy Institute (PHAI) contends that food industry processing and marketing practices have encouraged excessive food consumption. The PHAI Law and Obesity Project considers the existing state of regulation, legislation, and litigation related to the food industry's contribution to obesity, and the potential for new legal strategies to effectively reduce this contribution.

In 2005 California scored a legislative victory that prohibits the state's public elementary and middle schools from selling soda from vending machines and bans the sale of soda in the state's public high schools. John Graham, a former Harvard University professor of public health who serves as the administrator of the Office of Information and Regulatory Affairs, Office of Management and Budget, successfully campaigned to require food manufacturers to disclose the trans-fat content of their products on nutrition labels. (Trans fats are formed by the partial hydrogenation of vegetable oil—the process used to make vegetable oil more solid. Trans fats raise low-density lipoprotein cholesterol levels and may lower high-density lipoprotein cholesterol.) As of January 1, 2006, the U.S. Food and Drug Administration (FDA) required the disclosure of the trans-fat content of food that is sold in the United States.

The legal professor Richard Banzhaf, who campaigned against tobacco, advocates using the legal system to create change in Americans' diets. He exhorts attorneys to bring lawsuits against fast-food purveyors and manufacturers of junk food to increase consumer awareness of the role the food industry plays in promoting obesity.

Richard Berman, the executive director of the Center for Consumer Freedom, an advocacy group supported by restaurant and food companies and that represents major corporations such as RJR Nabisco, marshals lawyers, publicists, and lobbyists to respond to antiobesity crusaders. The center identifies itself as a nonprofit coalition that stands for "common sense and personal choice." It derides lawsuits and legislation aimed at limiting consumers' rights to choose the foods they want to consume, and it pokes fun at CSPI mandates to offer consumers nutritional data and the self-appointed "food police"—legislators, public health officials, and others—intent on modifying Americans' diets. The organization is credited with helping defeat a measure that would have required chain restaurants to offer nutritional data about their products. In "Study: Why Food and Drink Bans Won't Solve Childhood Obesity" (April 26, 2007, http://www.consumerfreedom.com/article _detail.cfm/article/182), it staunchly opposes banning or restricting food and drink sales in schools, asserting that

physical inactivity as opposed to the overconsumption of foods high in sugar, fat, and calories is the primary cause of childhood obesity.

## The American Obesity Association Action Plan

The American Obesity Association (May 2, 2005, http://obesity1.tempdomainname.com/subs/about.shtml) has an ambitious agenda for the government and private sector that enumerates specific funding priorities, programs, and services to prevent, treat, and educate Americans. It calls for:

- Recognizing obesity as a disease—to expand research and foster a national commitment to combating obesity comparable in scope and funding to those for cancer, human immunodeficiency syndrome/acquired immunodeficiency syndrome, and smoking.

- Making obesity a public health priority—working with the media and policy makers to improve understanding of obesity, to fight stigma and discrimination, and to effectively address the epidemic.

- Supporting the prevention of obesity—advocating for a federal Human Physical Activity Impact statement that enables policy makers to evaluate the impact of public projects on physical activity in communities.

- Preventing and treating childhood and adolescent obesity—support for research about the roles of the school in nutrition education and physical activity.

- Including coverage for obesity treatment in health insurance—including a Medicare prescription benefit to enable older adults and disabled people to gain access to antiobesity medications.

- Support for consumer protection agencies' efforts to identify and eliminate frauds and deceptive practices directed against people with obesity.

- Advancing new treatments of obesity—the AOA organized a coalition of fifteen pharmaceutical companies to work with the FDA on guidelines for developers of weight-loss drugs.

- Supporting the obesity community—providing a national, multidisciplinary Provider Directory that enables prospective patients to find health-care providers in their area and a Career Resource Center for prospective employers and job seekers.

At the close of 2007, at least two key AOA objectives had been realized. In October 2004 the AOA celebrated the decision by the Centers for Medicare and Medicaid Services to eliminate language from its policy that said obesity is not a disease. Laura Kann, Nancy D. Brener, and Howell Wechsler note in "Overview and Summary: School Health Policies and Programs Study 2006" (*Journal of School Health*, vol.77, no. 8, October 2007) that in 2006 the nation's schools made strides in terms of nutrition and

FIGURE 8.1

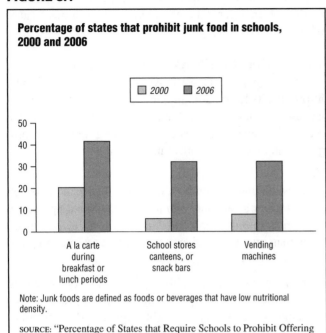

**Percentage of states that prohibit junk food in schools, 2000 and 2006**

Note: Junk foods are defined as foods or beverages that have low nutritional density.

SOURCE: "Percentage of States that Require Schools to Prohibit Offering Junk Foods in School Settings, 2000 and 2006," in *U.S. Schools Making Progress in Decreasing Availability of Junk Food and Promoting Physical Activity*, Centers for Disease Control and Prevention, National Center for Chronic Disease Prevention and Health Promotion, Division of Adolescent and Student Health, October 23, 2007, http://www.cdc.gov/DataStatistics/2007/shpps/ (accessed October 30, 2007)

fitness—a full 30% had banned junk food, compared to just 4% in 2000, and the percentage of school districts mandating elementary schools to offer physical education rose from 83% in 2000 to 93% in 2006. Figure 8.1 shows that the percentage of states that prohibit offering junk food in schools has grown between 2000 and 2006.

## OVERWEIGHT, OBESITY, AND THE LAW

Jeffrey Levi, Laura M. Segal, and Emily Gadola indicate in *F As in Fat: How Obesity Policies Are Failing in America, 2007* (August 2007, http://healthyamericans.org/reports/obesity2007/Obesity2007Report.pdf) that in 2006 eight states introduced legislation intended to offer or strengthen private insurance coverage for obesity prevention or treatment, especially for people who are clinically severe obese (body mass index [BMI] greater than 40). Indiana enacted legislation specifying confidentiality and reporting requirements for bariatric surgery complications and providing for insurance coverage for medically necessary expenses for clinically severe obesity treatment. Nine states—Connecticut, Indiana, Louisiana, Missouri, New Jersey, Oklahoma, South Carolina, Tennessee and Virginia—considered legislation governing obesity treatment in 2006.

During 2006 several states considered taxing foods and beverages with minimal nutritional value and using the revenues to finance school facilities or childhood obesity

prevention initiatives; however, no legislation to this effect was enacted. In "Patterns of Childhood Obesity Prevention Legislation in the United States" (*Preventing Chronic Diseases*, vol. 4, no. 3, July 2007), a review of state-level legislation aimed at combating childhood obesity, Tegan K. Boehmer et al. find that between 2003 and 2005, 717 bills and 134 resolutions were introduced. More than half (53%) of the resolutions were adopted, but just 17% of the bills became law. Most legislation addressed school nutrition standards and vending machines, followed by physical education and physical activity. Statewide initiatives (30%) were most often adopted, followed by model school policies (29%) and safe routes to school (28%).

The U.S. Senate's Healthy Students Act of 2007 and the U.S. House of Representatives' Stop Obesity in Schools Act of 2007 provide federal-level initiatives to promote children's health by encouraging increased physical activity and improved nutrition. Both bills would require the director of the Centers for Disease Control and Prevention to convene a Commission to Improve School Meals composed of nutrition and health experts to develop new nutritional standards for the School Lunch, Summer Food Service, Child and Adult Care Food, and School Breakfast programs. They would further require that the new standards ban foods of minimal nutritional value. At the close of 2007, neither bill had made it out of committee.

### Legislators Target School Programs

The National Conference of State Legislatures indicates in *2007 State Legislation Report* (2007, http://www.aap.org/advocacy/statelegrpt.pdf) that during 2006 state legislatures were actively considering policy options to address the obesity epidemic. Aiming to start early to prevent the onset of chronic conditions, legislators proposed a variety of policy approaches to create opportunities for a healthier diet and more exercise beginning in childhood. State legislatures in twenty-seven states considered or enacted legislation aimed at improving the nutritional quality of school foods and beverages.

Levi, Segal, and Gadola note that as of 2007 twelve states had some type of student BMI reporting or fitness screening in effect. All fifty states and the District of Columbia required physical education in schools, but the nature and extent of the requirement varied. Several states focused on increasing physical education requirements or encouraging positive physical activity programs for students during and after school. Some states and school districts cited the cost of physical education programs and an emphasis on academics as obstacles to increasing physical education programs.

### Lawsuits Attack Food Service Industry

A number of individuals and advocacy groups have brought lawsuits against the food service industry. Some claim they deserve compensation for the damage that fattening foods have done to their health. Others focus on advertising and marketing that they feel is deceptive and misleads people into eating unhealthy products. Many attorneys and public health professionals believe such lawsuits can serve as vehicles that reverse the obesity epidemic, in part because the media attention generated by such lawsuits motivates food companies to produce healthier products and to reconsider marketing and advertising practices.

The first class-action suit was the widely publicized case of Caesar Barber, a fifty-six-year-old New Yorker weighing 270 pounds, who claimed that four fast-food restaurants—McDonald's, Burger King, Wendy's, and KFC—jeopardized his health by promoting high-calorie, high-fat, and salty menu items. In "Whopper of a Lawsuit: Fast-Food Chains Blamed for Obesity, Illnesses" (ABC News.com, July 26, 2002), Geraldine Sealey reports that Barber filed the lawsuit in the New York State Supreme Court "on behalf of an unspecified number of other obese and ill New Yorkers who also feast on fast food." According to Sealey, Barber's suit alleged that the fast-food restaurants, where he ate "four or five times a week even after suffering a heart attack, did not properly disclose the ingredients of their food and the risks of eating too much." Even though Barber's suit was dismissed by two judges and he was barred from filing a third time, legal scholars assert that more cases like Barber's will be heard by the courts.

The legal community did not have long to wait. In January 2005 an appeals court ordered McDonald's to defend a 2004 lawsuit by the New York teenagers Ashley Pelman and Jazlen Bradley, who claimed the company hid the health risks of foods that made them obese. Samuel Hirsch, the lawyer who represented Caesar Barber, represented the teenagers. The suit was the first complaint accusing a fast-food chain of hiding health risks of its food to be considered by a judge. The teenagers said they ate at McDonald's restaurants three to five times a week over a fifteen-year period. Their suit claimed the company hid the health risks of Big Macs, Chicken McNuggets, and other foods high in fat and cholesterol in its 1987 advertisements in the United States and in brochures circulated in Great Britain. McDonald's defended the accuracy of its ads and asserted that there was no evidence that the teenage plaintiffs, one of whom was born in 1988, ever saw the ads. In early 2008 the suit remained unresolved.

### Legislation Protects Food Industry Interests

The food industry and others argue that Americans choose what they eat and should not be able to blame the food industry if their personal choices have unhealthy consequences. State and federal legislators who agree with this viewpoint have enacted or attempted to enact laws that protect the food industry from weight-related lawsuits.

On October 19, 2005, the House of Representatives passed a bill that would prevent most obesity or weight-related claims against the food industry and make it harder than ever before for consumers to sue restaurants and food retailers for serving fattening fare. By a vote of 307 to 119, lawmakers endorsed the Personal Responsibility in Food Consumption Act, which informs consumers that if they gain weight as a result of eating high-fat, high-calorie, and sugar-laden food, they have only themselves to blame. The legislation, however, did not receive a vote in the Senate, so it did not become law.

The restaurant and food-processing industries have also championed state measures such as the Idaho Commonsense Consumption Act, signed into law on April 2, 2004, which bans civil lawsuits for obesity and obesity-related health problems. The same month, Arizona (2004, http://azleg.state.az.us/alispdfs/46leg/2R/House/SummaryJUD.pdf) enacted legislation affirming that "there is no duty to warn a consumer that a non-defective food product may cause health problems if consumed excessively and provides an affirmative defense."

In *School Nutrition and Physical Education Legislation: An Overview of 2005 State Activity* (April 4, 2005, http://www.rwjf.org/files/research/NCSL%20-%20April%202005%20Quarterly%20Report.pdf), Carla I. Plaza notes that in 2005 Wyoming enacted a bill that limits an individual's ability to sue food and beverage companies. Specifically, Wyoming's Commonsense Consumption Act prohibits an individual from suing a manufacturer, seller, trade association, agricultural producer, wholesaler, broker, or retailer of a qualified product for injury or death based on the individual's weight gain, obesity, or a health condition related to weight gain or obesity. South Dakota and Utah passed comparable legislation in March 2004, followed by Colorado, Florida, Georgia, Missouri, and Tennessee in May 2004, Louisiana in June, Illinois in July, and Michigan in October. Besides Wyoming, Kansas, Kentucky, Maine, North Dakota, Ohio, Oregon, and Texas introduced and enacted legislation limiting obesity-related lawsuits in 2005.

Lianne S. Pinchuk states in "Are Fast Food Lawsuits Likely to Be the Next 'Big Tobacco'?" (*National Law Journal*, February 28, 2007) that by 2007 twenty-three states had enacted legislation to provide some measure of protection or even immunity for food companies threatened by obesity lawsuits.

## THE FOOD INDUSTRY RESPONDS TO PUBLIC OUTCRY

Mounting pressure on the food industry to change its marketing practices and offer healthier products has had some success. For example, in 2003 Coca-Cola withdrew from exclusive vending-machine contracts in schools and acquired Odwalla, an organic fruit-juice company, to enable the company to offer healthy beverages. Kraft announced intentions to eliminate in-school marketing to children, introduce smaller portions, and develop more nutritious products. Applebee's International offered Weight Watchers selections on its restaurant menus. McDonald's reduced the use of trans fats for cooking its French fries and introduced a line of salads as well as leaner versions of its Chicken McNuggets.

In March 2004 McDonald's responded to growing attention to the relationship between portion size and obesity by announcing that the corporation would discontinue its supersized products—French fries and soft drinks—in an effort to simplify its menu and appeal to consumers' heightened awareness about obesity. McDonald's also piloted a new "Go Active" meal for adults that included a salad, a pedometer to count steps, and a bottle of water in several test markets throughout the country. Industry observers applauded these moves, citing the corporation's shift from the "value" aspect of fast food— providing more food for less money—to a more health-conscious purveyor of salads and reasonable portion sizes that emphasize nutrition rather than value. They also expressed the hope that other fast-food chains would follow suit and offer more nutritional information and low-calorie fare.

In another effort to counter charges that its food is unhealthy and contributes to obesity, McDonald's began to display nutrition facts on the packaging of its menu items in 2006. Customers of the world's largest restaurant company can learn the amount of calories and fat, among other information, in a McDonald's product by looking at the wrapper instead of having to go to its Web site or ask for nutrition information at the counter.

Despite these positive changes, the Ethical Investment Research Services finds in *Obesity Concerns in the Food and Beverage Industry* (February 2006, http://www.eiris .org/files/research%20publications/seeriskobesityfeb06.pdf) that the efforts of six multinational food companies— Cadbury-Schweppes, Coca-Cola, Kraft Foods, McDonald's, PepsiCo, and Unilever—were uneven. McDonald's and Unilever were deemed the slowest to respond to the obesity epidemic, although both companies were lauded for acknowledging their responsibility for addressing the problem of childhood obesity. In terms of responsible advertising to children, the report gives the lowest score to McDonald's and the highest to Cadbury-Schweppes.

## WEIGHT-BASED DISCRIMINATION

Nearly everyone who is overweight or obese has suffered some form of bias, from disapproving glances and unsolicited advice about how to lose weight to the seemingly unending stream of "fat jokes" and the unflattering and even humiliating portrayal of overweight people in the media. Despite the pervasive anti-fat bias in American

culture, until recently there were anecdotal reports, but little evidence, demonstrating that negative attitudes toward obese individuals resulted in stigmatization and clear instances of discrimination.

In "Bias, Discrimination, and Obesity" (*Obesity Research*, vol. 9, no. 12, December 2001), Rebecca M. Puhl and Kelly D. Brownell review data revealing that systematic discrimination against obese individuals occurs in at least three areas: education, employment, and health care. They also acknowledge that evidence points to discrimination in adoption proceedings, jury selection, and housing.

Puhl and Brownell describe obese people as "the last acceptable targets of discrimination" and name rejection— teasing, taunts, derogatory comments, and derision—by peers as the first of many challenges overweight or obese youngsters will face. Some studies find distinct anti-fat bias in children as young as age three and increasingly negative stereotypic attitudes with age. Puhl and Brownell point to a landmark study conducted during the 1960s in which children were shown pictures of six children with various physical characteristics and disabilities, including use of crutches or wheelchair, amputations, or facial disfigurements, and were asked to rank them in order of whom they would be most likely to befriend. Most subjects ranked the picture of the obese child last. When this study was performed again in 2001, children in the fifth and sixth grade displayed the strongest bias against the obese child and expressed even more prejudice than their counterparts had forty years earlier. Teachers also revealed considerable bias, with nearly 30% in one survey describing becoming obese as "the worst possible thing that can happen to a person."

Puhl and Brownell observe that along with the psychological and social consequences of prejudice and exclusion, obese students suffered lower rates of college acceptance, with obese women gaining college admission less frequently (31%) than obese male applicants (42%). They also find that normal-weight college students received more financial support from their families than overweight students, and overweight women were least likely to receive financial support.

Overweight and obese job applicants and workers may be subjected to weight-based discrimination in employment. Many studies document discrimination in hiring practices, especially when the positions sought involved public contact, such as sales or direct customer service. Obese workers face inequities in wages, benefits, and promotions, and several studies confirm that the economic penalties are greater for women than for men. Overweight women earn less doing the same work as their normal-weight counterparts and have dimmer prospects for promotion. The courts have considered cases in which workers contended that their job terminations were weight-related. The outcomes of these cases indicate that termination can occur because of employer prejudice and arbitrary weight standards.

In "Gender, Body Mass, and Economic Status" (National Bureau of Economic Research Working Paper No. 11343, May 2005), Dalton Conley and Rebecca Glauber find that even though overweight women do not fare as well as their normal-weight peers in terms of income, overweight men are as successful, economically and in terms of job status, as normal-weight men. The researchers show that increased BMI significantly decreased women's family income as well as their "occupational prestige," a measure of the social status afforded to different jobs. A 1% increase in a woman's BMI reduced her family income by 0.6% and a 0.4% decrease in occupational prestige. Along with lower pay and less-prestigious jobs, heavier women's poorer socioeconomic outcomes were attributable to three factors:

- Overweight women tend to have lower chances of getting married

- When they do marry their spouses tend to have less earning power

- Overweight women have a higher risk for divorce

Consistent with past research, men experience no negative effects of body mass on economic outcomes. Overweight men are not less likely to marry, nor are they at increased risk for divorce, separation, or widowhood.

## Weight Bias among Health Professionals

Anti-fat bias among health-care professionals may discourage obese people from seeking medical care and compromise the care they receive. Even though research indicates that obese patients often delay or cancel medical appointments for a variety of reasons, including fear about being weighed or undressing in front of health professionals, speculation exists that presumed or real prejudice on the part of health professionals may also deter them from seeking medical care. When researchers asked more than four hundred physicians to name patient characteristics that provoked feelings of discomfort, reluctance, or dislike, one-third of the subjects mentioned obesity, making it the fourth most common condition named after drug addiction, alcoholism, and mental illness. The subjects also linked obesity to negative qualities such as poor hygiene, hostility, dishonesty, and noncompliance with prescribed treatment. Another survey of family physicians found that two-thirds said their obese patients lacked self-control, and nearly 40% characterized their obese patients as lazy. Nurses expressed similar attitudes—nearly half reported that they were uncomfortable caring for obese patients and 31% told surveyors they would prefer not to care for obese patients at all.

Puhl and Brownell note documented evidence that deeply held negative stereotypes adversely affect the clinical judgment of health professionals, including diagnosis and the quality of care delivered to obese patients. A survey of more than twelve hundred physicians revealed that most were ambivalent about caring for overweight and obese patients and did not treat them with the same determination they displayed toward normal-weight patients. Just 18% said they would refer an overweight patient to a weight-loss program, and less than half (42%) would refer a mildly obese patient to a weight-loss program.

Even health professionals who specialize in the medical treatment of obesity are not immune from anti-fat bias. Marlene B. Schwartz et al. administered a standardized test that measured bias to 389 health professionals—physicians, researchers, dieticians, nurses, psychologists, and others—who attended an international obesity conference in Quebec in 2001. The researchers reported the test results in "Weight Bias among Health Professionals Specializing in Obesity" (*Obesity Research*, vol. 11, no. 9, September 2003). Bias was assessed using the Implicit Associations Test (IAT), a timed test that analyzes the automatic associations respondents make about particular attributes. For example, the IAT helps identify whether or not test takers hold negative attitudes and stereotypical views about obese people, such as considering them to be lazy, unmotivated, sluggish, or worthless.

Schwartz et al. find that the health professionals they tested—one-third of whom provided direct clinical care to obese patients—exhibited significant anti-fat bias. They linked the stereotypes lazy, stupid, and worthless with obese people, with younger health professionals displaying more anti-fat bias than older health-care workers did. Schwartz et al. hypothesize that younger health professionals may be more strongly imprinted with societal pressures to be thin, which have intensified in recent decades. Another explanation may be that older health professionals, who have more maturity and experience, may have overcome some of their negative attitudes about obese patients. Despite the presence of bias, the researchers concede that even though it is intuitively appealing to assume that bias has an influence on treatment, their research does not demonstrate that bias resulted in poorer treatment of obese patients.

**OBESE AMERICANS RECEIVE FEWER PREVENTIVE HEALTH SERVICES.** Ironically, people who are obese and usually receive more medical care for chronic diseases related to obesity may also receive fewer preventive services. Does bias contribute to this disparity in preventive care? In "Associations between Obesity and Receipt of Screening Mammography, Papanicolaou Tests, and Influenza Vaccination: Results from the Health and Retirement Study (HRS) and the Asset and Health Dynamics among the Oldest Old (AHEAD) Study" (*American Journal of*

*Public Health*, vol. 95, no. 9, September 2005), Truls Østbye et al. examine the association between BMI and receipt of screening mammography and Papanicolaou tests (screening for cervical cancer) among middle-aged women and the association between BMI and receipt of influenza vaccination among older adults. The investigators analyzed data from the Health and Retirement Study (4,439 women aged fifty to sixty-one years) and the Asset and Health Dynamics among the Oldest Old Study (4,045 women and 2154 men aged seventy years or older).

Østbye et al. find significant differences in how often obese women were given mammograms and Pap smears to screen for cancer. They also note that obese men and women were also less likely to receive flu shots. Seventy-one percent of the obese women studied reported having mammograms, compared to 78% of those who were not obese. Similarly, 54% of the obese women reported having Pap smears, compared to 73% of the nonobese women. In addition, 57% of the obese men and women whose records were reviewed reported receiving flu shots, compared to 78% of the people who were of normal weight. Østbye et al. pose several potential explanations for the disparity in preventive services: obese patients' reluctance to undress for cancer screening tests, practitioners' difficulties in performing screening tests on obese women, and the observation that obese patients may require so many medical care services for chronic diseases that preventive care may be overlooked.

Nancy Klein Amy et al. examine in "Barriers to Routine Gynecological Cancer Screening for White and African-American Obese Women" (*International Journal of Obesity*, vol. 30, no. 1, 2006) the reasons obese women, who are at higher risk for gynecological cancers than nonobese women, are less likely to get cancer-screening tests. The investigators find that obese women delay cancer-screening tests and believe that their weight is a barrier to obtaining appropriate health care. The obese women reported disrespectful treatment, embarrassment at being weighed, negative attitudes of providers, unsolicited advice to lose weight, and medical equipment that was too small to be functional as reasons they delayed or avoided screening tests. The percentage of women who reported these barriers to seeking preventive care increased as the women's BMI increased.

## Airlines Weigh Their Options

In June 2002 Southwest Airlines became the center of a fiery debate when the airline decided to strengthen its enforcement of a policy established in 1980 of requesting and requiring passengers who, because of excessive girth, must occupy two airplane seats to purchase both seats. The policy allows passengers to be reimbursed for the additional seat if their flight is not full. The National Association to Advance Fat Acceptance (NAAFA), an

advocacy group, and other consumer groups called the move discriminatory. Southwest Airlines is not the only airline with this policy; Continental, Northwest, and other commercial carriers also require large-sized passengers to pay for two seats.

In 2003 the Federal Aviation Administration (FAA) proposed requiring all passengers on small airlines to be weighed in along with their luggage. The FAA asserted that before takeoff, the pilot must calculate the weight of the aircraft as well as that of its passengers, luggage, and crew to determine which seats passengers should occupy to ensure proper balance. For this reason it is vital to know exact passenger and luggage weights on small planes, where several people with a few extra pounds can tilt the plane away from its center of gravity. Even though operators of smaller commuter airlines acknowledged the safety issue, they were reluctant to support the FAA recommendation because they feared that weighing people would discourage them from using commuter airlines, many of which are already strapped financially.

In May 2003 the FAA ruled that airlines must assume that passengers weigh 190 or 195 pounds, depending on the season. At the same time, checked bags on domestic flights were adjusted from an estimated twenty-five pounds to thirty pounds. The thirty-pound estimate for checked bags on international flights remained unchanged. The requirement followed shortly after the crash of a commuter plane that killed all twenty-one people aboard. Investigators suspect the propeller plane was slightly above its maximum weight on takeoff, with most of the weight toward the tail. The weight distribution problem was compounded by a maintenance error that made it difficult to lower the nose with the control column. After the nineteen-seat plane rose above the ground, its nose pointed dangerously skyward; the pilots were unable to level it off, and the plane spun into the ground.

Andrew Dannenberg, Deron C. Burton, and Richard J. Jackson reveal in "Economic and Environmental Costs of Obesity: The Impact on Airlines" (*American Journal of Preventive Medicine*, vol. 27, no. 3, October 2004) that the average American gained ten pounds during the 1990s. The extra weight required an additional 350 million gallons of fuel used by airlines in 2000. This extra weight translated into about $275 million in excess costs in 2000 alone. The extra fuel represented 2.4% of the total volume of jet fuel used domestically that year, and along with the monetary cost, there was the environmental impact of burning all that extra jet fuel to transport what Dannenberg, Burton, and Jackson call "this additional adiposity." The article "Gov't Study: Obese Passengers Pushing up Cost of Flights" (Associated Press, November 4, 2004) notes that Jack Evans, the spokesperson for the Air Transport Association of America, which represents major U.S. airlines, agreed that weight is a real issue. He explained that weight considerations and fuel prices have prompted airlines to replace metal forks and spoons with plastic utensils and to forgo bulky magazines: "We're dealing in a world of small numbers—even though it has a very incremental impact. When you consider airlines are flying millions of miles, it adds up over time."

## Obese Female Shoppers Face Discrimination

In "The Stigma of Obesity in Customer Service: A Mechanism for Remediation and Bottom-Line Consequences of Interpersonal Discrimination" (*Journal of Applied Psychology*, vol. 91, no. 3, 2006), Eden B. King et al. report the results of a study that revealed widespread rude behavior and discrimination against obese female shoppers. The investigators discovered that when women aged nineteen to twenty-eight wore prosthetic suits designed to make them appear obese, they were treated more rudely and received less eye contact and fewer smiles from sales clerks at a Houston shopping mall than when they shopped without the fat suit. Even though nearly three-quarters of the sales clerks were women, they tended to interact less with the obese female shoppers, ending interactions abruptly and assuming more negative tones of voice with them.

Treatment of the obese shoppers was worse when they were dressed casually than when they wore professional attire, and their treatment improved when they shopped sipping diet soda and volunteered that they were trying to lose weight. A survey of shoppers conducted as part of this study found that survey respondents who were obese reported being subjected to more rude treatment from sales clerks, which prompted them to spend less time and money in the stores where they experienced discrimination. King et al. are optimistic that the financial and ethical implications of this study will provide powerful incentives for retailers to address size discrimination with employees.

## San Francisco Bans Weight-Based Discrimination and Hears Landmark Cases

The San Francisco Human Rights Commission reports in "Compliance Guidelines to Prohibit Weight and Height Discrimination" (http://www.sfgov.org/site/sfhumanrights_page.asp?id=5911) that on July 26, 2001, it unanimously approved historic guidelines for implementing a height/weight antidiscrimination law, and the city became the first jurisdiction in the United States to offer guidelines on how to prevent discrimination based on weight or height. Santa Cruz, California, Seattle, Washington, Washington, D.C., and Michigan have similar laws banning discrimination based on height or weight.

The strength of the ordinance was tested two years later when Jennifer Portnick, a 240-pound aerobics instructor, was refused a job at Jazzercise, Inc., an international dance-fitness organization based in Carlsbad, California, and brought her case before the San Francisco Human Rights Commission. She later reached an agreement with the company to drop a requirement about the appearance of instructors. It was the first case settled under the San Francisco ordinance, which has become known as the "Fat and Short Law."

Patricia Leigh Brown reports in "240 Pounds, Persistent and Jazzercise's Equal" (*New York Times*, May 8, 2002) that Portnick's attorney, Sondra Solovay, the author of *Tipping the Scales of Justice: Fighting Weight-Based Discrimination*, said Portnick was "geographically lucky" to have filed her case in one of just four jurisdictions in the country that outlawed weight-based discrimination.

## Weight Bias Influences Adoption Decision

Grant Slater reports in "Man Resorts to Surgery to Adopt Child" (Associated Press, August 25, 2007) the case of Gary Stocklaufer, a man weighing 558 pounds who was prevented from adopting a child because of his weight. Stocklaufer and his wife were ordered by a Missouri judge to give the four-month-old boy, a relative of the couple, whom they had raised since he was one week old, to another couple for possible adoption. Because the Stocklaufers serve as licensed foster parents and already have one adopted child, they and adoption activists alleged that weight was the deciding factor. In a desperate attempt to regain his son, Stocklaufer dieted to 480 pounds before he underwent bariatric surgery, which was intended to help him reduce his weight by more than 50%. Adoption experts consider this a landmark case because it is the first one in which a couple seeking to adopt has resorted to surgery to surmount the increasingly prevalent practice of denying adoptions on the basis of weight.

## The Origins of Stigma and Bias

Rebecca M. Puhl and Kelly D. Brownell observe in "Psychosocial Origins of Obesity Stigma: Toward Changing a Powerful and Pervasive Bias" (*Obesity Reviews*, vol. 4, no. 4, November 2003) that many people intensely dread the possibility of becoming obese. In one survey, 24% of women and 17% of men said they would sacrifice three or more years of their life to be thin. There are reports of women who choose not to become pregnant because they fear gaining weight and becoming fat. Others smoke cigarettes in an effort to remain thin or reject the advice that they quit smoking because they fear they will gain weight should they quit. This powerful fear of fat, coupled with widespread perceptions that overweight people lack competence, self-control, ambition, intelligence, and attractiveness, create a culture in which it is socially acceptable to hold negative stereotypes about obese individuals and to discriminate against them.

One explanation of the origin of weight stigma is that traditionally Americans believe in self-determination and individualism—people get what they deserve and are responsible for their circumstances. In this context, when overweight is viewed as resulting from controllable behaviors, it is easy to understand that if an individual believes overweight people are to blame for their weight, then they should be stigmatized. Other research findings—that many Americans view life as predictable, with effort and ability inevitably producing the desired outcomes, and the finding that attractive people are deemed good and believed to embody many positive qualities—support this theory. Interestingly, researchers find that in other countries the best predictors of anti-fat attitudes were cultural values that held both negative views about fatness and the belief that people are responsible for their life outcomes.

Several other theories about the origins of weight stigma have been proposed. Conflict theory suggests that prejudice arises from conflicts of interest between groups and struggles to acquire or retain resources or power. Social identity theory posits that groups develop their social identities by comparing themselves to other groups and designating other groups as inferior. Integrated threat theory proposes that stigmatized groups are perceived as a threat. Proponents of this theory suggest that overweight and obese people threaten deeply held cultural values of self-discipline, self-control, moderation, and thinness. Another theory, evolved dispositions theory, proposes that members of a group will be stigmatized if they threaten or undermine group functioning. This evolutionary adaptation may predispose people to shun obese individuals because they are at increased health risk and may not be able to make sufficient contributions to the group's welfare because of weight-related illness or disability.

**Reducing Weight Bias and Stigma**

In "Demonstrations of Implicit Anti-fat Bias: The Impact of Providing Causal Information and Evoking Empathy" (*Health Psychology*, vol. 22, no. 1, January 2003), Bethany A. Teachman et al. wondered if anti-fat bias would be reduced when people were told that an individual's obesity resulted largely from genetic factors rather than as the result of overeating and lack of exercise. The investigators assigned study participants to one of three groups. The first group received no information about the cause of obesity; the second group was given an article asserting that the principal cause of obesity was genetic; and the third group was given an article that attributed most obesity to overeating and lack of physical activity. As the researchers anticipated, the group told that obesity was controllable—resulting from overeating and inactivity—revealed the greatest amount of bias. However, to their surprise, Teachman et al. find that the group informed that

obesity was primarily genetic in origin did not have significantly lower levels of bias than either the control group that had received no prior information or the group informed that obesity was caused by overeating and inactivity.

Teachman et al. also wanted to find out whether eliciting empathy for obese people would significantly reduce negative attitudes. The researchers hypothesized that by sharing written stories about weight-based discrimination with study participants they would feel empathy with the subjects in the stories, which they would then generalize to the entire population of obese people. Even though some study participants in the group that read the stories displayed lower bias, the majority did not have lower bias than the control group that had not read the stories of discrimination. The investigators speculate that the stories describing negative evaluations of an obese person might actually have served to reinforce rather than diminish bias.

Puhl and Brownell note in "Psychosocial Origins of Obesity Stigma" that the increasing prevalence of obesity has not acted to reduce weight bias. They also refute the notion that stigma is necessary to motivate overweight and obese people to lose weight. They reiterate that dieting is not associated with long-term weight loss, regardless of the individual's motivation. Furthermore, stigma has led to discrimination and exerts a harmful influence on health and quality of life. These obesity experts contend that unless stigma is reduced, obese people will continue to contend with prejudice and discrimination.

Even though few studies have evaluated the effectiveness of strategies to reduce weight stigma, a variety of initiatives have produced varying degrees of attitudinal change. These approaches include:

- Educating participants about external uncontrollable causes such as the biological and genetic factors that contribute to obesity.

- Teaching and encouraging young children to practice size acceptance.

- Improving attitudes by combining efforts to elicit empathy with education about the uncontrollable causes of obesity.

- Encouraging direct personal contact with overweight and obese individuals to dispel negative stereotypes.

- Changing individuals' beliefs by exposing them to opposing attitudes and values held by a group that they consider important. This approach, which is called social consensus theory, relies on the observation that after learning that a group does not share the individuals' beliefs, they are more likely to modify their beliefs to be similar to those expressed by the group they respect or wish to join.

In "Psychosocial Origins of Obesity Stigma," Puhl and Brownell describe the results of their experiments using social consensus theory to modify attitudes toward obese people. They conducted experiments with university students in which participants reported their attitudes toward obese people before and after the researchers offered them varying consensus opinions of other students. In one experiment, participants who were told that other students held more favorable attitudes about obese people reported significantly fewer negative attitudes and more positive attitudes about obese people than they had before they learned about the opinions of other students. Furthermore, they also changed their ideas about the causes of obesity, favoring the uncontrollable causes after they were told the other students believed obesity was attributable to these causes.

A second experiment confirmed that the power to alter the participants' beliefs depended on whether the source of the opposing beliefs was an in-group or out-group. Not surprisingly, participants' attitudes toward obese people were more likely to change when the information they were given came from a source they valued—an in-group. In a third experiment the researchers compared attitudinal change produced by social consensus with other methods to reduce stigma, including one in which participants were given written material about the uncontrollable or controllable causes of obesity. Puhl and Brownell find that social consensus was as effective as or more effective than any of the other methods they applied. They state that social consensus theory also offers an explanation about why obese individuals themselves express negative stereotypes—they want to belong to the valued social group and choose to accept negative stereotypes to align themselves with current culture. Furthermore, by accepting prevailing cultural values and beliefs, they not only resemble the in-group more closely but also distance themselves from the out-group, where identity and membership are defined by being overweight or obese.

Even though Puhl and Brownell consider social consensus a promising approach to reducing weight bias and stigma, they caution that there are many unanswered questions about its widespread utility and effectiveness. They conclude that "an ideal and comprehensive theory of obesity stigma would identify the origins of weight bias, explain why stigma is elicited by obese body types, account for the association between certain negative traits and obesity, and suggest methods for reducing bias. Existing theories do not yet meet all these criteria."

## Advocacy Groups Promote Size and Weight Acceptance

*People get so many conflicting messages about what is healthy and what is attractive. The same thin celebrities who were being glamorized in recent years are now being airbrushed to look even thinner on magazine covers. That sends a terrible message, both to the celebrities and to the public. Love your body, it's the only one you have. You have to take care of yourself—and that starts with self-esteem.*

—Allen Steadham, director of the International Size Acceptance Association, press release (July 2003)

There is a growing consumer movement that advocates size and weight acceptance with the overarching goal of assisting people to have positive body images at any weight and to achieve health at any size. Nearly all organizations that champion size acceptance characterize the preoccupation with dieting and weight loss as unhealthy and unproductive, citing statistics about diet failures, the dangers of weight cycling (the repeated loss and regain of body weight), as well as frustration and low self-esteem. The size acceptance movement proposes that it is possible to be fit and fat and that health and beauty are attainable at all weights. It also works to reduce "fat phobia," anti-fat bias, and weight-based discrimination.

The International Size Acceptance Association (ISAA) is an organization that promotes size acceptance and aims to end size discrimination throughout the world by means of advocacy and visible, lawful actions. In "Healthy Body Esteem: Love Your Body It's the Only One You Have" (2003, http://www.size-acceptance.org/downloads/ Healthy_Body_Esteem.pdf), the ISAA discusses its Respect Fitness Health Initiative and Healthy Body Esteem campaigns, which provide an alternative to the "diet-of-the-day" pressures and gloom-and-doom predictions about size and weight that assault people every day. The ISAA asserts that people of all sizes can become more fit, and the organization is committed to helping people of all sizes strive for higher levels of fitness and improvements in their overall quality of life. Similarly, the ISAA observes that everyone could benefit from healthier food choices and is committed to helping inform the public about healthy nutrition.

Another group, the Council on Size and Weight Discrimination, Inc., a nonprofit consumer advocacy organization working to end "sizism," bigotry, and discrimination against people who are heavier than average, focuses its advocacy efforts on affecting changes in medical treatment, job discrimination, and media images. The council's basic principles were derived from "Tenets of the Nondiet Approach" (Karin Kratina, Dayle Hayes, and Nancy King, eds., *Moving away from Diets: Healing Eating Problems and Exercise Resistance*, 2003) and focus on:

- Total health enhancement and well-being, rather than on weight loss or achieving a specific "ideal weight"
- Self-acceptance and respect for the diversity of bodies that come in a wide variety of shapes and sizes, rather than on the pursuit of an idealized weight at all costs

- The pleasure of eating well, based on internal cues of hunger and satiety (the feeling of fullness or satisfaction after eating), rather than on external food plans or diets

- The joy of movement, encouraging all physical activities, rather than on prescribing a specific routine of regimented exercise

The National Association to Advance Fat Acceptance (NAAFA; September 2007, http://www.naafa.org/documents/brochures/naafa-info.html#whatis) is a non-profit human rights organization dedicated to eliminating discrimination based on body size and providing people with the "tools for self-empowerment through public education, advocacy, and member support." The NAAFA has assumed a proactive role in protesting social prejudice, bias, and discrimination, as well as working with the Federal Trade Commission to stop diet fraud. The organization also seeks to improve legal protection for people who are overweight and obese by educating lawmakers and serving as a national legal clearinghouse for attorneys challenging size discrimination.

# CHAPTER 9
# DIET AND WEIGHT-LOSS LORE, MYTHS,
# AND CONTROVERSIES

One of the challenges facing public-health professionals as they seek to combat obesity among Americans is helping consumers to distinguish myths, lore, legends, and outright fraud from accurate, usable information about nutrition, diet, exercise, and weight loss. Some of these inaccuracies are so long-standing and deeply rooted in American culture that even the most educated consumers unquestioningly accept them as facts. Others began with a kernel of truth but have been so wildly distorted or misinterpreted that they are confusing, misleading, or entirely erroneous. The rapid influx and dissemination of information about the origins of overweight and obesity and conflicting accounts of how best to treat these problems compound the challenge. With media reports and advertisements trumpeting different diets nearly every week, it is no wonder that Americans are confused about diet and weight loss.

The fiction that people who are overweight or obese are lazy and weak-willed is among the most harmful myths because it serves to promote stigma, bias, and discrimination. Another common misconception is that it is equally easy or difficult for all people to lose weight. There are biological and behavioral factors that affect an individual's body weight, and people vary in terms of genetic propensity to become overweight, basal metabolic rate (BMR), and the number of fat cells. BMR, often referred to simply as the metabolic rate, is the number of calories an individual expends at rest to maintain normal body functions. BMR changes with age, weight, height, gender, diet, and exercise habits and has been found to vary by as much as one thousand calories per day. Differences in metabolic rate explain, in part, why not all people who adhere to the same diet achieve the same results in terms of pounds lost or rate of weight loss. Another factor that produces variation in weight loss is the number of fat cells in the dieter's body. Even though fat cells do not determine body weight, they are affected by weight gain

and act to limit weight loss because their number cannot be decreased. For example, a normal-weight person has about 40 billion fat cells, whereas an individual who weighs 250 pounds with a body mass index (BMI) of 40 may have as many as 100 billion fat cells. Weight loss causes fat cells to shrink in size but does not decrease their number. As a result, individuals with twice as many fat cells as normal-weight people may be able to shrink their fat cells to a normal size but even when they have attained a healthy weight they will still have twice as many fat cells.

## DIET AND WEIGHT-LOSS MYTHS

It is impossible to recount all the fantastic and improbable claims that have been made in recent years. This section considers some of the most persistent myths about diet, exercise, and weight loss.

### Low-Carbohydrate Diets

MYTH. A low-carbohydrate diet is the fastest, healthiest, and best way to lose weight.

FACT. Low-carbohydrate diets may initially produce more rapid weight loss than other diets; however, most of the loss is water weight rather than fat. The water loss occurs as the kidneys flush out the excess waste products resulting from the digestion of protein and fat. Many low-carbohydrate diets encourage the consumption of high-fat foods, such as butter, heavy cream, bacon, and cheese. Long-term, high-fat diets may raise blood cholesterol levels, and low-carbohydrate, high-protein diets produce a state of ketosis (the accumulation of ketones from partly digested fats as a result of inadequate carbohydrate intake), which may increase the risk of gout (a severe arthritis attack that occurs in one joint—typically the big toe, ankle, or knee—caused by defects in uric acid metabolism) and kidney stones. Furthermore, most nutritionists and researchers concur that even though some weight-loss diets are nutritionally inadequate and others are even dangerously insufficient, nearly all diets can affect weight loss, and currently no compelling evidence exists to proclaim that one diet is vastly superior to another. A key factor in the success of any weight-loss diet is adherence—whether dieters can remain faithful to the regimen they have chosen, and to date low-carbohydrate diets have not demonstrated superiority in terms of adherence. Boredom and frustration with a low-carbohydrate regimen may occur when dieters crave the carbohydrates that they are forbidden or can eat only in small amounts.

Still, there is one unanswered question about diet and weigh loss: Why do some dieters successfully lose weight using low-carbohydrate or low-fat diets, whereas others on the same diets are unsuccessful? Cara B. Ebbeling et al. assert in "Effects of a Low-Glycemic Load vs Low-Fat Diet in Obese Young Adults" (*Journal of the American Medical Association*, vol. 297, no. 19, May 16, 2007) that which diet will be the most effective for each individual depends in part on the dieter's hormonal profile—specifically on differences in insulin secretion as measured by serum insulin concentration. The researchers compared seventy-three subjects following a low-glycemic load diet or low-fat diet and measured their body weight, body fat, and insulin concentration before and after six months of dieting and during a twelve-month follow-up period. During the six months of dieting, high insulin secretors lost more weight (2.2 pounds per month) on the low-glycemic-load diet, than on the low-fat diet (0.9 pounds per month). After eighteen months, the high insulin secretors had lost a total of 12.8 pounds on the low-glycemic-load diet, compared to just 2.6 pounds on the low-fat diet. The low-glycemic-load dieters also lost more body fat than the low-fat dieters and were more successful at maintaining their weight losses. In contrast, dieters who were considered low insulin secretors fared equally well on both diets. Ebbeling et al. also observed that independent of insulin secretion status, the low-glycemic-load diet had beneficial effects—high-density lipoprotein increased and triglycerides decreased. Subjects on the low-fat diet did not realize these benefits, but did experience reductions in low-density lipoprotein.

## Calorie Reduction

MYTH. The dieter needs to cut calories drastically to lose weight.

FACT. Weight loss may be accomplished with modest reductions in calorie consumption. Low-calorie diets often result in metabolic adaptations, such as a significant reduction in resting metabolic rate, which may produce weight maintenance or even weight gain rather than the desired weight loss. Many nutritionists and diet plans advise simultaneously reducing total caloric-intake and modifying the balance of macronutrients (nutrients that the body uses in relatively large amounts: carbohydrates,

fats, and proteins)—some weight-loss diets reduce fat intake, others reduce carbohydrates.

## Negative-Calorie Foods

MYTH. It takes more calories to eat and digest some foods such as celery or cabbage than these foods contain, so eating them causes or speeds weight loss.

FACT. There are no foods that when eaten cause weight loss. Foods containing caffeine may temporarily boost metabolism but they do not cause weight loss. However, some recent evidence suggests that eating grapefruit or drinking grapefruit juice may help people who are obese to lose weight. Ken Fujioka et al. at the Scripps Clinic in San Diego, California, compared weight loss over a twelve-week period among one hundred obese individuals. One-third of the subjects ate half a grapefruit before each meal three times per day, whereas another drank a glass of grapefruit juice before every meal. The third group did not include grapefruit in their meals. According to Marina Murphy, in "Grapefruit Diet Works and May Prevent Diabetes" (*Chemistry and Industry*, no. 3, February 2, 2004), Fujioka et al. reported that after twelve weeks subjects who ate grapefruit lost an average of 3.6 pounds, and those who drank grapefruit juice lost an average of 3.3 pounds, whereas those in the control group who consumed no grapefruit lost an average of 0.5 pounds. Fujioka et al. attributed the weight loss to lowered levels of insulin, which were confirmed by measurements of blood glucose and insulin levels. They posited that the more efficiently sugar is metabolized, the less likely it is to be stored as fat. Furthermore, lowering insulin levels reduces feelings of hunger—elevated insulin levels stimulate the brain's hypothalamus, producing feelings of hunger.

## Eating at Night

MYTH. Eating after 8:00 PM causes weight gain.

FACT. Weight gain or loss does not depend on the time of day food is consumed—excess calories will be stored as fat whether they are consumed midmorning or just before bedtime. In general, weight is governed by the amount of food consumed measured in total calorie count, and the amount of physical activity expended during the day.

There is, however, evidence that eating early in the day and eating breakfast are habits associated with maintaining a healthy weight. In "Make It an Early Bird" (*New York Times*, November 21, 2007), Jennifer Ackerman indicates that research reveals that people who eat breakfast tend to consume fewer calories throughout the day, compared to those who make dinner their biggest meal. This may be because the system in the brain that signals satiety (the feeling of fullness or satisfaction after eating) is more effective early in the day—at night an individual may be more prone to succumb to overeating.

## Natural Weight-Loss Products

MYTH. Organic, natural, or herbal weight-loss products are safer than synthetic (produced in the laboratory) over-the-counter (nonprescription) or prescription drugs.

FACT. Simply because products are organic or naturally occurring does not necessarily mean that they are risk-free or safe. For example, according to the press release "FTC Charges Direct Marketers of Ephedra Weight Loss Products with Making Deceptive Efficacy and Safety Claims" (July 1, 2003, http://www.ftc.gov/opa/2003/07/ephedra.shtm), in July 2003 the Federal Trade Commission (FTC) took action against the marketers of weight-loss products containing ephedra, which is derived from a leafless desert shrub, and hydroxycitric acid, which is an extract from brindall berries. The actions targeted deceptive effectiveness, safety, and side-effect claims for weight-loss supplements containing these dietary supplements. The FTC challenged advertising claims that ephedra and other natural supplements caused rapid, substantial, and permanent weight loss without diet or exercise, as well as the claims that these weight-loss products are "100% safe," "perfectly safe," or have "no side effects."

## Low-Fat and Low-Carbohydrate Foods

MYTH. Low fat or nonfat means few or no calories.

FACT. A low-fat or nonfat food is usually lower in calories than the same size—as measured by weight—portion of the full-fat food; however, a food product can contain zero grams of fat and still have a high calorie content. Many fat-free foods replace the fat with sugar and contain just as many or more calories as full-fat versions. Even though most fruits and vegetables are naturally low in fat and calories, processed low-fat or nonfat foods may be high in calories because extra sugar, flour, or starch thickeners have been added to enhance the low-fat foods' taste or texture.

Similarly, low-carbohydrate foods are often higher in calories than their "regular" counterparts because their fat content is higher. Many foods that are naturally low in carbohydrates such as meat, butter, and cheese are also calorie-dense. Many nutritionists suggest limiting the consumption of low-carbohydrate versions of foods, such as low-carbohydrate frozen desserts, because they not only contain as many or more calories per serving than ice cream but also are often sweetened with artificial sweeteners that lack any nutrients.

## Eliminating Starchy Foods

MYTH. Pasta, potatoes, and bread are fattening foods and should be eliminated or sharply limited when trying to lose weight.

FACT. Potatoes, rice, pasta, bread, beans, and some starchy vegetables such as squash, yams, sweet potatoes, turnips, beets, and carrots are not innately fattening. They are rich in complex carbohydrates, which are important sources of energy. Furthermore, foods that are high in complex carbohydrates are often low in fat and calories because carbohydrates contain only four calories per gram, compared to the nine calories per gram contained by fats. In "Effects of an Ad Libitum Low-Fat, High-Carbohydrate Diet on Body Weight, Body Composition, and Fat Distribution in Older Men and Women" (Archives of Internal Medicine, vol. 164, no. 2, January 26, 2004), Nicholas P. Hays et al. report the results of a small study in which dieters lost substantial amounts of weight on a high-carbohydrate, low-fat regimen. Meals were prepared for the subjects, who were told to eat as much as they wanted and to return any uneaten food, which enabled the researchers to calculate the subjects' caloric intake. Surprisingly, subjects who consumed a high-carbohydrate, low-fat diet with no quantity or caloric restrictions lost significant amounts of weight. The researchers speculate that low-fat, high-carbohydrate diets may reduce body weight via reduced food intake, because complex carbohydrate-rich foods are more satiating and less energy-dense than higher-fat foods and conclude that their "data support the alteration of dietary macronutrient composition without emphasis on caloric restriction as an effective means of promoting weight loss."

## Genetic Destiny

MYTH. People from families where many members are overweight or obese are destined to become overweight.

FACT. It is true that studies of families find similarities in body weight and that immediate relatives of obese people are at an increased risk for overweight and obesity, compared to people with normal-weight family members. Even though it is generally accepted that genetic susceptibility or predisposition to overweight or obesity is a factor, researchers believe that environmental and behavioral factors make equally strong, if not stronger, contributions to the development of obesity. As a result, people from overweight or obese families may have to make concerted efforts to maintain healthy body weight and prevent weight gain, but they are not destined to become obese simply by virtue of the genes they inherited.

## Exercise Alone

MYTH. Exercise is a better way to lose weight than dieting.

FACT. Even though there are many health benefits from exercise, weight loss is not generally considered a direct benefit. Research has consistently demonstrated that for weight loss, diet trumps exercise because it is simpler to reduce caloric intake significantly through diet than to increase caloric expenditure significantly through exercise. For example, if a 155-pound person wished to reduce his or her consumption by 400 calories per day, it might be achieved by simply eliminating dessert and reducing portion sizes. In contrast, expending four hundred calories requires considerable effort. To burn 400 calories, a 155-pound person would have to spend an hour bicycling at about ten miles an hour; hiking cross country, mowing the lawn, or ice skating at nine miles per hour; or water skiing or walking uphill at about 3.5 miles per hour. However, many studies demonstrate that exercise is an important way to prevent overweight and maintain weight loss.

Cris A. Slentz et al. find in "Effects of the Amount of Exercise on Body Weight, Body Composition, and Measures of Central Obesity: STRRIDE—A Randomized Controlled Study" (*Archives of Internal Medicine*, vol. 164, no. 1, January 12, 2004) that as little as thirty minutes of walking daily is enough exercise to prevent weight gain for most sedentary people, and that exertion above that may even cause weight and fat loss. The investigators randomly assigned 182 overweight, inactive adults aged forty to sixty-five to one of three programs of escalating exercise or to a control group that did not exercise for eight months. One group did the equivalent of twelve miles of walking per week, another completed the equivalent of twelve miles of jogging per week, and the most intense exercise group performed exercise comparable to jogging twenty miles per week. All the exercise was performed on treadmills, elliptical trainers, or stationary bicycles in supervised settings. The subjects were encouraged not to change their diets during the study.

Subjects in the two low-level exercise groups lost weight and fat, and those in the most intense exercise program lost more weight and fat than the others. The vigorous exercise group had a 3.5% weight loss and the two low-dose groups had slightly more than a 1% weight loss, whereas the control group had a 1.1% weight gain. Exercise dose and intensity also determined changes in waist circumference—subjects who did not exercise had a 0.8% increase in waist circumference. The two groups doing the lower amounts of exercise had decreases of about 1.5%, and the most intensely exercising group reported a waist decline of 3.4%. Slentz et al. determine that a modest amount of exercise—thirty minutes per day—can prevent weight gain without changes in diet.

## Eating Disorders

MYTH. Eating disorders occur exclusively among middle- and upper-class white females.

FACT. Like many myths about diet, weight, and nutrition, this one is based on fact: an estimated 90% of people with anorexia nervosa or bulimia nervosa are female; however, according to Susan Z. Yanovski of the National Institutes of Health in "Eating Disorders, Race, and Mythology" (*Archives of Family Medicine*, vol. 9, no. 1, January 2000), binge-eating disorder occurs in both genders and across all socioeconomic classes. Yanovski attributes the myth that eating disorders are limited to middle- and upper-class white women to the fact that many studies were conducted on college campuses where few minority students were enrolled, and other research looked at people seeking treatment, often at referral centers. Yanovski observes that "studies done on such populations, which may be more likely to be white and of higher socioeconomic status, have limited generalizability." She also cites research that finds that minorities are substantially affected by eating disorders—one study found that African-American women were as likely as white women to report binge eating. Another revealed that the prevalence of binge eating was comparable among Hispanic, non-Hispanic white, and African-American women, but that binge-eating symptoms were more severe among the Hispanic group. Yanovski concludes that the "recognition that eating disorders are color-blind can ensure that appropriate recognition and treatment are available to all patients at risk."

Anna Keski-Rahkonen et al. indicate in "Epidemiology and Course of Anorexia Nervosa in the Community" (*American Journal of Psychiatry*, vol. 164, no. 8, August 2007) that there is a substantially higher lifetime prevalence of anorexia nervosa than reported in previous studies—as high as 270 cases per 100,000 among women between the ages of fifteen to nineteen. Keski-Rahkonen et al. also offer a hopeful finding: most young women recovered within five years and usually progressed to full recovery.

## WHY DIETS FAIL

Historically, diets have been considered to have "failed" when lost weight is regained. Many nutritionists and obesity researchers believe that diets fail because most are not sustainable. The more restrictive the diet, the less likely an individual will be to remain faithful to it because, in general, people cannot endure extended periods of hunger and deprivation. Another reason diets may fail is that they neglect to teach dieters new eating habits to assist them to maintain their weight loss. Most overweight people gained their excess weight by consuming more calories per day than they needed. Dieting creates a temporary deficit of calories or specific macronutrients such as carbohydrates or fat. Because the weight-loss diet is viewed as a temporary measure with a beginning and an end, at its conclusion most dieters return to their previous eating habits and often regain the lost weight or even more weight. Many nutritionists and dieticians who work with people who are overweight or obese assert that diets do not fail; instead, dieters fail to learn how to eat properly to prevent weight regain.

Consumers are not the only ones who believe that diets are doomed to failure; many health professionals and researchers cite the statistic that 95% of diets fail. This oft-cited statistic has been attributed to Albert Stunkard of the University of Pennsylvania and the director emeritus of the American Obesity Association. Stunkard put forth the 95% failure rate in an account of research he performed in 1959, which involved advising one hundred overweight patients to diet, with no follow-up or support to increase their adherence to the diet. In "Whether Obesity Should Be Treated?" (*Health Psychology*, vol. 12, no.5, September 1993), Kelly D. Brownell observes that this statistic has been widely applied even though it is quite dated, was not confirmed by subsequent studies, and involved only subjects in university-based research programs.

The article "New Diet Winners: We Rate the Diet Books and Plans" (*Consumer Reports*, June 2007) observes that only recently have successful dieters been studied to learn from their successes and incorporate them into more effective, and ideally sustainable, weight-loss plans. It cites as an example the new emphasis on achieving satiety without consuming too many calories by consuming low-density foods. This article may help dispel the myth that dieters are doomed to failure.

## Improving Long-Term Weight Loss

More recent research demonstrates that dieters find it challenging to maintain weight loss; however, it refutes the 95% failure rate. In "Successful Weight Loss Maintenance" (*Annual Review of Nutrition*, vol. 21, 2001), Rena R. Wing and James O. Hill propose defining "successful long-term weight loss maintenance as intentionally losing at least 10% of initial body weight and keeping it off for at least one year." Using this definition, the investigators offer more favorable outcomes of weight-loss efforts. Wing and Hill report that more than 20% of overweight or obese people can and do lose 10% or more of body weight and maintain the weight loss for more than a year. Analyzing data from the National Weight Control Registry, they also find that people who successfully maintained long-term weight loss—an average weight loss of 66.1 pounds for an average of 5.5 years—shared common behaviors that promoted weight loss and weight maintenance. These behavioral strategies included eating a diet low in fat, frequent self-monitoring of body weight and food intake, and high levels of regular physical activity. Wing and Hill also posit that weight-loss maintenance may become easier over time because they observe that once weight loss had been maintained for two to five years, the chances of longer-term success were greatly increased.

Even though Wing and Hill offer more optimistic estimates of successful weight loss and weight maintenance than what Stunkard reported, there is obviously considerable room for improvement. In "Long-Term Maintenance of Weight Loss: Current Status" (*Health Psychology*, vol. 19, no. 1, supplement, January 2000), Robert W. Jeffery et al. identify areas of investigation that might produce strategies to assist more people to control their weight effectively. The researchers assert that despite high rates of dieting and the possibility of long-term success in voluntary weight loss overall, successful weight losses are being offset by failures. Jeffery et al. speculate that the reason for this overall lack of success is that improvements in long-term weight loss have thus far lagged behind improvements in short-term weight loss.

Jeffery et al. describe the typical course of weight loss and regain among people participating in behavioral treatment for obesity as rapid initial weight loss that slows, with maximum weight loss achieved approximately six months after treatment began. Thereafter, weight regain begins and continues until weight stabilizes at or slightly below the starting weight. The investigators speculate that the behavior changes that are prescribed are sufficient for weight loss, and failure to maintain behavior changes may be due to loss of knowledge and skills, loss of motivation, or unpleasant side effects of behavior change such as hunger, psychological stress, or social pressure. Historically, researchers favor either a biological interpretation of the challenge of weight maintenance—the importance of biological determinants of body weight—or a behavioral explanation. Behavioral scientists interpret the weight loss–weight regain pattern as evidence of how difficult it is to achieve lasting change in given the environmental factors that influence behaviors.

Jeffery et al. classify efforts to improve long-term maintenance of weight loss as attempts to increase the intensity of initial treatment, extend the length of treatment, alter dietary and exercise prescriptions, enhance motivation, and teach maintenance-specific behavioral skills. An example of high-intensity obesity treatment is the use of very-low-calorie diets (VLCDs). VLCDs restrict food intake for periods of two to three months to six hundred to eight hundred calories per day, substantially lower than conventional low-calorie diets, which range from one thousand to twelve hundred calories per day. VLCDs consistently produce larger initial weight losses than conventional low-calorie diets. However, they have not proven successful in improving long-term weight loss. The larger, rapid weight losses generated by severe calorie restriction are followed by larger and more rapid regains, which offset the initial losses. Two or more years after treatment, people who were placed on VLCDs fared no better than those who lost weight using less intense regimens.

Treating obesity like chronic diseases such as diabetes and high blood pressure that require ongoing management appears to help; however, attendance at treatment sessions declines over time and is associated with weight regain. Efforts to modify dietary and exercise prescriptions have focused on emphasizing exercise instead of focusing solely on dietary changes. Even though some studies show that the addition of exercise improved short-term weight loss and weight loss at eighteen-month follow-up visits, exercise was found to slow but not prevent weight regain.

Nutrigenetics (using genetic information to custom-tailor a weight-loss diet) may help improve the success of weight-loss and weight-maintenance efforts. In "Improved Weight Management Using Genetic Information to Personalize a Calorie Controlled Diet" (*Nutrition Journal*, vol. 6, no.29, October 18, 2007), Ioannis Arkadianos et al. indicate that they offered nutrigenetic testing and developed individual diets to people who historically had failed to lose weight. They compare the results these dieters achieved to a control group that did not receive nutrigenetic screening or a personalized diet and find that subjects in the nutrigenetic group fared better in terms of adherence to their diets, weight loss and maintenance, and improvements in blood glucose levels.

Approaches to enhance motivation focus on two areas: improved social supports and tangible financial incentives. Strategies to improve social supports emphasize including spouses or significant others in the weight-loss process to teach them to provide social support for their partners' weight-loss efforts. Such strategies demonstrate modest success as do contracts in which groups agree to aim for individual or group weight loss.

In "A Pilot Study Testing the Effect of Different Levels of Financial Incentives on Weight Loss among Overweight Employees" (*Journal of Occupational and Environmental Medicine*, vol. 49, no. 9, September 2007), Eric A. Finkelstein et al. find that financial incentives may be effective inducements to lose weight. The researchers followed two hundred overweight workers in North Carolina, who were randomly assigned to one of three groups. One group received no incentives, whereas the other two groups received $7 or $14 for each percentage point of weight lost. For example, a two-hundred-pound subject in the group receiving $7 for each percentage point group who lost ten pounds, or 5% of his or her weight, received $35. Finkelstein et al. find that workers who received the most money and other incentives such as time off lost the most weight. At three months, subjects with no financial incentive lost 2 pounds, those in the $7 group lost approximately 3 pounds, and those in the $14 group lost 4.7 pounds.

Teaching patients skills that are useful for weight maintenance as opposed to weight loss emphasizes that there are two distinctly different sets of strategies: one set focuses on weight loss and the other on maintaining a stable energy balance around a lower weight. The most commonly used model for teaching maintenance-specific skills is relapse prevention, which involves teaching people to identify situations in which lapses in behavioral adherence are likely to occur, to plan strategies in advance to prevent lapses, and to get back on track should they occur. Relapse prevention is based on the idea that breaking the so-called rules in terms of remaining faithful to diet and exercise programs may often lead to negative psychological reactions that in turn prompt reversion to pre-weight-loss behaviors. To date, only one study—Bas Verplanken and Wendy Wood's "Interventions to Break and Create Consumer Habits" (*Journal of Public Policy and Marketing*, vol. 25, no. 1, spring 2006)—has examined the effectiveness of this approach. Verplanken and Wood hypothesize that learning and practicing a well-defined, positive response to relapses might help people sustain weight loss. However, their findings do not support this hypothesis.

Jeffery et al. acknowledge that weight management is a continuing source of fascination and frustration for researchers as well as for dieters. They recommend that research consider additional areas such as:

- Considering obesity as a chronic disorder requiring continuous care, with the aim of developing cost-effective methods for delivering care indefinitely.

- Examining psychological, behavioral, biological, and environmental factors that relate to weight loss, maintenance of weight loss, and weight regain to identify the key factors associated with successful long-term weight loss.

- Improving the assessment of energy intake and expenditure and of behavior patterns associated with change in energy intake and expenditure.

- Examining the role of behavioral preferences in obesity and its treatment in an effort to answer questions such as: Can behavioral preferences or reinforcement values be changed in ways that would facilitate long-term weight loss? Do they change spontaneously after behavior changes?

- Researching why long-term outcomes of behavior treatment for obesity in children and adolescents have been more successful than treatment for obesity in adults.

- Learning more about the role of physical activity and social support in relationship to long-term weight loss.

- Discovering safer and more effective medications to treat obesity and developing new ways to integrate medications into effective programs of weight control.

## WEIGHT-LOSS SCHEMES DEFRAUD CONSUMERS

There is a long history of marketing so-called miraculous, fat-burning pills, potions, and products to Americans seeking effortless weight loss. Peter N. Stearns, in *Fat History: Bodies and Beauty in the Modern West* (1997), and Laura Fraser, in *Losing It: False Hopes and Fat Profits in the Diet Industry* (1998), offer detailed histories of magical cures and weight-loss fads. At the turn of the twentieth century products such as obesity belts and chairs that delivered electrical stimulation, as well as corsets, tonics, and mineral waters, claimed to cause weight loss.

Diet pills arrived on the scene in 1910 with the introduction of weight-loss tablets that contained arsenic (a poisonous metallic element), strychnine (a plant toxin formerly used as a stimulant), caffeine, and pokeberries (formerly used as a laxative). In the 1920s cigarette makers promoted their product as a diet aid, urging Americans to smoke rather than eat. During the 1930s diet pills containing dinitrophenol, a chemical used to manufacture explosives, dyes, and insecticides, enjoyed brief popularity after it was observed that factory workers making munitions lost weight. Their popularity was short-lived, as cases of temporary blindness and death were attributed to their use.

The second half of the twentieth century saw the proliferation of questionable, and often entirely worthless, weight-loss devices and gimmicks, including inflatable suits to "sweat off pounds," diet drinks and cookies, and slimming creams, patches, shoe inserts, and wraps to reduce fat thighs and abdomens. Even though the claims made for many of these products sounded too good to be true, unsuspecting Americans spent billions of dollars in the hope of achieving quick, easy, and permanent weight loss.

### Weighing the Claims

In May 2000 the Partnership for Healthy Weight Management, a coalition of scientific, academic, health-care, government, commercial, and public interest representatives, initiated consumer and media education programs that not only aimed to increase public awareness of the obesity epidemic in the United States but also to promote responsible marketing of weight-loss products and programs. The partnership also published the consumer guide *Finding a Weight Loss Plan That Works for You* (2005, http://www.ftc.gov/bcp/edu/pubs/consumer/health/hea05 .pdf), which was designed to help overweight or obese consumers find weight-loss solutions to meet their needs. The guide contains a checklist that enables consumers to compare weight-loss plans based on a variety of criteria. (See Table 9.1.) It also advises consumers about how to select weight-loss programs and services based on specific information from potential providers. The coalition

also launched the Ad Nauseam (2006, http://www .consumer.gov/weightloss/adnauseum.pdf) campaign to encourage the media to demand proof before accepting advertising copy that contains unbelievable, dubious, or extravagant promises of weight-loss success.

In *Weight-Loss Advertising: An Analysis of Current Trends* (September 2002, http://www.ftc.gov/bcp/reports/ weightloss.pdf), Richard L. Cleland et al. of the FTC report that as much as 55% of advertising for weight-loss products and services contains false or unsupported effectiveness claims. Nearly 40% of the three hundred advertisements reviewed by Cleland et al. made at least one assertion that was most likely false, and an additional 15% made at least one representation that was very likely false, or in the best cases, lacked adequate substantiation. Table 9.2 shows the frequency of claims made by various types of weight-loss products and services. Cleland et al. also observe that despite an unprecedented law enforcement effort in the decade preceding their study, the incidence of false and deceptive weight-loss advertising claims appeared to have increased.

On November 19, 2002, the FTC convened a workshop attended by researchers, scholars, media experts, and medical professionals from the government, academia, and private industry that aimed to evaluate claims and develop new and more effective ways to combat false and deceitful weight-loss advertising claims. The FTC summarized the workshop proceedings, including attendees' assessments of eight broad categories of advertising claims, in *Deception in Weight-Loss Advertising Workshop: Seizing Opportunities and Building Partnerships to Stop Weight-Loss Fraud* (December 2003, http://www .ftc.gov/os/2003/12/031209weightlossrpt.pdf). The following section considers the advertising claims and summarizes the attendees' assessments of these claims. It also draws on an analysis of the FTC report by Stephen Barrett in "Impossible Weight-Loss Claims: Summary of an FTC Report" (December 16, 2003, http://www.quackwatch.org/ 01QuackeryRelatedTopics/PhonyAds/weightlossfraud .html).

### No Diet or Exercise Required

CLAIM. The advertised product causes substantial weight loss without exercise or diet.

EXAMPLES. "U.S. patent reveals weight loss of as much as 28 pounds in 4 weeks.... Eat all your favorite foods and still lose weight. The pill does all the work," and "Lose up to 2 pounds daily without diet or exercise." Table 9.3 contains other examples of comparable claims.

ASSESSMENT. The consensus was that products purporting to cause weight loss without diet or exercise would either need to cause malabsorption (impair the absorption) of calories or to increase metabolism. Because the number of calories that can be malabsorbed is limited to twelve

**TABLE 9.1**

## Checklist for evaluating weight loss products and services

Use this checklist to gather and compare information from all weight loss programs you're considering.

Make several copies of the blank form so you can fill out one for each program. A provider's willingness to give you this information is an important factor in choosing a program. If you need help to evaluate the information you gather, talk with your primary health care provider or a registered dietitian.

Program name _____

Address _____

Phone number _____

In this program, my daily caloric intake will be: _____

My daily caloric intake is determined by: _____

I ☐ will ☐ will not be evaluated initially by program staff.

The evaluation will be made by (check all that apply):
☐ Physician          ☐ Nurse          ☐ Registered dietitian          ☐ Other company-trained employee

My progress is supervised by (check all that apply):
☐ Physician          ☐ Nurse          ☐ Licensed psychologist
☐ Registered dietitian          ☐ Company-trained employee

I ☐ will          ☐ will not be evaluated by a physician during the course of my treatment.

During the first month, my progress will be monitored:
☐ Weekly          ☐ Biweekly          ☐ Monthly          ☐ Other _____

After the first month, my progress will be monitored:
☐ Weekly          ☐ Biweekly          ☐ Monthly          ☐ Other _____

My weight loss plan includes (check all that apply):
☐ Nutrition information about healthy eating          ☐ At least 1,200 calories/day for women or 1,400 calories/day for men
☐ Suggested menus and recipes          ☐ Keeping food diaries or other monitoring activities
☐ Portion control          ☐ Liquid meal replacements
☐ Prepackaged meals          ☐ Dietary supplements (vitamins, minerals, botanicals, herbals)
☐ Prescription weight loss drugs          ☐ Help with weight maintenance and lifestyle changes
☐ Surgery

My plan includes regular physical activity that is (check both if both apply):
☐ Supervised (at the program site)          _____ times per week, _____ minutes per session.
☐ Unsupervised (on my own time)          _____ times per week, _____ minutes per session.

The physical activity includes (check all that apply):
☐ Walking          ☐ Swimming          ☐ Stationary cycling
☐ Strength training          ☐ Aerobic dancing          ☐ Other _____

The weight loss plan includes (check all that apply):
☐ Family counseling          ☐ Group support          ☐ Lifestyle modification advice
☐ Weight maintenance advice          ☐ Weight maintenance counseling
The staff explained the risks associated with this weight loss progam. They are:

_____
_____

The staff explained the costs of this program. (Check all that apply and fill in the blanks.)
☐ I will be charged a one-time entry fee of $ ___.
☐ I will be charged $ ___ per visit.
☐ Food replacements will cost about $ ___ per month.
☐ Prescription weight loss drugs will cost about $ ___ per month.
☐ Vitamins and other dietary supplements will cost about $ ___ per month.
☐ Diagnostic tests are required and will cost about $ ___ .
☐ Other costs include _____ at $ ___.

**Total cost for this program $____**

The program gave me information about:
☐ The health risks of being overweight.          ☐ The difficulty many people have maintaining weight loss.
☐ The health benefits of weight loss.          ☐ How to improve my chances at maintaining my weight.

Other information to ask for:
Participants in this program have lost an average of ___ lbs. over ___ months/years.
Participants in this program have kept off ___ % of their weight loss for ___ years.

This information is based on the following (check one):
☐ All participants.
☐ Participants who completed the program.
☐ Other _____

Notes: _____
_____

SOURCE: "Checklist for Evaluating Weight Loss Products and Services," in *Finding a Weight Loss Program that Works for You*, Federal Trade Commission, The Partnership for Healthy Weight Management, 2005, http://www.ftc.gov/bcp/edu/pubs/consumer/health/hea05.pdf (accessed November 12, 2007)

hundred to thirteen hundred calories per week, or about one-third of a pound per week, malabsorption alone is unlikely to lead to substantial weight loss. Similarly, there is no thermogenic (heat producing) agent, such as ephe-drine combined with caffeine, able to boost metabolism enough to produce weight loss without diet or exercise. In fact, the mechanism by which ephedrine products appear to assist weight loss is by suppressing appetite rather than

TABLE 9.2

**Frequency of claims by product category**

| | Dietary supplements (157) | Hypnosis (27) | Meal replacement (33) | Food (15) | Plans/programs/ diet centers (21) | Wraps (10) | Transdermal products (11) | Other (24) |
|---|---|---|---|---|---|---|---|---|
| Testimonials | 59% | 96% | 70% | 80% | 76% | 50% | 45% | 63% |
| Fast results | 60% | 59% | 58% | 33% | 43% | 90% | 73% | 42% |
| Guaranteed results | 59% | 93% | 36% | 7% | 24% | 60% | 55% | 42% |
| Natural | 56% | 11% | 42% | 47% | 14% | 50% | 27% | 33% |
| No diet or exercise | 55% | 56% | 12% | 20% | 14% | 20% | 27% | 33% |
| Long-term/permanent | 38% | 100% | 18% | 27% | 33% | 60% | 18% | 50% |
| Safe/no side effects | 55% | 30% | 27% | 20% | 24% | 50% | 36% | 25% |
| Before-and-after | 33% | 85% | 36% | 60% | 76% | 40% | 36% | 25% |
| Clinically proven | 53% | 0 | 36% | 7% | 10% | 40% | 82% | 33% |
| No more failure | 32% | 89% | 27% | 20% | 24% | 20% | 9% | 38% |
| Medical approval | 34% | 7% | 12% | 7% | 10% | 0 | 36% | 38% |
| Excessive wt. loss warning | 12% | 4% | 0 | 0 | 5% | 0 | 9% | 0 |

SOURCE: Richard L. Cleland et al., "Table 2. Frequency of Claims by Product Category," in *Weight Loss Advertising: An Analysis of Current Trends*, Federal Trade Commission, September 2002, http://www.ftc.gov/bcp/reports/weightloss.pdf (accessed November 12, 2007)

TABLE 9.3

**Examples of claims that promise weight loss without diet or exercise**

"Awesome attack on bulging fatty deposits … has virtually eliminated the need to diet." (Konjac root pill)

"They said it was impossible, but tests prove [that] my astounding diet-free discovery melts away. . . 5, 6, even 7 pounds of fat a day." (ingredients not disclosed)

"The most powerful diet pill ever discovered! No diet or workout required. The secret weight-loss pill behind Fitness models, Show Biz and Entertainment professionals! No prescription required to order." (ingredients not disclosed)

"Lose up to 30 lbs . . . No impossible exercise! No missed meals! No boring foods or small portions!" (plant extract fucus vesiculosus)

"Lose up to 8 to 10 pounds per week … [n]o dieting, no strenuous exercise." (elixir purportedly containing 16 plant extracts)

"My 52 lbs of unwanted fat relaxed away without dieting or grueling exercise." (hypnosis seminar)

"No exercise… [a]nd eat as much as you want—the more you eat, the more you lose, we'll show you how." (meal replacement)

SOURCE: Richard L. Cleland et al., "Table 5. Lose Weight without Diet or Exercise Claims," in *Weight Loss Advertising: An Analysis of Current Trends*, Federal Trade Commission, September 2002, http://www.ftc.gov/bcp/reports/weightloss.pdf (accessed November 12, 2007)

speeding metabolism. Furthermore, even though green tea extract was found to increase metabolism, it was by a scant 4%.

## No Restrictions on Eating

**CLAIM.** Users can lose weight while still enjoying unlimited amounts of high-calorie foods.

**EXAMPLE.** "Eat All the Foods You Love and Still Lose Weight (Pill Does All the Work)."

**ASSESSMENT.** This claim was viewed as a variation of the assertion that dieters can lose weight without reducing caloric intake or increasing exercise, because this claim states that users not only can lose weight without reducing

caloric intake but also may increase caloric intake and still lose weight. The assembled experts concurred that if this claim was true, it would defy the laws of physics.

## Permanent Weight Loss

**CLAIM.** The advertised product causes permanent weight loss.

**EXAMPLES.** "Take it off and keep it off. You won't gain the weight back afterwards because your weight will have reached an equilibrium," and "People who use this product say that even when they stop using the product, their weight does not jump up again."

**ASSESSMENT.** Even if a product caused weight loss through a reduction of calories, appetite suppression, or malabsorption, weight would be regained once use of the product stopped and calorie consumption returned to previous levels. Researchers and health professionals have repeatedly observed that dieters tend to regain weight lost over time once the diet, intervention, or other treatment ends. According to the National Academy of Science, Food, and Nutrition Board, "Many programs and services exist to help individuals achieve weight control. But the limited studies paint a grim picture: those who complete weight-loss programs lose approximately 10 percent of their body weight only to regain two-thirds of it back within 1 year and almost all of it back within 5 years." Furthermore, there are no published scientific studies supporting the claim that a nonprescription drug, dietary supplement, cream, wrap, device, or patch can cause permanent weight loss.

## Fat Blockers

**CLAIM.** The advertised product causes substantial weight loss through the blockage or absorption of fat or calories.

**EXAMPLES.** "[The named ingredient] can ingest up to 900 times its own weight in fat, that's why it's a fantastic fat blocker," and "The Super Fat Fighting Formula inhibits fats, sugars and starches from being absorbed in the intestines and turning into excess weight, so that you can lose pounds and inches easily."

**ASSESSMENT.** Science does not support the possibility that sufficient malabsorption of fat or calories can occur to cause substantial weight loss. To lose even one pound per week requires malabsorption of about five hundred calories per day or about fifty-five grams of fat. To lose two pounds per day, as promised in some advertisements, would require the malabsorption of seven thousand calories per day, which is impossible given that it is several times the total calories that most people consume daily, let alone the number of calories consumed from fat. The FTC has challenged deceptive fat-blocker claims for some of the most popular diet products on the market. The evidence supports the position that consumers cannot lose substantial weight through the blockage of the absorption of fat. It is not scientifically feasible for a nonprescription drug, dietary supplement, cream, wrap, device, or patch to cause substantial weight loss through the blockage of absorption of fat or calories.

## Quick Weight Loss

**CLAIM.** The user of the advertised product can safely lose more than three pounds a week for time periods exceeding four weeks. Table 9.4 shows claims that promise unbelievably rapid results.

**EXAMPLES.** "Lose three pounds per week, naturally and without side effects."

**ASSESSMENT.** Significant health risks are associated with medically unsupervised, rapid weight loss over extended periods of time. In general, "the more restrictive the diet, the greater are the risks of adverse effects associated with weight loss." One of the best documented risks is the increased incidence of gallstones. The claim that consumers using products such as these can safely lose more than three pounds per week for a period of more than four weeks is not scientifically feasible.

## Weight-Loss Creams and Patches

**CLAIM.** The advertised product that is worn on the body or rubbed into the skin causes substantial weight loss.

**EXAMPLES.** "Lose two to four pounds daily with the Diet Patch," and "Thigh Cream drops pounds and inches from your thighs."

**ASSESSMENT.** Diet patches and creams that are worn or applied to the skin have not been proven to be safe or effective. Furthermore, their alleged mechanisms of action are not scientifically credible.

**TABLE 9.4**

### Examples of claims that promise fast results

"This combination of plant extracts constitutes a weight-loss plan that facilitates what is probably the fastest weight loss ever observed from an entirely natural treatment." (elixir purportedly containing 16 plant extracts)

"Just fast and easy, effective weight loss!" (fucus vesiculosus)

"Lose 10 lbs. in 8 Days!" (apple cider vinegar)

"Rapid weight loss in 28 days!" (ephedra)

"Knock off your unwanted weight and fat deposits at warp speeds! You can lose 18 pounds in one week!" (ingredients not disclosed)

"Clinically proven to cause rapid loss of excess body fat." (phosphosterine)

"Two clinically proven fat burning formulations that are guaranteed to get you there fast or it costs you absolutely nothing." (ingredients not disclosed)

SOURCE: Richard L. Cleland et al., "Table 4. Representative Claims that Promise Fast Results," in *Weight Loss Advertising: An Analysis of Current Trends*, Federal Trade Commission, September 2002, http://www.ftc.gov/bcp/reports/weightloss.pdf (accessed November 12, 2007)

## Guaranteed Success

**CLAIM.** The advertised product causes substantial weight loss for all users.

**EXAMPLES.** "Lose excess body fat. No willpower required. Works for everyone no matter how many times you've tried and failed before."

**ASSESSMENT.** This claim assumes that overweight and obesity arise from a single cause or are amenable to a single solution. Because the causes of overweight and obesity are thought to be genetic factors and environmental conditions, and contributing factors such as diet, metabolic rate, level of physical activity, and adherence to weight-loss treatment vary, it is unlikely that one product would be effective for all users. Even U.S. Food and Drug Administration–approved, prescription drugs for weight loss have a high level of nonresponders, and surgical treatment for obesity is not successful 100% of the time. The claim that a nonprescription drug, dietary supplement, cream, wrap, device, or patch will cause substantial weight loss for all users is not scientifically feasible.

## Targeted Weight-Loss Products

**CLAIM.** Users of the advertised product can lose weight from only those parts of the body where they wish to lose weight.

**EXAMPLES.** Testimonial advertising included claims such as "And it has taken off quite some inches from my butt (5 inches) and thighs (4 inches), my hips now measure 35 inches. I still wear the same bra size though. The fat has disappeared from exactly the right places."

**ASSESSMENT.** Small published studies of aminophylline cream indicate that its use may cause the redistribution of fat from the thighs to other fat stores; however, it has not been shown to cause fat loss. Even if some products were capable of causing more weight loss from certain areas of the body, no part would be spared completely—fat is lost from all fat stores throughout the body.

### Red Flag Campaign and Big Fat Lie Initiative Target Phony Weight-Loss Claims

Another outcome of the November 2002 workshop was the design of an education initiative to assist the media to voluntarily screen weight-loss product ads containing claims that are "too good to be true." The media were targeted for intensive education not only because broad-based public education has proven largely inadequate to protect consumers from persuasive messages trumpeting easy weight loss but also to acknowledge the media's powerful ability to reduce weight-loss fraud by sharply reducing the dissemination of obviously false weight-loss advertising. On December 9, 2003, the FTC launched its Red Flag campaign to more effectively assist the media to reduce deceptive weight-loss advertising and promote positive, reliable advertising messages about weight loss.

In April 2004 the FTC filed claims against seven companies for making false weight-loss claims, and in November 2004 the FTC announced six new cases against advertisers using bogus weight-loss claims. In each of these cases, the FTC sought to stop the bogus ads and to secure reparation for consumers. The FTC also launched in November 2004 Operation Big Fat Lie, a nationwide law enforcement action against the six companies making false weight-loss claims in national advertisements. Operation Big Fat Lie aims to stop deceptive advertising and provide refunds to consumers harmed by unscrupulous weight-loss advertisers; encourage the media not to carry advertisements containing bogus weight-loss claims; and educate consumers to be wary of companies promising miraculous weight loss without diet or exercise. The FTC also launched a Web site (http://wemarket4u.net/fatfoe/) to help consumers identify false weight-loss claims.

## DO VLCDs INCREASE LONGEVITY?

Even though most Americans are overweight, some people are experimenting with extremely low-calorie diets in the hope that by remaining extremely thin they will stave off disease and live longer. Advocates of extreme caloric restriction (CR) contend that sharply reducing caloric intake creates biochemical changes that slow the aging process, which theoretically should increase life expectancy.

Most people would find it impossible to adhere to semistarvation diets, but there is sound scientific evidence—such as Luigi Fontana and Samuel Klein's "Aging, Adiposity, and Calorie Restriction" (*Journal of the American Medical Association*, vol. 297, no. 9, March 7, 2007) and Arthur V. Everitt and David G. Le Couteur's "Life Extension by Calorie Restriction in Humans" (*Annals of the New York Academy of Sciences*, vol. 1114, no. 1, October 2007)—that subsistence diets increase the life span of fruit flies, worms, spiders, guppies, mice, and hamsters by between 10% and 40%. In theory, semistarvation prolongs life by reducing metabolism—how quickly glucose is used for energy—in an evolutionary adaptation to conserve calories during periods of famine. Dieters are familiar with this process—they know from experience that as they eat less, their metabolic rate drops, which makes losing weight increasingly more difficult. CR adherents experience comparable drops in metabolic rate—one study found that their body temperature dropped by a full degree. Proponents of CR assert that even though metabolism is vital for life, it is also destructive because it produces unstable molecules known as free radicals that can damage cells through a process called oxidation.

Animal studies find that CR inhibits the growth of cancerous tumors, possibly because at lower body temperatures the body may be better able to repair damaged deoxyribonucleic acid, which provides the genetic information necessary for the organization and functioning of most living cells and controls the inheritance of traits and characteristics. Animals on CR diets have reduced levels of blood sugar and insulin and greater insulin sensitivity, all of which reduces their risk for diabetes and cardiovascular disease. There is even evidence that CR boosts brain function. Mice with the tendency to develop neurological conditions such as Alzheimer's or Parkinson's disease developed these conditions later and more slowly when they were placed on CR diets, and rodents on CR diets displayed better memory and learning than those on normal diets. There is also evidence that CR influences patterns of gene expression. As animals age, certain genes tend to turn off and become inactive, whereas others are activated. In "Genomic Profiling of Short- and Long-Term Caloric Restriction Effects in the Liver of Aging Mice" (*Proceedings of the National Academy of Sciences*, vol. 98, no. 19, September 11, 2001), Shelley X. Cao et al. indicate that CR prevents 70% of change in gene expression in mice.

In 2004 the National Institutes of Health began a seven-year study to explore the effect of CR on human metabolism. The study is exploring the benefits and risks associated with CR. CR adherents report immediate health benefits including increased mental acuity, reduced need for sleep, sharply reduced cholesterol and fasting blood sugar levels, weight loss, and reduced blood pressure. The regimen is clearly not easy, and even its staunchest advocates, such as members of the Caloric Restriction Society, concede that many people who practice CR experience constant hunger, obsessions with food, mood disorders such as irritability and depression, and lowered libido. CR can also cause people to feel cold, and even with adequate vitamin and mineral supplementation it can cause some people to suffer from osteoporosis (decreased bone mass) and hair loss.

In "Why Dietary Restriction Substantially Increases Longevity in Animal Models but Won't in Humans" (*Ageing Research Reviews*, vol. 4, no. 3, August 2005), John P. Phelan and Michael R. Rose challenge the notion that CR will increase longevity. The scientists conclude that severely restricting calories over decades may add a few years to a human life span, but will not enable humans to live to 125 years or older. The investigators developed a mathematical model based on the known effects of calorie intake and life span that showed that people who consume the most calories have a shorter life span, and that if people severely restrict their calories over their lifetime, their life span increases by between 3% and 7%—far less than the twenty-plus years some hoped could be achieved by drastic CR. Phelan and Rose opine that "longevity is not a trait that exists in isolation; it evolves as part of a complex life history, with a wide range of underpinning physiological mechanisms involving, among other things, chronic disease processes." They advise Americans to "try to maintain a healthy body weight, but don't deprive yourself of all pleasure. Moderation appears to be a more sensible solution."

Everitt and Le Couteur confirm that even though short-term CR does improve specific markers associated with longevity such as deep body temperature and plasma insulin levels, CR is unlikely to offer markedly increased longevity. The authors cite as evidence the Okinawans, the longest-lived people on earth, who consume 40% fewer calories than the average American and live just four years longer. Everitt and Le Couteur opine that "the effects of CR on human life extension are probably much smaller than those achieved by medical and public health interventions, which have extended life by about 30 years in developed countries in the 20th century, by greatly reducing deaths from infections, accidents, and cardiovascular disease."

# CHAPTER 10
# PREVENTING OVERWEIGHT AND OBESITY

*If and when the public chooses to use government power to offset the factors that promote obesity, we can do so. A day may come when we decide to limit advertising of unhealthy food, strengthen lifestyle teaching in schools, and create stronger financial incentives to adhere to lifestyle recommendations. The more eager we the people are to fight the obesogenic environment, the more responsive and effective our governments will become.*

—Michael Dansinger of the Tufts–New England Medical Center, "Beating Obesity Is Not Mission Impossible" (*Medscape General Medicine*, vol. 9, no. 4, 2007)

Many obesity researchers and health professionals believe that the most effective way to win the war on obesity is to intensify efforts to prevent overweight and obesity among children, adolescents, and adults. They assert that over time prevention is far more cost effective than the expenditures associated with weight-loss efforts and medical treatment of obesity-related diseases. They also observe that prevention is a preferable strategy because to date no universally effective long-term treatment consistently produces and maintains weight loss.

The *Surgeon General's Call to Action to Prevent and Decrease Overweight and Obesity, 2001* (2001, http://www.surgeongeneral.gov/topics/obesity/calltoaction/ CalltoAction.pdf) calls for the design and implementation of interventions to prevent and decrease overweight and obesity, both individually and collectively. It asserts that effective actions must occur at many levels and acknowledges that even though individual behavioral change is at the core of all strategies to reduce overweight and obesity, to be optimally effective, efforts must not be limited to individual behavioral change.

The *Call to Action* recommends actions to modify group influences by initiating prevention programs targeting families, communities, employers and workers, the health-care delivery system, and the media, as well as changes in public policy. Furthermore, the report calls for concerted efforts and predicts that actions to prevent and reduce overweight and obesity will fail unless changes are made at every level of American society. Characterizing these problems as societal rather than individual, the report observes that individual behavioral change is possible only in "a supportive environment with accessible and affordable healthy food choices and opportunities for regular physical activity." The report also warns that actions aimed exclusively at individual behavioral change that did not consider social, cultural, economic, and environmental influences would be counterproductive, serving only to reinforce negative stereotypes, bias, and stigmatization of people who are overweight or obese.

The *Call to Action* promises to abide by five overarching principles to guide its recommendations about how to prevent and decrease overweight and obesity:

- Promote the recognition of overweight and obesity as major public health problems.

- Assist Americans in balancing healthy eating with regular physical activity to achieve and maintain a healthy or healthier body weight.

- Identify effective and culturally appropriate interventions to prevent and treat overweight and obesity.

- Encourage environmental changes that help prevent overweight and obesity.

- Develop and enhance public-private partnerships to help implement this vision.

Many public health professionals believe that environmental change and policy interventions are the most promising strategies for generating and maintaining healthy nutrition and physical activity behaviors at a population level. Environmental interventions are those actions that modify availability of, access to, pricing of, or education about foods at the places where they are purchased. Policy interventions legislate, regulate, or, through formal or informal rules, serve to guide individual and collective behavior. Examples of environmental and policy initiatives that have met with success include:

- Increasing the availability of fruits and vegetables at school and workplace cafeterias, and the addition of fresh fruit to refrigerated vending machines.

- Replacing soft drinks in school vending machines with fruit juices and water.

- Instituting daily physical education requirements for students.

- Providing point-of-purchase nutrition information at restaurants and grocery stores to encourage healthy food choices.

- Allowing workers adequate break time and a location where nursing mothers can express milk so their babies can continue to accrue the health benefits of breastfeeding even after their mothers return to work.

## PREVENTION EFFORTS TARGET FAMILIES, COMMUNITIES, AND SCHOOLS

Public health education, communication, and other programs aimed at families and communities are identified as the cornerstone of prevention efforts. The *Call to Action* puts forth communication strategies and corresponding actions that may be taken to promote awareness about the effects of overweight on health and support healthy eating and physical activity. For example, the communication strategy of educating expectant parents and other community members about the protective effect of breastfeeding against the development of obesity was translated into the action of creating community environments that promote and support breastfeeding. (The Centers for Disease Control and Prevention notes in *Does Breastfeeding Reduce the Risk of Pediatric Overweight?* [July 2007, http://www.cdc.gov/nccdphp/dnpa/nutrition/pdf/breastfeeding_r2p.pdf] that children who are breastfed for nine months are 30% less likely to become overweight.) Similarly, the communication objective to heighten consumer awareness about reasonable food and beverage portion sizes was coupled with action to encourage the food industry to provide sensible food and beverage portion sizes.

Prevention efforts were not only directed to families and communities but also to policy makers, whose actions to establish social and environmental policy could support communities and families to be more physically active and consume healthier diets. Policy makers were exhorted to create more community-based obesity prevention and treatment programs for children and adults and provide demonstration grants to improve access to, and availability of, healthy affordable foods in inner cities. They were also advised to enact public policy to

create and maintain safe and accessible sidewalks, walking and bicycle paths, and stairs.

In the community, schools offer ideal settings and multiple opportunities for preventing overweight and obesity by educating children about, and engaging them in, healthy eating and physical activity. To reinforce their messages concerning the importance of school physical activity and nutrition programs, schools can ensure that breakfast and lunch programs meet nutrition standards and provide food options that are low in fat, calories, and added sugars. Other ways to enact this communication strategy include offering healthy snacks in vending machines and school stores and providing all students with quality daily physical education to cultivate the knowledge, attitudes, skills, behaviors, and confidence needed to be physically active for life.

### Population-Based Prevention Programs Target Racial and Ethnic Minority Groups

In "Population-Based Interventions Engaging Communities of Color in Healthy Eating and Active Living: A Review" (*Preventing Chronic Disease*, vol. 1, no. 1, January 2004), Antronette K. Yancey et al. review studies of population-based interventions targeting communities composed primarily of members of racial and ethnic minorities. They identify twenty-three interventions intended to promote healthy eating and active lifestyles aimed at African-American, Hispanic, Native American, Alaskan Native, Asian-American, Native Hawaiian, and other Pacific Islander populations that met specific study criteria and were implemented between 1972 and 2000.

Yancey et al. describe several initiatives instituted between the 1970s and early 1990s that produced modest but measurable improvement in diet. For example, one program promoted reducing cholesterol and saturated fat intake via targeted print and electronic media in three semirural northern California towns with substantial Latino populations. Another intervention sought to engage African-American residents of public housing communities in Birmingham, Alabama, in group exercise programs. A third program cultivated regional coalitions of community-based organizations to develop fitness promotion activities such as walking clubs, cooking demonstrations and classes, aerobic exercise classes, walking trails, and health fairs.

### Federally Funded National Nutrition Education

Together, the U.S. Department of Health and Human Services (HHS) and the U.S. Department of Agriculture (USDA) update *Nutrition and Your Health: Dietary Guidelines for Americans* every five years. First published in 1980, the guidelines serve as the basis for federal food and nutrition education programs. Historically, some public health professionals believed that the USDA food pyramid was flawed because its composition was unduly influenced by pressure from the food industry, whose members know that even subtle changes to the guidelines can affect a food manufacturer's sales. Furthermore, these public health professionals asserted that the guidelines should not be expected to represent objective scientific evidence because they are developed by the U.S. government agency responsible for agriculture, rather than for health.

In January 2004 members of the Dietary Guidelines Advisory Committee met to discuss the sixth version of the dietary guidelines, *Dietary Guidelines for Americans, 2005* (January 2005, http://www.health.gov/dietaryguidelines/dga2005/document/pdf/DGA2005.pdf). Among the issues the committee considered were a reassessment of the food pyramid, the components of a healthy American diet, and energy balance. In preparation for the meeting, the thirteen committee members reviewed recent scientific research, including the Institute of Medicine's *Dietary Reference Intakes for Energy, Carbohydrate, Fiber, Fat, Fatty Acids, Cholesterol, Protein, and Amino Acids* (2002) and the World Health Organization's (WHO) *Diet, Nutrition, and the Prevention of Chronic Diseases* (2003, http://whqlibdoc.who.int/trs/WHO_TRS_916.pdf).

The Institute of Medicine report asserts that to meet daily energy and nutritional needs while minimizing the risk for chronic disease, adults should get 45% to 65% of their calories from carbohydrates, 20% to 35% from fat, and 10% to 35% from protein. Earlier guidelines advised diets with 50% or more of carbohydrates and 30% or less of fat, with comparable protein-intake recommendations in previous and current guidelines. The guidelines for children are similar to those for adults, except that infants and younger children are advised a slightly higher proportion of fat—25% to 40% of their caloric intake. The report also emphasizes balancing diet with physical activity and recommends total daily calorie consumption for individuals based on height, weight, gender, and four different levels of physical activity. Its recommendation of an hour per day of physical activity was derived from studies of average daily energy expended by people who maintain a healthy weight.

The WHO report calls on a team of global experts to identify new recommendations for governments on diet and exercise to combat obesity and related chronic diseases. The report advises changing daily nutritional intake and increasing energy expenditure by:

- Reducing consumption of foods high in saturated fat and sugar

- Sharply reducing the amount of salt in the diet

- Increasing the amount of fresh fruit and vegetables in the diet

- Engaging in moderate-intensity physical activity for at least one hour per day

The WHO report specifically recommends limiting fat to between 15% and 30% of total daily intake and saturated fats to less than 10% of this total. It suggests that between 55% and 75% of daily intake should be carbohydrates but that added sugars (refined or simple sugars as opposed to those naturally occurring in fruit and complex carbohydrates) should be limited to 10% or less. Protein should make up 10% to 15% of calorie intake and salt should be restricted to less than five grams per day (about one teaspoon).

## New Food Pyramids Debuted in 2005

The new dietary guidelines offered a number of recommendations that made the food pyramid outdated. For example, the guidelines emphasized choosing complex carbohydrates over simple ones, such as by choosing bread and pasta made from whole-grain flour instead of white flour. They also stipulated the amount of saturated fat Americans should consume to keep saturated fat below 10% of their total calorie intake. Furthermore, they advised sharply limiting added sugars and choosing and preparing foods with little salt (sodium chloride) so that daily intake totaled less than twenty-three hundred milligrams (approximately one teaspoon of salt) of sodium. Table 10.1 shows the sodium content for selected foods. It is interesting to note that processed foods that do not necessarily taste salty, such as tomato soup, nevertheless contain significant amounts of added sodium.

In April 2005 the government replaced the single, one-size-fits-all triangular pyramid with twelve individually tailored food pyramids and a new guide, called MyPyramid (http://www.mypyramid.gov/), to help Americans improve their eating habits. Each food pyramid is intended to meet the varying nutritional needs of people based on their age and level of physical activity. Table 10.2 displays the

**TABLE 10.1**

**Range of sodium content for selected foods**

| Food group | Serving size | Range (mg) |
|---|---|---|
| Breads, all types | 1 oz | 95–210 |
| Frozen pizza, plain, cheese | 4 oz | 450–1200 |
| Frozen vegetables, all types | 1/2 c | 2–160 |
| Salad dressing, regular fat, all types | 2 Tbsp | 110–505 |
| Salsa | 2 Tbsp | 150–240 |
| Soup (tomato), reconstituted | 8 oz | 700–1260 |
| Tomato juice | 8 oz (~1 c) | 340–1040 |
| Potato chips* | 1 oz (28.4 g) | 120–180 |
| Tortilla chips* | 1 oz (28.4 g) | 105–160 |
| Pretzels* | 1 oz (28.4 g) | 290–560 |

*All snack foods are regular flavor, salted.
Note: None of the examples provided were labeled low-sodium products. Serving sizes were standardized to be comparable among brands within a food. Pizza and bread slices vary in size and weight across brands.

SOURCE: "Table 15. Range of Sodium Content for Selected Foods," in *Dietary Guidelines for Americans, 2005*, 6th ed., U.S. Department of Health and Human Services and U.S. Department of Agriculture, January 2005, http://www.health.gov/dietaryguidelines/dga2005/document/pdf/Chapter8.pdf (accessed November 12, 2007)

**TABLE 10.2**

**Estimated calorie requirements for each gender and age group at three levels of physical activity**

[Estimates are rounded to the nearest 200 calories]

| Gender | Age (years) | Activity level[a, b, c] | | |
| | | Sedentary[a] | Moderately active[b] | Active[c] |
|---|---|---|---|---|
| Child | 2–3 | 1,000 | 1,000–1,400[d] | 1,000–1,400[d] |
| Female | 4–8 | 1,200 | 1,400–1,600 | 1,400–1,800 |
| | 9–13 | 1,600 | 1,600–2,000 | 1,800–2,200 |
| | 14–18 | 1,800 | 2,000 | 2,400 |
| | 19–30 | 2,000 | 2,000–2,200 | 2,400 |
| | 31–50 | 1,800 | 2,000 | 2,200 |
| | 51+ | 1,600 | 1,800 | 2,000–2,200 |
| Male | 4–8 | 1,400 | 1,400–1,600 | 1,600–2,000 |
| | 9–13 | 1,800 | 1,800–2,200 | 2,000–2,600 |
| | 14–18 | 2,200 | 2,400–2,800 | 2,800–3,200 |
| | 19–30 | 2,400 | 2,600–2,800 | 3,000 |
| | 31–50 | 2,200 | 2,400–2,600 | 2,800–3,000 |
| | 51+ | 2,000 | 2,200–2,400 | 2,400–2,800 |

Note: These levels are based on estimated energy requirements (EER) from the Institute of Medicine Dietary Reference Intakes Macronutrients Report, 2002, calculated by gender, age, and activity level for reference-sized individuals. "Reference size," as determined by the Institute of Medicine, is based on median height and weight for ages up to age 18 years of age and median height and weight for that height to give a body mass index (BMI) of 21.5 for adult females and 22.5 for adult males.
[a]Sedentary means a lifestyle that includes only the light physical activity associated with typical day-to-day life.
[b]Moderately active means a lifestyle that includes physical activity equivalent to walking about 1.5 to 3 miles per day at 3 to 4 miles per hour, in addition to the light physical activity associated with typical day-to-day life.
[c]Active means a lifestyle that includes physical activity equivalent to walking more than 3 miles per day at 3 to 4 miles per hour, in addition to the light physical activity associated with typical day-to-day life.
[d]The calorie ranges shown are to accommodate needs of different ages within the group. For children and adolescents, more calories are needed at older ages. For adults, fewer calories are needed at older ages.

SOURCE: "Table 3. Estimated Calorie Requirements (in Kilocalories) for Each Gender and Age Group at Three Levels of Physical Activity," in *Dietary Guidelines for Americans, 2005*, 6th ed, U.S. Department of Health and Human Services and U.S. Department of Agriculture, January 2005, http://ww.health.gov/dietaryguidelines/dga2005/document/pdf/Chapter2.pdf (accessed November 12, 2007)

calorie requirements based on age, gender, and physical activity level calculated by the Institute of Medicine and included in the new dietary guidelines.

In contrast to the old food pyramid, which featured horizontal bands representing food groups, the new food pyramids contain rainbow-colored bands that run vertically from the tip of the pyramid to the base. The new pyramids also have staircases climbing up one side, exhorting chronically sedentary Americans to become more active. Figure 10.1 shows the cornerstones of the new pyramid food guidance system: physical activity, variety, proportionality, moderation, gradual improvement, and personalization. Furthermore, unlike the old food pyramid, the new food pyramids aim to help people control their portion sizes. The old food pyramid used serving sizes, whereas the new ones offer standardized measures such as cups and ounces.

The new pyramid food guidance system delivers basic messages about healthy eating and physical activity that are nearly universally applicable. Americans should:

- Eat at least three ounces of whole-grain cereals, rice, or pasta every day.

- Choose low-fat or fat-free milk, yogurt, and other dairy products.

- Choose food and beverages low in added sugars.

Mypyramid.gov offers consumers the opportunity to receive customized food plans based on age, gender, and activity level. Consumers may then print personalized posters, worksheets, and other information to help them start on their healthy eating plan. The revised pyramid food guidance system offers a wealth of detailed information about food intake including the number of calories people of various ages and activity levels should consume, the amounts of food to consume from different food groups, and portion size information for twelve different calorie levels, ranging from one thousand calories to thirty-two hundred calories per day. (See Table 10.3 and Table 10.4.)

The 2005 pyramid education plan includes sample menus such as the two-thousand-calorie food plan shown in Table 10.5 and Table 10.6. It also provides a food tracking chart that helps consumers monitor their intake and assess the quality of their food choices in terms of meeting their nutritional needs within the allotted calorie level.

Fans of the revised dietary guidelines and pyramids assert that even if the new pyramids and online educational resources fail to motivate Americans to change their diet and increase their activity level, the guidelines will still have a salutary effect in terms of their influence on the food industry. For example, they suggest food manufacturers will be prompted to eliminate trans fats from their products and add whole-grain products in response to the guidelines.

**New Pyramid for Children**

In September 2005 the USDA released MyPyramid for Kids (http://www.mypyramid.gov/kids/index.html), a new food pyramid for children aged six to eleven that replaced the 1999 version. Like the adult pyramid, the children's pyramid displays each of the six major food groups using familiar images such as apples for fruits and bread for grains to encourage children to choose healthy foods rather than burgers and fries. The children's version of the pyramid features a girl running up the steps to the top and kids playing soccer, baseball, and basketball, walking a dog, riding a bike, stretching, picnicking, and doing yoga. (See Figure 10.2.)

Children are encouraged to play hard and be more physically active to meet the government's recommended sixty minutes of exercise per day. The USDA suggests that children set up home gyms, substituting items such

**FIGURE 10.1**

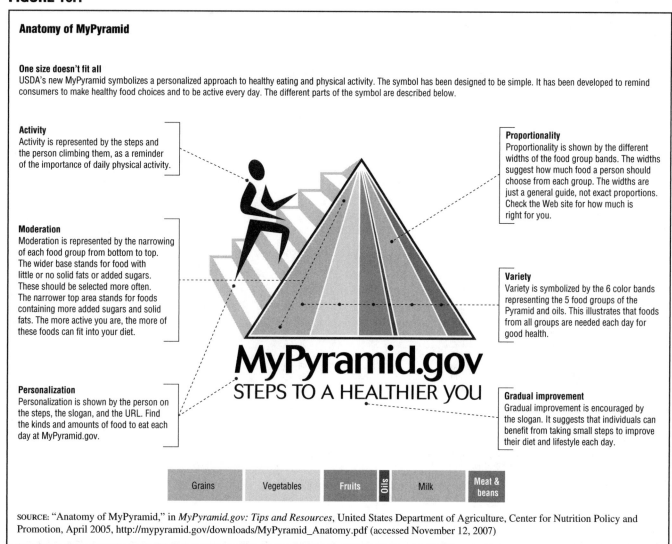

**Anatomy of MyPyramid**

**One size doesn't fit all**
USDA's new MyPyramid symbolizes a personalized approach to healthy eating and physical activity. The symbol has been designed to be simple. It has been developed to remind consumers to make healthy food choices and to be active every day. The different parts of the symbol are described below.

**Activity**
Activity is represented by the steps and the person climbing them, as a reminder of the importance of daily physical activity.

**Moderation**
Moderation is represented by the narrowing of each food group from bottom to top. The wider base stands for food with little or no solid fats or added sugars. These should be selected more often. The narrower top area stands for foods containing more added sugars and solid fats. The more active you are, the more of these foods can fit into your diet.

**Personalization**
Personalization is shown by the person on the steps, the slogan, and the URL. Find the kinds and amounts of food to eat each day at MyPyramid.gov.

**Proportionality**
Proportionality is shown by the different widths of the food group bands. The widths suggest how much food a person should choose from each group. The widths are just a general guide, not exact proportions. Check the Web site for how much is right for you.

**Variety**
Variety is symbolized by the 6 color bands representing the 5 food groups of the Pyramid and oils. This illustrates that foods from all groups are needed each day for good health.

**Gradual improvement**
Gradual improvement is encouraged by the slogan. It suggests that individuals can benefit from taking small steps to improve their diet and lifestyle each day.

**MyPyramid.gov**
**STEPS TO A HEALTHIER YOU**

Grains | Vegetables | Fruits | Oils | Milk | Meat & beans

SOURCE: "Anatomy of MyPyramid," in *MyPyramid.gov: Tips and Resources*, United States Department of Agriculture, Center for Nutrition Policy and Promotion, April 2005, http://mypyramid.gov/downloads/MyPyramid_Anatomy.pdf (accessed November 12, 2007)

as soup cans and stairs for weights and stair machines. Just as it does for adults, the USDA Web site provides a worksheet that allows kids to record and track their food consumption and physical activity.

To attract, inform, and entertain children, the USDA Web site features an interactive computer spaceship game. Players who balance food and exercise properly can blast off in an electronic spaceship to Planet Power. Choosing too many foods high in fat and sugar will cause the ship to sputter on the launch pad and release black smoke.

The MyPyramid for Kids program also employs an array of teaching materials for parents, childcare providers, and educators, including tip sheets and posters, lesson plans, CDs, Go Fish game cards, coloring books, and songs. The USDA Web site "Team Nutrition" (http://www.fns.usda.gov/tn/) offers school success stories and provides initiatives ranging from planting a school garden and organizing a school health fair to nutrition education programs, cooking classes, and poster contests.

Like its adult counterpart, MyPyramid for Kids garnered praise and criticism. According to Libby Quaid, in "Government Unveils Food Pyramid for Kids" (Associated Press, September 29, 2005), Michael Jacobson, the executive director of the Center for Science in the Public Interest, said in a news release that the guidelines did not go far enough: "If the government really wanted to improve kids' eating habits, it would get junk food out of schools, it would ban junk food advertising on television, it would require calorie counts on fast-food menu boards and sponsor hard-hitting educational materials." Other consumer groups said they believed federal funds would be better spent on a mass media campaign to promote eating fruits and vegetables.

**5 a Day for Better Health Program**

The 5 a Day for Better Health program is the nation's largest public-private nutrition education initiative. The

**TABLE 10.3**

**MyPyramid food intake pattern calorie levels**

| Activity level | Males | | | Activity level | Females | | |
|---|---|---|---|---|---|---|---|
| | Sedentary* | Mod. active* | Active* | | Sedentary* | Mod. active* | Active* |
| Age | | | | Age | | | |
| 2 | 1,000 | 1,000 | 1,000 | 2 | 1,000 | 1,000 | 1,000 |
| 3 | 1,000 | 1,400 | 1,400 | 3 | 1,000 | 1,200 | 1,400 |
| 4 | 1,200 | 1,400 | 1,600 | 4 | 1,200 | 1,400 | 1,400 |
| 5 | 1,200 | 1,400 | 1,600 | 5 | 1,200 | 1,400 | 1,600 |
| 6 | 1,400 | 1,600 | 1,800 | 6 | 1,200 | 1,400 | 1,600 |
| 7 | 1,400 | 1,600 | 1,800 | 7 | 1,200 | 1,600 | 1,800 |
| 8 | 1,400 | 1,600 | 2,000 | 8 | 1,400 | 1,600 | 1,800 |
| 9 | 1,600 | 1,800 | 2,000 | 9 | 1,400 | 1,600 | 1,800 |
| 10 | 1,600 | 1,800 | 2,200 | 10 | 1,400 | 1,800 | 2,000 |
| 11 | 1,800 | 2,000 | 2,200 | 11 | 1,600 | 1,800 | 2,000 |
| 12 | 1,800 | 2,200 | 2,400 | 12 | 1,600 | 2,000 | 2,200 |
| 13 | 2,000 | 2,200 | 2,600 | 13 | 1,600 | 2,000 | 2,200 |
| 14 | 2,000 | 2,400 | 2,800 | 14 | 1,800 | 2,000 | 2,400 |
| 15 | 2,200 | 2,600 | 3,000 | 15 | 1,800 | 2,000 | 2,400 |
| 16 | 2,400 | 2,800 | 3,200 | 16 | 1,800 | 2,000 | 2,400 |
| 17 | 2,400 | 2,800 | 3,200 | 17 | 1,800 | 2,000 | 2,400 |
| 18 | 2,400 | 2,800 | 3,200 | 18 | 1,800 | 2,000 | 2,400 |
| 19–20 | 2,600 | 2,800 | 3,000 | 19–20 | 2,000 | 2,200 | 2,400 |
| 21–25 | 2,400 | 2,800 | 3,000 | 21–25 | 2,000 | 2,200 | 2,400 |
| 26–30 | 2,400 | 2,600 | 3,000 | 26–30 | 1,800 | 2,000 | 2,400 |
| 31–35 | 2,400 | 2,600 | 3,000 | 31–35 | 1,800 | 2,000 | 2,200 |
| 36–40 | 2,400 | 2,600 | 2,800 | 36–40 | 1,800 | 2,000 | 2,200 |
| 41–45 | 2,200 | 2,600 | 2,800 | 41–45 | 1,800 | 2,000 | 2,200 |
| 46–50 | 2,200 | 2,400 | 2,800 | 46–50 | 1,800 | 2,000 | 2,200 |
| 51–55 | 2,200 | 2,400 | 2,800 | 51–55 | 1,600 | 1,800 | 2,200 |
| 56–60 | 2,200 | 2,400 | 2,600 | 56–60 | 1,600 | 1,800 | 2,200 |
| 61–65 | 2,000 | 2,400 | 2,600 | 61–65 | 1,600 | 1,800 | 2,000 |
| 66–70 | 2,000 | 2,200 | 2,600 | 66–70 | 1,600 | 1,800 | 2,000 |
| 71–75 | 2,000 | 2,200 | 2,600 | 71–75 | 1,600 | 1,800 | 2,000 |
| 76 and up | 2,000 | 2,200 | 2,400 | 76 and up | 1,600 | 1,800 | 2,000 |

Note: Calorie levels are provided for each year of childhood, from 2–18 years, and for adults in 5-year increments.
*Calorie levels are based on the estimated energy requirements (EER) and activity levels from the Institute of Medicine Dietary Reference Intakes Macronutrients Report, 2002.
Sedentary=less than 30 minutes a day of moderate physical activity in addition to daily activities.
Mod. active=at least 30 minutes up to 60 minutes a day of moderate physical activity in addition to daily activities.
Active=60 or more minutes a day of moderate physical activity in addition to daily activities.

SOURCE: "MyPyramid Food Intake Pattern Calorie Levels," in *MyPyramid.gov: Tips and Resources*, United States Department of Agriculture, Center for Nutrition Policy and Promotion, April 2005, http://mypyramid.gov/downloads/MyPyramid_Calorie_Levels.pdf (accessed November 12, 2007)

program originated in the California Department of Health Services in 1988 and is jointly sponsored by the National Cancer Institute and the Produce for Better Health Foundation (PBH), a nonprofit consumer-education foundation representing the fruit and vegetable industry. In 2001 the national 5 a Day partnership expanded to include other voluntary health organizations and produce associations. Besides the National Cancer Institute and the PBH, the partnership now includes representatives from the USDA, the Centers for Disease Control and Prevention (CDC), the American Cancer Society, the Produce Marketing Association, the United Fresh Fruit and Vegetable Association, the National Alliance for Nutrition and Activity, and the Association of State and Territorial Directors of Health Promotion and Public Health Education.

The 5 a Day for Better Health program (January 14, 2007, http://www.keepkidshealthy.com/nutrition/5_a_day .html) aims to increase fruit and vegetable consumption.

Its objectives are "to increase public awareness of the importance of eating five or more servings of fruits and vegetables every day for better health; and to provide consumers with specific information about how to include more servings of fruits and vegetables into daily eating patterns." Data from the CDC reveal that in 2005 less than one-quarter (23%) of Americans ate the recommended five servings per day of fruit and vegetables. (See Figure 10.3.) Biing-Hwan Lin, Jane Reed, and Gary Lucier of the Economic Research Service reveal in *U.S. Fruit and Vegetable Consumption: Who, What, Where, and How Much* (October 2004, http://www.ers.usda.gov/publications/ aib792/aib792-2/aib792-2.pdf) that fruit and vegetable consumption vary by geography, age, race, and ethnicity as well as by whether the food is eaten at home or away from home. The researchers find that:

• Seniors eat fewer French fries and potato chips than toddlers; however, seniors eat more fresh and canned potatoes than younger consumers.

**TABLE 10.4**

**MyPyramid food intake patterns**

| Calorie level[a] | 1,000 | 1,200 | 1,400 | 1,600 | 1,800 | 2,000 | 2,200 | 2,400 | 2,600 | 2,800 | 3,000 | 3,200 |
|---|---|---|---|---|---|---|---|---|---|---|---|---|
| **Daily amount of food from each group** | | | | | | | | | | | | |
| Fruits[b] | 1 cup | 1 cup | 1.5 cups | 1.5 cups | 1.5 cups | 2 cups | 2 cups | 2 cups | 2 cups | 2.5 cups | 2.5 cups | 2.5 cups |
| Vegetables[c] | 1 cup | 1.5 cups | 1.5 cups | 2 cups | 2.5 cups | 2.5 cups | 3 cups | 3 cups | 3.5 cups | 3.5 cups | 4 cups | 4 cups |
| Grains[d] | 3 oz-eq | 4 oz-eq | 5 oz-eq | 5 oz-eq | 6 oz-eq | 6 oz-eq | 7 oz-eq | 8 oz-eq | 9 oz-eq | 10 oz-eq | 10 oz-eq | 10 oz-eq |
| Meat and beans[e] | 2 oz-eq | 3 oz-eq | 4 oz-eq | 5 oz-eq | 5 oz-eq | 5.5 oz-eq | 6 oz-eq | 6.5 oz-eq | 6.5 oz-eq | 7 oz-eq | 7 oz-eq | 7 oz-eq |
| Milk[f] | 2 cups | 2 cups | 2 cups | 3 cups | 3 cups | 3 cups | 3 cups | 3 cups | 3 cups | 3 cups | 3 cups | 3 cups |
| Oils[g] | 3 tsp | 4 tsp | 4 tsp | 5 tsp | 5 tsp | 6 tsp | 6 tsp | 7 tsp | 8 tsp | 8 tsp | 10 tsp | 11 tsp |
| Discretionary calorie allowance[h] | 165 | 171 | 171 | 132 | 195 | 267 | 290 | 362 | 410 | 426 | 512 | 648 |
| **Vegetable subgroup amounts are per week** | | | | | | | | | | | | |
| Dark green vegetables | 1 c/wk | 1.5 c/wk | 1.5 c/wk | 2 c/wk | 3 c/wk | 3 c/wk | 3 c/wk | 3 c/wk | 3 c/wk | 3 c/wk | 3 c/wk | 3 c/wk |
| Orange vegetables | .5 c/wk | 1 c/wk | 1 c/wk | 1.5 c/wk | 2 c/wk | 2 c/wk | 2 c/wk | 2 c/wk | 2.5 c/wk | 2.5 c/wk | 2.5 c/wk | 2.5 c/wk |
| Legumes | .5 c/wk | 1 c/wk | 1 c/wk | 2.5 c/wk | 3 c/wk | 3 c/wk | 3 c/wk | 3 c/wk | 3.5 c/wk | 3.5 c/wk | 3.5 c/wk | 3.5 c/wk |
| Starchy vegetables | 1.5 c/wk | 2.5 c/wk | 2.5 c/wk | 2.5 c/wk | 3 c/wk | 3 c/wk | 6 c/wk | 6 c/wk | 7 c/wk | 7 c/wk | 9 c/wk | 9 c/wk |
| Other vegetables | 3.5 c/wk | 4.5 c/wk | 4.5 c/wk | 5.5 c/wk | 6.5 c/wk | 6.5 c/wk | 7 c/wk | 7 c/wk | 8.5 c/wk | 8.5 c/wk | 10 c/wk | 10 c/wk |

Notes: The suggested amounts of food to consume from the basic food groups, subgroups, and oils to meet recommended nutrient intakes at 12 different calorie levels. Nutrient and energy contributions from each group are calculated according to the nutrient-dense forms of foods in each group (e.g., lean meats and fat-free milk). The table also shows the discretionary calorie allowance that can be accommodated within each calorie level, in addition to the suggested amounts of nutrient-dense forms of foods in each group.

**Estimated daily calorie needs** To determine which food intake pattern to use for an individual, the following chart gives an estimate of individual calorie needs. The calorie range for each age/sex group is based on physical activity level, from sedentary to active.

| | Calorie range | | | | Calorie range | | |
|---|---|---|---|---|---|---|---|
| **Children** | **Sedentary[i]** | → | **Active[j]** | **Children** | **Sedentary[i]** | → | **Active[j]** |
| 2–3 years | 1,000 | → | 1,400 | 2–3 years | 1,000 | → | 1,400 |
| **Females** | | | | **Males** | | | |
| 4–8 years | 1,200 | → | 1,800 | 4–8 years | 1,400 | → | 2,000 |
| 9–13 | 1,600 | → | 2,200 | 9–13 | 1,800 | → | 2,600 |
| 14–18 | 1,800 | → | 2,400 | 14–18 | 2,200 | → | 3,200 |
| 19–30 | 2,000 | → | 2,400 | 19–30 | 2,400 | → | 3,000 |
| 31–50 | 1,800 | → | 2,200 | 31–50 | 2,200 | → | 3,000 |
| 51+ | 1,600 | → | 2,200 | 51+ | 2,000 | → | 2,800 |

[a]Calorie levels are set across a wide range to accommodate the needs of different individuals. The calorie range chart can be used to help assign individuals to the food intake pattern at a particular calorie level.

[b]Fruit group includes all fresh, frozen, canned, and dried fruits and fruit juices. In general, 1 cup of fruit or 100% fruit juice, or 1/2 cup of dried fruit can be considered as 1 cup from the fruit group.

[c]Vegetable group includes all fresh, frozen, canned, and dried vegetables and vegetable juices. In general, 1 cup of raw or cooked vegetables or vegetable juice, or 2 cups of raw leafy greens can be considered as 1 cup from the vegetable group.

[d]Grains group includes all foods made from wheat, rice, oats, cornmeal, barley, such as bread, pasta, oatmeal, breakfast cereals, tortillas, and grits. In general, 1 slice of bread, 1 cup of ready-to-eat cereal, or 1/2 cup of cooked rice, pasta, or cooked cereal can be considered as 1 ounce equivalent from the grains group. At least half of all grains consumed should be whole grains.

[e]Meat & beans group in general, 1 ounce of lean meat, poultry, or fish, 1 egg, 1 Tbsp. peanut butter, 1/4 cup cooked dry beans, or 1/2 ounce of nuts or seeds can be considered as 1 ounce equivalent from the meat and beans group.

[f]Milk group includes all fluid milk products and foods made from milk that retain their calcium content, such as yogurt and cheese. Foods made from milk that have little to no calcium, such as cream cheese, cream, and butter, are not part of the group. Most milk group choices should be fat-free or low-fat. In general, 1 cup of milk or yogurt, 1 1/2 ounces of natural cheese, or 2 ounces of processed cheese can be considered as 1 cup from the milk group.

[g]Oils include fats, from many different plants and from fish, that are liquid at room temperature, such as canola, corn, olive, soybean, and sunflower oil. Some foods are naturally high in oils, like nuts, olives, some fish, and avocados. Foods that are mainly oil include mayonnaise, certain salad dressings, and soft margarine.

[h]Discretionary calorie allowance is the remaining amount of calories in a food intake pattern after accounting for the calories needed for all food groups—using forms of foods that are fat-free or low-fat and with no added sugars.

[i]Sedentary means a lifestyle that includes only the light physical activity associated with typical day-to-day life.

[j]Active means a lifestyle that includes physical activity equivalent to walking more than 3 miles per day at 3 to 4 miles per hour, in addition to the light physical activity associated with typical day-to-day life.

SOURCE: "MyPyramid Food Intake Patterns," in *MyPyramid.gov: Tips and Resources*, United States Department of Agriculture, Center for Nutrition Policy and Promotion, April 2005, http://mypyramid.gov/downloads/MyPyramid_Food_Intake_Patterns.pdf (accessed November 12, 2007)

- Toddlers also like to eat apples, fresh as well as processed, whereas adults age twenty to fifty-nine eat the fewest apples.
- Women forty and older eat the most spinach, while teenage girls eat the least.
- High-income consumers drink more orange juice, whereas low-income consumers drink more orange drinks (less than 10% juice).
- Consumption of French fries does not vary by income.
- Compared to low-income consumers, high-income consumers eat more of many vegetables, including fresh celery, garlic, cucumbers, bell peppers, mushrooms, and tomatoes.
- Eighty-eight percent of French fries are eaten away from home; fast-food establishments account for 67%.

**TABLE 10.5**

## Sample menus for a 2000-calorie food plan

| Day 1 | Day 2 | Day 3 | Day 4 | Day 5 |
|-------|-------|-------|-------|-------|
| **Breakfast** | **Breakfast** | **Breakfast** | **Breakfast** | **Breakfast** |
| Breakfast burrito | Hot cereal | Cold cereal | 1 whole wheat English muffin | Cold cereal |
| 1 flour tortilla (7" diameter) | 1/2 cup cooked oatmeal | 1 cup bran flakes | 2 tsp soft margarine | 1 cup puffed wheat cereal |
| 1 scrambled egg (in 1 tsp | 2 tbsp raisins | 1 cup fat-free milk | 1 tbsp jam or preserves | 1 tbsp raisins |
| soft margarine) | 1 tsp soft margarine | 1 small banana | 1 medium grapefruit | 1 cup fat-free milk |
| 1/3 cup black beans* | 1/2 cup fat-free milk | 1 slice whole wheat toast | 1 hard-cooked egg | 1 small banana |
| 2 tbsp salsa | 1 cup orange juice | 1 tsp soft margarine | 1 unsweetened beverage | 1 slice whole wheat toast |
| 1 cup orange juice | **Lunch** | 1 cup prune juice | **Lunch** | 1 tsp soft margarine |
| 1 cup fat-free milk | Taco salad | **Lunch** | White bean–vegetable soup | 1 tsp jelly |
| **Lunch** | 2 ounces tortilla chips | Tuna fish sandwich | 1 1/4 cup chunky vegetable soup | **Lunch** |
| Roast beef sandwich | 2 ounces ground turkey, sauteed | 2 slices rye bread | 1/2 cup white beans | Smoked turkey sandwich |
| 1 whole grain sandwich bun | in 2 tsp sunflower oil | 3 ounces tuna (packed in water, drained) | 2 ounce breadstick | 2 ounces whole wheat pita bread |
| 3 ounces lean roast beef | 1/2 cup black beans* | 2 tsp mayonnaise | 8 baby carrots | 1/4 cup romaine lettuce |
| 2 slices tomato | 1/2 cup iceberg lettuce | 1 tbsp diced celery | 1 cup fat-free milk | 2 slices tomato |
| 1/4 cup shredded romaine lettuce | 2 slices tomato | 1/4 cup shredded romaine lettuce | **Dinner** | 3 ounces sliced smoked turkey breast* |
| 1/8 cup sauteed mushrooms (in 1 tsp oil) | 1 ounce low-fat cheddar cheese | 2 slices tomato | Rigatoni with meat sauce | 1 tbsp mayo-type salad dressing |
| 1 1/2 ounce part-skim mozzarella cheese | 2 tbsp salsa | 1 medium pear | 1 cup rigatoni pasta (2 ounces dry) | 1 tsp yellow mustard |
| 1 tsp yellow mustard | 1/2 cup avocado | 1 cup fat-free milk | 1/2 cup tomato sauce tomato bits* | 1/2 cup apple slices |
| 3/4 cup baked potato wedges* | 1 tsp lime juice | **Dinner** | 2 ounces extra lean cooked ground | 1 cup tomato juice* |
| 1 tbsp ketchup | 1 unsweetened beverage | Roasted chicken breast | beef (sauteed in 2 tsp vegetable oil) | **Dinner** |
| 1 unsweetened beverage | **Dinner** | 3 ounces boneless skinless | 3 tbsp grated Parmesan cheese | Grilled top loin steak |
| **Dinner** | Spinach lasagna | chicken breast* | Spinach salad | 5 ounces grilled top loin steak |
| Stuffed broiled salmon | 1 cup lasagna noodles, cooked (2 oz dry) | 1 large baked sweet potato | 1 cup baby spinach leaves | 3/4 cup mashed potatoes |
| 5 ounce salmon filet | 2/3 cup cooked spinach | 1/2 cup peas and onions | 1/2 cup tangerine slices | 2 tsp soft margarine |
| 1 ounce bread stuffing mix | 1/2 cup ricotta cheese | 1 tsp soft margarine | 1/2 ounce chopped walnuts | 1/2 cup steamed carrots |
| 1 tbsp chopped onions | 1/2 cup tomato sauce tomato bits* | 1 ounce whole wheat dinner roll | 3 tsp sunflower oil and vinegar dressing | 1 tbsp honey |
| 1 tbsp diced celery | 1 ounce part-skim mozzarella cheese | 1 tsp soft margarine | 1 cup fat-free milk | 2 ounces whole wheat dinner roll |
| 2 tsp canola oil | 1 ounce whole wheat dinner roll | 1 cup leafy greens salad | **Snacks** | 1 tsp soft margarine |
| 1/2 cup saffron (white) rice | 1 cup fat-free milk | 3 tsp sunflower oil and vinegar dressing | 1 cup low-fat fruited yogurt | 1 cup fat-free milk |
| 1/2 cup steamed broccoli | **Snacks** | **Snacks** | | **Snacks** |
| 1 tsp soft margarine | 1/2 ounce dry-roasted almonds* | 1/4 cup dried apricots | | 1 cup low-fat fruited yogurt |
| 1 cup fat-free milk | 1/4 cup pineapple | 1 cup low-fat fruited yogurt | | |
| **Snacks** | 2 tbsp raisins | | | |
| 1 cup cantaloupe | | | | |

# TABLE 10.5

## Sample menus for a 2000-calorie food plan [CONTINUED]

| Day 6 | Day 7 |
|---|---|
| **Breakfast** | **Breakfast** |
| French toast | Pancakes |
| 2 slices whole wheat French toast | 3 buckwheat pancakes |
| 2 tsp soft margarine | 2 tsp soft margarine |
| 2 tbsp maple syrup | 3 tbsp maple syrup |
| 1/2 medium grape fruit | 1/2 cup strawberries |
| 1 cup fat-free milk | 3/4 cup honeydew melon |
| | 1/2 cup fat-free milk |
| **Lunch** | **Lunch** |
| Vegetarian chili on baked potato | Manhattan clam chowder |
| 1 cup kidney beans* | 3 ounces canned clams (drained) |
| 1/2 cup tomato sauce w/tomato tidbits* | 3/4 cup mixed vegetables |
| 3 tbsp chopped onions | 1 cup canned tomatoes* |
| 1 ounce low fat cheddar cheese | 10 whole wheat crackers* |
| 1 tsp vegetable oil | 1 medium orange |
| 1 medium baked potato | 1 cup fat-free milk |
| 1/2 cup cantaloupe | **Dinner** |
| 3/4 cup lemonade | Vegetable stir-fry |
| **Dinner** | 4 ounces tofu (firm) |
| Hawaiian pizza | 1/4 cup green and red bell peppers |
| 2 slices cheese pizza | 1/2 cup bok choy |
| 1 ounce canadian bacon | 2 tbsp vegetable oil |
| 1/4 cup pineapple | 1 cup brown rice |
| 2 tbsp mushrooms | 1 cup lemon-flavored iced tea |
| 2 tbsp chopped onions | **Snacks** |
| Green salad | 1 ounce sunflower seeds* |
| 1 cup leafy greens | 1 large banana |
| 3 tsp sunflower oil and vinegar dressing | 1 cup low-fat fruited yogurt |
| 1 cup fat-free milk | |
| **Snacks** | |
| 5 whole wheat crackers* | |
| 1/8 cup hummus | |
| 1/2 cup fruit cocktail (in water or juice) | |

| Food group | | Daily average over one week |
|---|---|---|
| Grains | Total grains (oz eq) | 6.0 |
| | Whole grains | 3.4 |
| | Refined grains | 2.6 |
| Vegetables* | Total vegetables (cups) | 2.6 |
| Fruits | Fruits (cups) | 2.1 |
| Milk | Milk (cups) | 3.1 |
| Meat & beans | Meat/beans (oz eg) | 5.6 |
| Oils | Oils (tsp/grams) | 7.2 tsp/32.4 g |
| *Vegetable subgroups | | (Weekly totals) |
| | Dark-green vegetables (cups) | 3.3 |
| | Orange vegetables (cups) | 2.3 |
| | Beans/peas (cups) | 3.0 |
| | Starchy vegetables (cups) | 3.4 |
| | Other vegetables (cups) | 6.6 |

| Nutrient | Daily average over one week |
|---|---|
| Calories | 1,994 |
| Protein, g | 98 |
| Protein, % kcal | 20 |
| Carbohydrate, g | 264 |
| Carbohydrate % kcal | 53 |
| Total fat, g | 67 |
| Total fat, % kcal | 30 |
| Saturated fat, g | 16 |
| Saturated fat, % kcal | 7 |
| Monounsaturated fat, g | 23 |
| Polyunsaturated fat, g | 23 |
| Linoleic acid, g | 21 |
| Alpha-linolenic acid, mg | 1.1 |
| Cholesterol, mg | 207 |
| Total dietary fiber, g | 31 |
| Potassium, mg | 4,715 |
| Sodium, mg* | 1,948 |
| Calcium, mg | 1,389 |
| Magnesium, mg | 432 |
| Copper, mg | 1.9 |
| Iron, mg | 2.5 |
| Phosphorus, mg | 1,830 |
| Zinc, mg | 14 |
| Thiamin, mg | 1.9 |
| Riboflavin, mg | 21 |
| Niacin equivalents, mg | 24 |
| Vitamin in B6, mg | 2.9 |
| Vitamin in B12, mcg | 18.4 |
| Vitamin in C, mg | 190 |
| Vitamin in E, mg (AT) | 18.9 |
| Vitamin in A, mcg (RAE) | 1,430 |
| Dietary folate equivalents, mcg | 558 |

*Starred items are foods that are labelled as no-salt-added, low-sodium, or low-salt versions of the foods. They can also be prepared from scratch with little or no salt. All other foods are regular commercial products which contain variable levels of sodium. Average sodium level of the 7 day menu assumes no-salt-added in cooking or at the table.

Notes: Averaged over a week, this seven day menu provides all of the recommended amounts of nutrients and food from each food group.

SOURCE: "Sample Menus for a 2000-Calorie Food Plan," in *MyPyramid.gov: Tips and Resources*, United States Department of Agriculture, Center for Nutrition Policy and Promotion, April 2005, http://www.mypyramid.gov/downloads/sample_menu.pdf (accessed November 12, 2007)

**TABLE 10.6**

**MyPyramid food choices based on 2,000 calories per day**

[Based on the information you provided, this is your daily recommended amount from each food group.]

| Grains<br>6 ounces | Vegetables<br>2 1/2 cups | Fruits<br>2 cups | Milk<br>3 cups | Meat & beans<br>5 1/2 ounces |
|---|---|---|---|---|
| Make half your grains whole<br>Aim for at least 3 ounces<br>   of whole grains a day | Vary your veggies<br>Aim for these amounts each week:<br>Dark green veggies=3 cups<br>Orange veggies=2 cups<br>Dry beans & peas=3 cups<br>Starchy veggies=3 cups<br>Other veggies=6 1/2 cups | Focus on fruits<br>Eat a variety of fruit<br>Go easy on fruit juices | Get your calcium-rich foods<br>Go low-fat or fat-free when you<br>   choose milk, yogurt, or cheese | Go lean with protein<br>Choose low-fat or lean meats and<br>   poultry<br>Vary your protein routine-choose more<br>   fish, beans, peas, nuts, and seeds |

Find your balance between food and physical activity.
Be physically active for at least 30 minutes most days of the week.
Know your limits on fats, sugars, and sodium.
Your allowance for oils is 6 teaspoons a day.
Limit extras—solid fats and sugars—to 265 calories a day.

Your results are based on a 2,000 calorie pattern.                    Name:_____

Note: This calorie level is only an estimate of your needs. Monitor your body weight to see if you need to adjust your calorie intake.

SOURCE: "MyPyramid Steps to a Healthier You," in *MyPyramid.gov: Tips and Resources*, United States Department of Agriculture, Center for Nutrition Policy and Promotion, April 2005, http://mypyramid.gov/downloads/results/results_2000_18.pdf (accessed November 12, 2007)

- About 60% of catsup is consumed away from home, and fast-food establishments account for one-third.

- African-American consumers, though just 13% of the U.S. population, account for 21% of sweet potato consumption. Puerto Rican Hispanics consume as many sweet potatoes as African-American consumers.

- White and Hispanic consumers eat more bell peppers than African-Americans and Asian-Americans. African-Americans eat one-third less per capita than others.

- Per capita spinach consumption is highest among Asian-Americans.

- Asian-Americans like to eat mushrooms. African-Americans eat only one-third as much as Asian-Americans per capita.

- Hispanics consume three times more dry beans per capita than the national average.

- Consumers in the South eat more fresh cabbage than consumers in other regions. As for sauerkraut, three-fourths is eaten in the Midwest and East.

- Watermelon consumption is greatest in the West. Vegetable consumption varies by where consumers live.

- Consumers in suburban and rural areas eat about 40% fewer fresh snap beans than those living in cities.

- Suburban consumers eat more cucumbers than other consumers do.

- Consumers purchase sweet corn as fresh, frozen, or canned in nearly equal proportions.

- Processed tomato products account for 80% of total tomato consumption. The largest processed use of tomatoes is for sauces, followed by tomato paste, canned whole tomato products, and catsup and juice.

The 5 a Day for Better Health program provides customized, age-appropriate health education materials for children such as lesson plans and activity sheets. "There's a Rainbow on My Plate" (February 24, 2003, http://www.dole5aday.com/Media/Press/RecentReleases/240203.jsp?topmenu=5) is a comprehensive nutrition education curriculum that encourages kindergarten through sixth-grade students to develop healthy eating habits. "There's a Rainbow on My Plate" debuted in 2003 in three thousand supermarkets and twelve thousand elementary schools in the United States. Lessons featured information about fresh, dried, frozen, and canned fruits and vegetables and 100% fruit juices as well as the locations of these foods in the supermarket, and definitions of serving sizes. Participating schools received teacher's guides with lesson plans and activity sheets, coloring books, packs of crayons, and take-home flyers for parents.

Nutrition research reveals that active men should consume even more than the five servings of fruit and vegetables the program has promoted since its inception. In 2000 the 5 a Day for Better Health program launched the Men Shoot for 9 program to encourage men to eat nine servings of fruits and vegetables every day. Along with reducing the risk for heart disease, high blood pressure, stroke, many cancers, and diabetes, diets rich in fruits and vegetables can help prevent overweight and obesity. Fruits and vegetables are naturally low in calories and fat, and their high water and fiber content produce feelings of satiety (the feeling of fullness or

**FIGURE 10.2**

**MyPyramid for kids**

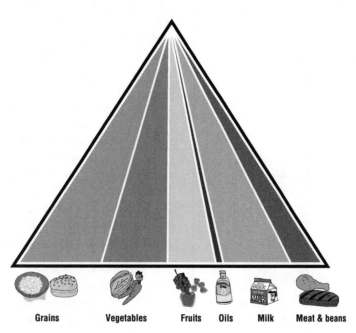

| Grains | Vegetables | Fruits | Oils | Milk | Meat & beans |

**Grains**
Make half your
grains whole

Start smart with breakfast. Look for whole-grain cereals. Just because bread is brown doesn't mean it's whole-grain. Search the ingredients list to make sure the first word is "whole" (like "whole wheat").

**Vegetables**
Vary your veggies

Color your plate with all kinds of great-tasting veggies. What's green and orange and tastes good? Veggies! Go dark green with broccoli and spinach, or try orange ones like carrots and sweet potatoes.

**Fruits**
Focus on fruits

Fruits are nature's treats—sweet and delicious. Go easy on juice and make sure it's 100%.

**Oils**

Oils are not a food group, but you need some for good health. Get your oils from fish, nuts, and liquid oils such as corn oil, soybean oil, and canola oil.

**Milk**
Get your
calcium-rich foods

Move to the milk group to get your calcium. Calcum builds strong bones. Look at the carton or container to make sure your milk, yogurt, or cheese is lowfat or fat-free.

**Meat & beans**
Go lean with protein

Eat lean or lowfat meat, chicken, turkey, and fish. Ask for it baked, broiled, or grilled—not fried. It's nutty, but true. Nuts, seeds, peas, and beans are all great sources of protein, too.

For an 1,800–calorie diet, you need amounts below from each food group. To find the amounts that are right for you, go to MyPrymaid.gov.
**Grains** Eat 6 oz. every day; at least half should be whole
**Vegetables** Eat 2 1/2 cups every day
**Fruits** Eat 1 1/2 cups every day
**Milk** Get 3 cups every day; for kids ages 2 to 8, it's 2 cups
**Meat & beans** Eat 5 oz. every day

**Find your balance between food and fun**
Move more. Aim for at least 60 minutes everyday, or most days.
Walk, dance, bike, rollerblade—it all counts. How great is that!

**Fats and sugars—know your limits**
Get your fat facts and sugar smarts from the nutrition facts label.
Limit solid fats as well as foods that contain them.
Choose food and beverages low in added sugars and other caloric sweeteners.

SOURCE: "MyPyramid for Kids," in *MyPyramid.gov: For Kids*, United States Department of Agriculture, Center for Nutrition Policy and Promotion, September 2005, http://teamnutrition.usda.gov/Resources/mpk_poster2.pdf (accessed November 12, 2007)

satisfaction after eating). Combined with an active life-style and low-fat diet, eating greater amounts of fruits and vegetables and fewer high-calorie foods at meals can help control weight. The Men Shoot for 9 program teaches men that they can feel full and consume fewer calories when they substitute vegetables for foods that contain more fat and calories.

**State Funding for Prevention Efforts**

The CDC Nutrition and Physical Activity Program to Prevent Obesity and Other Chronic Diseases aims to help states prevent obesity and other chronic diseases by focusing on poor nutrition and inadequate physical activity. The program supports states to develop and implement nutrition and physical activity interventions, and sponsors initiatives

**FIGURE 10.3**

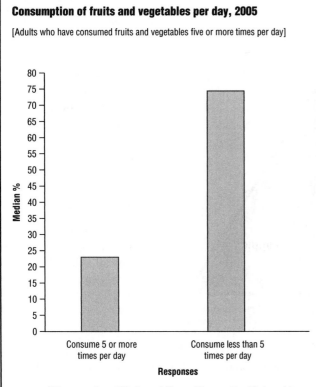

**Consumption of fruits and vegetables per day, 2005**

[Adults who have consumed fruits and vegetables five or more times per day]

SOURCE: "Consumption of Fruits and Vegetables per Day Nationwide (States, DC, and Territories)—2005," in *Behavioral Risk Factor Surveillance System: Prevalence Data*, Centers for Disease Control and Prevention, National Center for Chronic Disease Prevention and Health Promotion, April 17, 2007, http://apps.nccd.cdc.gov/brfss/display.asp? cat=FV&yr=2005&qkey=4415&state=US (accessed November 12, 2007)

to help populations balance caloric intake and expenditure, increase physical activity, improve nutrition by increasing consumption of fruits and vegetables, reduce television time, and increase breastfeeding.

In fiscal year 2007 twenty-one states received between $400,000 and $450,000 for capacity building, in which state health departments gather data, build partnerships, and create statewide health plans. Essentially, capacity building lays the necessary groundwork on which to institute nutrition and physical activity interventions. In addition, seven states received between $750,000 and $1.3 million for basic implementation, bringing the total number of states that received funding to twenty-eight. Figure 10.4 shows the states funded for basic implementation and those that received funds for capacity building.

## IS NUTRITION EDUCATION WORKING TO IMPROVE AMERICANS' DIETS?

The Healthy Eating Index (HEI) is a measure developed in 1990 by the USDA to assess the overall health value of Americans' diets. It captures the type and quantity of foods people eat and the degree to which diets comply

with specific recommendations in the USDA dietary guidelines and the food pyramids. The HEI assigns points for eating consistently within USDA guidelines. It assesses ten dietary components—grains, vegetables, fruits, milk, meat, total fat, total saturated fat, cholesterol, sodium, and a varied diet—on a scale of zero to ten. Individuals who eat grains, vegetables, fruits, milk, meat (including chicken and fish), as well as a variety of foods at or above the USDA recommended levels receive a maximum score of ten. A score of zero is assigned when the recommended amount of those components is not eaten. For fat, saturated fat, cholesterol, and sodium, a score of ten is awarded for eating the recommended amount or less. The highest possible score is a hundred; a score of eighty or above is considered a healthy diet, scores between fifty-one and eighty show a need for dietary improvement, and scores below fifty indicate poor diets. Table 10.7 shows the components and standards for scoring used in the 2005 HEI.

The market research firm NPD Group reports in *21st Annual Eating Patterns in America* (2006) that Americans' want to change their food purchasing and eating habits. Adults surveyed in 2006 said they wanted to:

- Increase whole grains in their diet (64%)

- Consume more dietary fiber (58%)

- Increase calcium (58%) and vitamin C (55%) consumption

- Reduce fat in their diet (71%)

- Reduce calories (62%), cholesterol (62%), and sugar (59%)

According to the report, Harry Balzer, the vice president of the NPD Group, named convenience the driving force behind consumers' food choices and cited the resurgence of sandwiches as dinner fare as evidence of American's desire for quick, easy meals. More than half of the households surveyed chose evening meals that were easy to prepare and required little or no planning. Nearly 40% prepared their meals from ingredients they had on hand, 35% said meal choices were made to accommodate the preferences of all family members, and 34% said quick, easy cleanup was a determining factor when choosing what to make for dinner. Less than half (47%) of dinners included at least one fresh ingredient, down from 56% in 1985.

The NPD report also observes that rates of dieting in the United States have declined. In 1990, 54% of women and 26% of men were dieting. In 2006, even though 60% of adults said they wanted to lose twenty pounds, just 26% of women and 19% of men said they were on a diet—the lowest rate of dieting in at least sixteen years. According to the NPD Group, the top ten diets Americans followed in 2006 were:

FIGURE 10.4

**States funded to prevent obesity and other chronic diseases, fiscal year 2007**

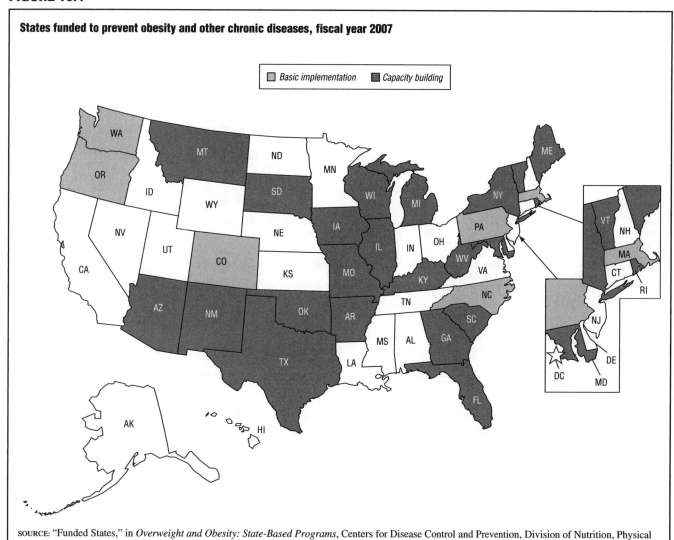

SOURCE: "Funded States," in *Overweight and Obesity: State-Based Programs*, Centers for Disease Control and Prevention, Division of Nutrition, Physical Activity, and Obesity, National Center for Chronic Disease Prevention and Health Promotion, May 22, 2007, http://www.cdc.gov/nccdphp/dnpa/obesity/state_programs/funded_states/index.htm (accessed November 12, 2007)

- A diet of their own
- A diet prescribed by a physician
- Weight Watchers
- A low-fat diet
- A low-calorie diet
- Atkins (a low-carbohydrate diet)
- South Beach (a low-carbohydrate diet)
- A sugar-free diet
- Slim-Fast (a brand of shakes, bars, snacks, packaged meals, and other dietary supplement foods)

### Americans' Snack Food Choices

In 2004 the NPD Group released its first-ever study about Americans' snack food choices: *Snacking in America*. The report finds that salty snacks such as pretzels, potato chips, and tortilla chips accounted for about a quarter of the convenience foods Americans choose for snacks. Even though children and teens snack on sugary treats such as candy, gum, chewy fruit snacks, and breath mints, which accounted for an additional 14% of snack food choices, some of the fastest-growing snack foods among children and adolescents aged two to eighteen were healthier choices. Yogurt was the fastest-growing snack food in terms of consumption frequency among children under thirteen. On average, children under thirteen ate yogurt eleven more times in the year ending June 2003 than they did three years earlier. Children aged two to seven ate yogurt as a snack nearly fourteen times more often in 2003 than they did in 1999, and children aged eight to twelve snacked on yogurt eight and a half times more in 2003 than five years earlier. Children may be making some healthier choices when it comes to snacks, but they are also snacking more frequently—in 2003 children and teens consumed about twenty-two more snacks per person per year than reported in 1999.

**TABLE 10.7**

**Healthy eating index, 2005**

| Component | Maximum points | Standard for maximum score | Standard for minimum score of zero |
|---|---|---|---|
| Total fruit (includes 100% juice) | 5 | ≥0.8 cup equiv. per 1,000 kcal | No fruit |
| Whole fruit (not juice) | 5 | ≥0.4 cup equiv. per 1,000 kcal | No whole fruit |
| Total vegetables | 5 | ≥1.1 cup equiv. per 1,000 kcal | No vegetables |
| Dark green and orange vegetables and legumes[a] | 5 | ≥0.4 cup equiv. per 1,000 kcal | No dark green or orange vegetables or legumes |
| Total grains | 5 | ≥3.0 oz equiv. per 1,000 kcal | No grains |
| Whole grains | 5 | ≥1.5 oz equiv. per 1,000 kcal | No whole grains |
| Milk[b] | 10 | ≥1.3 cup equiv. per 1,000 kcal | No milk |
| Meat and beans | 10 | ≥2.5 oz equiv. per 1,000 kcal | No meat or beans |
| Oils[c] | 10 | ≥12 grams per 1,000 kcal | No oil |
| Saturated fat | 10 | ≤7% of energy[d] | ≥15% of energy |
| Sodium | 10 | ≤0.7 gram pr 1,000 kcal[d] | ≥2.0 grams per 1,000 kcal |
| Calories from solid fat, alcohol, and added sugar (SoFAAS) | 20 | ≤20% of energy | ≥50% of energy |

Note: Intakes between the minimum and maximum levels are scored proportionately, except for saturated fat and sodium (see footnote d).
[a]Legumes counted as vegetables only after meat and beans standard is met.
[b]Includes all milk products, such as fluid milk, yogurt, and cheese.
[c]Includes nonhydrogenated vegetable oils and oils in fish, nuts, and seeds.
[d]Saturated fat and sodium get a score of 8 for the intake levels that reflect the 2005 Dietary Guidelines, <10% of calories from saturated fat and 1.1 grams of sodium/1,000 kcal, respectively.

SOURCE: Patricia M. Guenther et al., "Healthy Eating Index—2005 Components and Standards for Scoring," in *Healthy Eating Index—2005*, USDA, Center for Nutrition Policy and Promotion, December 2006, http://www.cnpp.usda.gov/Publications/HEI/healthyeatingindex2005factsheet.pdf (accessed November 12, 2007)

In June 2005 the NPD Group's SnackTrack reported that fruit was the number-one snack food consumed by children aged two to twelve. Among boys aged two to seven, fresh fruit was consumed more often than any other snack food, followed by yogurt, potato chips, chocolate candy, and cookies. Girls the same age ranked fresh fruit first, followed by yogurt, gum, potato chips, and chocolate candy. Boys aged eight to twelve named fresh fruit, potato chips, gum, ice cream, and chocolate candy as their top-five picks, and girls aged eight to twelve said fresh fruit, gum, potato chips, ice cream, and chocolate candy were their favorite snack foods.

Chewing gum was the number-one snack food named by adults aged eighteen to fifty-four in 2005. The NPD Group's SnackTrack found that age was a factor in the gum-chewing habits of Americans. Even though gum was a popular snack among children under thirteen years old, it did not rank as number one until the teenage years. After gum, the top-ranking snack foods among men in 2005 were chocolate candy, fresh fruit, potato chips, breath/candy mints, ice cream, nuts, cookies, tortilla chips, and candy bars. Women said chocolate candy, fresh fruit, potato chips, breath/candy mints, ice cream, cookies, nuts, yogurt, and crackers were their choices.

Packaged Facts notes in *On-the-Go Eating in the US: Consumer, Foodservice, Retailing and Marketing Trends* (2007) that one-third of Americans regularly skip meals and graze on snack foods, perhaps in part inspired by reports about the wisdom of eating small meals throughout the day, and the availability of smaller portion-controlled snack food offerings from the food-service industry. The report also attributes frequent snacking as well as increasing reliance of fast foods and take-out meals to time constraints—a full 33% of Americans felt they did not have the time to prepare or eat healthy meals.

**Interventions to Promote Healthy Weight**

In *The Guide to Community Preventive Services* (2005), the CDC reviews the effectiveness of community, population, and health-care system strategies to address a variety of public health challenges, including interventions to prevent obesity and promote healthy eating and physical activity. The guide considers the effectiveness of population-based interventions that promote healthy growth and development of children and adolescents and that support healthy weights among adults, and focuses on school-based strategies, worksite programs, health-care system interventions, and community-wide initiatives.

The CDC review concludes that worksite programs that combine nutrition and physical activity are effective and recommends them as strategies to control overweight and obesity. The review also determines that more evidence is needed to determine the effectiveness of school-based programs to control overweight and obesity.

**PREVENTION PROGRAMS AT THE WORK SITE**

Along with school-based nutrition programs and education initiatives aimed at the public at large, several notable obesity prevention efforts involved developing and implementing strategies to integrate physical activity and healthy food choices into routine worksite activities. Examples of such activities include incorporating planned activity breaks with music into long meetings, offering healthy food choices at meetings and during breaks as well as at employee cafeterias, and hosting walking meetings.

Because more than one hundred million Americans spend a large number of their waking hours at work, the worksite presents another opportunity for prevention programs. The *Call to Action* advises moving beyond traditional workplace health education programs. It recommends more intensive and comprehensive efforts such as modifying physical and social environments, instituting policies consistent with the objective of preventing overweight and obesity, and extending worksite prevention efforts not only to employees but also to the families of employees and their communities.

Examples of worksite obesity prevention strategies include:

- Instituting flexible work hours and schedules to create opportunities for regular physical activity during the workday
- Ensuring that healthy food options are available
- Establishing worksite exercise facilities or creating incentives for employees to join local fitness centers
- Developing incentives for workers to achieve and maintain a healthy body weight
- Encouraging employers to require weight management and physical activity counseling as covered benefits in health insurance contracts
- Creating work environments that promote and support breastfeeding
- Instituting federal worksite programs promoting healthy eating and physical activity that not only can serve as models but also may be easily adapted for use in the private sector

Recent research suggests that obesity may begin at the office. W. Kerry Mummery et al. examine in "Occupational Sitting Time and Overweight and Obesity in Australian Workers" (*American Journal of Preventive Medicine*, vol. 29, no. 2, August 2005) the role of the workplace in the problem of overweight and obesity by studying the association between occupational sitting time and overweight and obesity in a sample of adults employed full time. The investigators find that the more time workers sit at their desks, the more likely they are to be overweight. Higher total daily sitting time was associated with a 68% increased risk of being overweight or obese.

Overall, men sat an average of 209 minutes while at work, 20 minutes more than the average for women. Mummery et al. suggest that the extra twenty minutes might make a difference because they find a significant association between sitting time and overweight and obesity in male workers, but not in female workers.

Mummery et al. assert that encouraging workers to exercise may favorably influence a company's bottom line. They conclude, "Time and productivity lost due to chronic diseases associated with overweight and obesity may make it financially worthwhile for employers to be more proactive in the health of their employees by promoting physical activity at work."

### Offices of the Future May Improve, Rather Than Imperil, Health and Fitness

Steve Karnowski notes in "Researcher Sees Future Where People Walk at Work" (Associated Press, June 7, 2005) that James Levine, a Mayo clinic obesity researcher who studies nonexercise activity thermogenesis (NEAT; the calories people burn during everyday activities such as standing, walking, or even fidgeting), redesigned his office in 2005 to encourage physical activity to burn calories. Levine explains that because it is metabolically more effective and probably easier for most people to put more NEAT into their lives to achieve and maintain healthy body weights than to seek organized exercise, the physically active office would be a natural outgrowth of NEAT research.

Levine's office of the future holds meetings while walking laps on a track rather than sitting around a conference table eating donuts. Workers at computers walk on a treadmill rather than sit, and presentations are made standing at magnetic marker boards rather than sitting at desks or conference tables.

Levine's retrofitted office even appeals to his colleagues who already exercise regularly because they assert that standing and moving keeps them alert and focused throughout the day. Levine admits that there is pressure in his office to work while standing and to keep moving throughout the day, but he contends that this positive peer pressure is preferable to the pressure to bring unhealthy snack foods to the worksite.

## INTENSIFYING THE PREVENTION AGENDA IN THE HEALTH-CARE SYSTEM

Interactions with health-care professionals are important opportunities to deliver powerful prevention messages. Physicians' and other health professionals' prescriptions and recurring advice to prevent weight gain in order to prevent disease or reduce symptoms of existing disease are often powerful inducements for behavioral change. Most Americans have at least annual contact with a health-care professional, and if this contact includes information about the importance of weight management, then it may reinforce prevention messages received in other settings such as schools and worksites. Furthermore, health-care professionals are instrumental in shaping public policy and can leverage their expertise and credibility to present accurate messages in the media and catalyze sweeping changes in the community at large.

Examples of strategies to expand on prevention efforts in the health-care delivery system include:

- Training health-care providers and health profession students to use effective techniques to prevent and treat overweight and obesity
- Cultivating partnerships between health-care providers, schools, faith-based groups, and other community organizations to target social and environmental causes of overweight and obesity
- Classifying obesity as a disease to enable reimbursement for prevention efforts
- Partially or fully covering weight-management services including nutrition education and physical activity programs as health plan benefits

## USING THE MEDIA TO COMMUNICATE THE PREVENTION MESSAGE

The *Call to Action* underscores the pivotal role of the media in prevention efforts. The media can communicate and educate the public about healthy behaviors and health risks associated with overweight and obesity. It can introduce and reinforce prevention messages from health-care professionals and can assist to alter attitudes and perceptions by celebrating healthy eating and physical activity.

Since 1995 the International Food Information Council (IFIC) has tracked media coverage of diet, nutrition, and food safety. In the first such report, *Food for Thought* (1995), the IFIC noted that the leading nutrition and food issues receiving newspaper, television, and other media coverage during the previous twelve months were reducing fat intake; the impact of diet on disease risks; and discussions of foodborne illnesses, vitamin and mineral intake, disease causation, caloric intake, antioxidants, cholesterol intake, sugar intake, and fiber intake. As obesity became a more prominent issue in the late 1990s, the IFIC reports in the "Executive Summary" (*Food for Thought VI* (December 2005, http://www.ific.org/research/upload/ExecSummaryFFTVI.pdf) that the number of stories about diet, weight loss, nutrition, and obesity skyrocketed from 1,270 in 1995, to 2,412 in 2005. This increase reflected both a rising volume of coverage and an escalation in the number of media outlets reporting about diet, overweight, and obesity. In *Food for Thought VI*, the IFIC reports that obesity was the leading topic in food and nutrition media stories during 2004, followed by disease prevention, physical activity, weight management, disease causation, vitamin and mineral intake, fat intake, functional foods, mad cow disease, calorie intake, and biotechnology.

The WHO cautions against the judicious use of the media to combat the obesity epidemic. In the press release

"WHO Encourages Media to Put Obesity in Perspective" (June 26, 2003, http://www.nacsonline.com/), Derek Yach, the WHO executive director for noncommunicable diseases and mental health, asserts that the media's fixation on obesity threatens to overshadow efforts to improve global health. Yach states, "Of course obesity is important but it isn't the only issue, and we wouldn't want that to be seen as the only issue." He believes the WHO would oppose measures such as "fat taxes" intended to discourage consumption of high-fat foods. Yach offers that food manufacturers have expressed to the WHO their willingness to produce more healthy products, and he explains that based on recommendations from the World Bank, the WHO does not feel that manipulating taxes to modify consumption is advisable and that it could have undesirable effects.

In 2007 some observers expressed dismay with Small Steps (http://www.adcouncil.org/default.aspx?id=54), a media campaign targeting obesity that was created by the Ad Council and the U.S. Department of Health and Human Services. The government-funded campaign, which costs more than $1.5 million a year, features television spots intended to encourage people to make changes—such as eating healthy snacks and taking stairs instead of elevators—to improve their health.

According to the article "U.S. Obesity Ads Called 'Namby-Pamby'" (Associated Press, October 22, 2007), critics describe the ads, one of which features people finding blobs of fat on the floor and observing that it must be fat lost by someone choosing healthy snacks, as lacking the dramatic impact of antismoking campaigns and as too tame to be effective. They also question whether it is appropriate to tackle the urgent health consequences of an obesity epidemic with a campaign that emphasizes such small lifestyle changes. Describing the campaign, Michael Jacobson, the director of the Center for Science in the Public Interest, said, "It's so namby-pamby I think people will shrug it off," and Kelly Brownell, the director of Yale University's Center for Eating and Weight Disorders, opined, "I think 'Small Steps' is a euphemism for small vision." Jacobson also contends that the campaign fails to acknowledge one of the root causes of the obesity epidemic—ready access to inexpensive high-fat and high-calorie food—that the government should address. He said, "The U.S. government doesn't have the guts to go after junk food producers."

## EXPERTS TARGET CHILDHOOD OBESITY AND WOMEN

In 2005 the American Heart Association (AHA) launched a new initiative aimed at combating childhood obesity. The AHA initiative provides recommendations specifically directed at the promotion of physical activity in schools. The AHA is also acting to enhance existing programs such as Choose to Move, an Internet-based program that helps women add activity to their daily life and provides nutrition education. In "Nine out of 10 Women Attempt Exercise Goals after Initial Failure, Survey Shows" (October 16, 2007, http://www.choosetomove.org/pdf/CTM%20Exercise%20Survey%202007%20FINAL.pdf), the AHA notes that it added a new feature, the Choose to Move Countdown, a downloadable desktop tool that offers daily exercise tips, motivation, and nutrition information for twelve weeks to help women jumpstart healthy lifestyle changes.

In *A Nation at Risk: Obesity in the United States Statistical Sourcebook* (May 2005, http://www.americanheart.org/downloadable/heart/1114880987205NationAtRisk.pdf), the AHA provides solid science about nutrition, physical activity, and weight to the public, health-care professionals, and policy makers.

## ECONOMIC INCENTIVES FOR PREVENTION AND TREATMENT

Richard Hyer indicates in "Government-Funded Weight-Loss Programs Recommended for Low-Income Population" (*Medscape Medical News*, October 8, 2007) that many clinicians and obesity researchers believe that government-funded weight-loss programs and state-funded programs for communities for which commercial diet and exercise programs are cost-prohibitive would increase participation in weight-loss programs.

In anticipation of the 2008 presidential elections, during 2007 discussions about universal health-care coverage resurfaced and public health professionals hoped for a national health-care plan that would cover obesity prevention efforts as well as treatment. In "Preventing Obesity in America: Is It Achievable?" (*American Journal of Lifestyle Medicine*, vol. 1, no. 6, 2007), Clinton L. Greenstone of the University of Michigan Medical School calls for "more available and affordable weight loss programs, community- and work-based weight loss programs and broader insurance coverage for weight loss programs."

# CHAPTER 11

# PUBLIC OPINION AND ACTION ABOUT DIET, WEIGHT, NUTRITION, AND PHYSICAL ACTIVITY

*Americans are not confused about the facts. They know obesity is a serious health threat and that being overweight can lead to diabetes, heart attacks and cancer. As the survey shows, people also know they should be getting more physical activity each week and eating more fruits and vegetables, and many are doing so. That's terrific news. The survey shows that people are knowledgeable, and most of them are motivated, and we hope that people will take even more steps to achieve a healthy weight and physical fitness for themselves and their families.*

—Julie Gerberding, the director of the Centers for Disease Control and Prevention, commenting on the results of the Harvard School of Public Health Poll, July 14, 2005

A July 2007 Gallup Poll revealed that most Americans (85%) described their diet as "very/somewhat healthy." (See Figure 11.1.) Just 15% described their diet as "not too/not at all healthy." It appears that despite rising rates of overweight and obesity and evidence that many Americans are consuming too many calories, most believe their diets are healthy.

Similarly, even though the Centers for Disease Control and Prevention (CDC) indicates that two-thirds of Americans are overweight or obese, the July 2007 Gallup Poll found that more than half (52%) of Americans felt their weight is "about right," 42% described themselves as "overweight," and 5% considered themselves "underweight." (See Figure 11.2.) This self-assessment of weight is surprising in view of survey respondents' self-report of weight, which has been increasing steadily since 1990. Table 11.1 shows that the average weight of adults rose from 161 pounds in 1990 to 175 pounds in 2006. There has also been a rise in the percentage of people who weigh two hundred pounds or more—from just 15% in 1990 to 23% of adults in 2006.

An analysis of body mass index (BMI, which is a numerical expression that describes the relationship between weight and height that is associated with body fat and health risk) reveals that the Gallup Poll respondents are closer in weight status to CDC reports than to their own self-assessments. Table 11.2 shows that in November 2006, 38% of respondents had a BMI of 26 to 30, meaning they were classified as overweight, and an additional 20% had a BMI over 30, which indicates obesity.

Despite the increasing prevalence of obesity and its relationship to many chronic conditions, only 28% of survey respondents said obesity had caused a serious health problem for a relative; more than two-thirds (71%) denied having a family member with an obesity-linked health problem. (See Table 11.3.)

## CONSUMER KNOWLEDGE OF NUTRIENTS AND THEIR HEALTH BENEFITS

In *Experimental Study of Health Claims on Food Packages* (May 2007, http://www.cfsan.fda.gov/~comm/crnutri4.html), Chung-Tung Jordan Lin of the U.S. Food and Drug Administration reports the findings from research that considered consumer perceptions of health and other claims (nutrient content claims, structure/function claims, and dietary guidance statements) on food packages. Because the study was based on survey results from an Internet consumer panel, it only represents respondents' knowledge, attitudes, and behavior, rather than the beliefs of all Americans. Still, it offers insight into consumer understanding of health claims that do not name the specific nutrients that are involved in the diet-disease relationship (e.g., "Yogurt may reduce the risk of osteoporosis") and health claims that name the nutrient (e.g., "Calcium-rich foods, such as yogurt, may reduce the risk of osteoporosis").

The study had two phases. The first phase collected information about the awareness of foods and nutrients and their possible health benefits. The second phase assessed consumer understanding of various health

## FIGURE 11.1

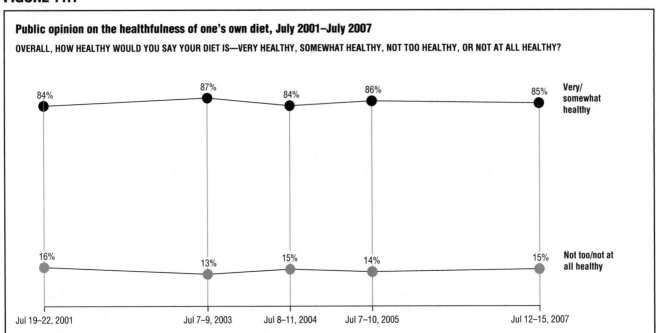

**Public opinion on the healthfulness of one's own diet, July 2001–July 2007**

OVERALL, HOW HEALTHY WOULD YOU SAY YOUR DIET IS—VERY HEALTHY, SOMEWHAT HEALTHY, NOT TOO HEALTHY, OR NOT AT ALL HEALTHY?

SOURCE: "Overall, how healthy would you say your diet is—very healthy, somewhat healthy, not too healthy, or not at all healthy?" in *Nutrition and Food*, The Gallup Organization, 2007, http://www.gallup.com/poll/6424/Nutrition-Food.aspx (accessed November 14, 2007). Copyright © 2007 by The Gallup Organization. Reproduced by permission of The Gallup Organization.

## FIGURE 11.2

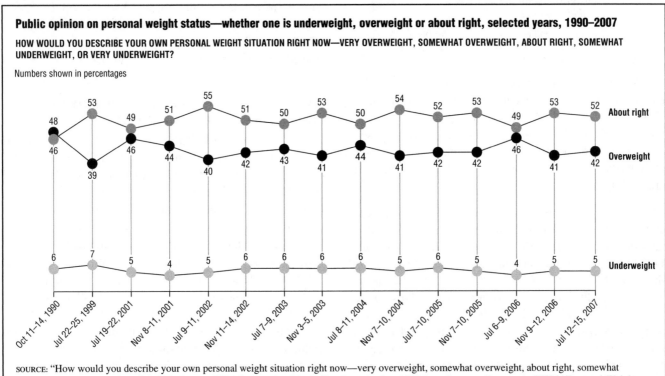

**Public opinion on personal weight status—whether one is underweight, overweight or about right, selected years, 1990–2007**

HOW WOULD YOU DESCRIBE YOUR OWN PERSONAL WEIGHT SITUATION RIGHT NOW—VERY OVERWEIGHT, SOMEWHAT OVERWEIGHT, ABOUT RIGHT, SOMEWHAT UNDERWEIGHT, OR VERY UNDERWEIGHT?

Numbers shown in percentages

SOURCE: "How would you describe your own personal weight situation right now—very overweight, somewhat overweight, about right, somewhat underweight, or very underweight?" in *Personal Weight Situation*, The Gallup Organization, 2007, http://www.gallup.com/poll/7264/Personal-Weight-Situation.aspx#1 (accessed November 14, 2007). Copyright © 2007 by The Gallup Organization. Reproduced by permission of The Gallup Organization.

claims and messages. Lin was especially eager to learn whether consumers were able to identify the nutrient linked to a specific health benefit, and whether they knew the food sources from which these nutrients might be obtained. In Phase 1, respondents were asked about their awareness of three foods (yogurt, orange juice, and

TABLE 11.1

## Poll respondents' report of their own current weight, selected years 1990–2006

WHAT IS YOUR APPROXIMATE CURRENT WEIGHT?

| | 124 lbs. or less % | 125–149 lbs. % | 150–174 lbs. % | 175–199 lbs. % | 200 lbs. and over % | No opinion % | Average weight |
|---|---|---|---|---|---|---|---|
| **National adults** | | | | | | | |
| 2006 Nov 9–12 | 8 | 19 | 23 | 22 | 23 | 5 | 175 |
| 2005 Nov 7–10 | 9 | 19 | 24 | 19 | 25 | 4 | 173 |
| 2004 Nov 7–10 | 8 | 19 | 25 | 20 | 24 | 4 | 173 |
| 2003 Nov 3–5 | 8 | 21 | 21 | 22 | 24 | 4 | 174 |
| 2002 Nov 11–14 | 9 | 22 | 22 | 18 | 25 | 4 | 173 |
| 2001 Nov 8–11 | 8 | 20 | 24 | 22 | 22 | 4 | 171 |
| 2001 Jul 19–22 | 8 | 19 | 26 | 20 | 22 | 5 | 173 |
| 1999 Jul 22–25 | 11 | 19 | 25 | 20 | 20 | 4 | 170 |
| 1990 Oct 11–14 | 12 | 27 | 27 | 16 | 15 | 3 | 161 |
| **Men** | | | | | | | |
| 2006 Nov 9–12 | 1 | 6 | 21 | 32 | 38 | 2 | 194 |
| 2005 Nov 7–10 | 1 | 7 | 24 | 27 | 40 | 1 | 193 |
| 2004 Nov 7–10 | * | 7 | 23 | 30 | 39 | 1 | 191 |
| 2003 Nov 3–5 | 1 | 7 | 19 | 34 | 38 | 1 | 195 |
| 2002 Nov 11–14 | 1 | 8 | 25 | 24 | 41 | 1 | 193 |
| 2001 Nov 8–11 | 1 | 8 | 22 | 33 | 35 | 1 | 189 |
| 2001 Jul 19–22 | 3 | 8 | 26 | 28 | 34 | 1 | 188 |
| 1999 Jul 22–25 | 1 | 6 | 27 | 31 | 33 | 2 | 190 |
| 1990 Oct 11–14 | 2 | 11 | 35 | 27 | 25 | * | 180 |
| **Women** | | | | | | | |
| 2006 Nov 9–12 | 14 | 31 | 26 | 13 | 9 | 6 | 155 |
| 2005 Nov 7–10 | 17 | 30 | 24 | 11 | 12 | 6 | 154 |
| 2004 Nov 7–10 | 16 | 29 | 27 | 11 | 11 | 6 | 156 |
| 2003 Nov 3–5 | 15 | 34 | 23 | 10 | 11 | 6 | 153 |
| 2002 Nov 11–14 | 15 | 35 | 21 | 12 | 10 | 7 | 153 |
| 2001 Nov 8–11 | 15 | 31 | 26 | 11 | 11 | 6 | 153 |
| 2001 Jul 19–22 | 13 | 29 | 26 | 12 | 12 | 8 | 158 |
| 1999 Jul 22–25 | 20 | 32 | 24 | 9 | 8 | 7 | 150 |
| 1990 Oct 11–14 | 21 | 42 | 20 | 7 | 5 | 5 | 142 |

SOURCE: "What is your approximate current weight?" in *Personal Weight Situation*, The Gallup Organization, July 2007, http://www.gallup.com/poll/7264/Personal-Weight-Situation.aspx#1 (accessed November 14, 2007). Copyright © 2007 by The Gallup Organization. Reproduced by permission of The Gallup Organization.

TABLE 11.2

## BMI (body mass index) distribution based on self-report of height and weight, 2001–06

[BMI distribution among adult population; in percent.]

| | <20 (under weight) | 20–25 (normal) | 26–30 (over weight) | 30+ (obese) | Missing |
|---|---|---|---|---|---|
| **National adults** | | | | | |
| 2006 Nov 9–12 | 6 | 31 | 38 | 20 | 5 |
| 2005 Nov 7–10 | 7 | 32 | 36 | 21 | 4 |
| 2004 Nov 7–10 | 6 | 32 | 39 | 20 | 3 |
| 2003 Nov 3–5 | 4 | 36 | 35 | 20 | 5 |
| 2001 Nov 8–11 | 5 | 36 | 36 | 18 | 5 |
| **Men** | | | | | |
| 2006 Nov 9–12 | 3 | 27 | 45 | 22 | 3 |
| 2005 Nov 7–10 | 2 | 26 | 47 | 23 | 2 |
| 2004 Nov 7–10 | 2 | 28 | 50 | 18 | 2 |
| 2003 Nov 3–5 | 3 | 28 | 45 | 23 | 1 |
| 2001 Nov 8–11 | 3 | 32 | 46 | 17 | 2 |
| **Women** | | | | | |
| 2006 Nov 9–12 | 9 | 36 | 32 | 17 | 7 |
| 2005 Nov 7–10 | 11 | 39 | 26 | 18 | 6 |
| 2004 Nov 7–10 | 9 | 35 | 29 | 21 | 6 |
| 2003 Nov 3–5 | 6 | 43 | 26 | 17 | 8 |
| 2001 Nov 8–11 | 6 | 39 | 28 | 19 | 8 |

Note: Body mass index is a ratio of one's weight to their height. It is calculated as a person's weight (in kilograms) divided by their height (in meters) squared.

SOURCE: "[BODY MASS INDEX]," in *Personal Weight Situation*, The Gallup Organization, July 2007, http://www.gallup.com/poll/7264/Personal-Weight-Situation.aspx#1 (accessed November 14, 2007). Copyright © 2007 by The Gallup Organization. Reproduced by permission of The Gallup Organization.

## TABLE 11.3

**Public opinion on whether obesity has caused a serious health problem for a family member, July 2007**

HAS OBESITY EVER BEEN A CAUSE OF SERIOUS HEALTH PROBLEMS IN YOUR FAMILY?

| | Yes | No | No opinion |
|---|---|---|---|
| 2007 Jul 12–15 | 28% | 71 | 1 |

SOURCE: "Has obesity ever been a cause of serious health problems in your family?" in *Personal Weight Situation*, The Gallup Organization, 2007, http://www.gallup.com/poll/7264/Personal-Weight-Situation.aspx#1 (accessed November 14, 2007). Copyright © 2007 by The Gallup Organization. Reproduced by permission of The Gallup Organization.

## TABLE 11.4

**Consumer assessment of the healthfulness of yogurt, orange juice, and pasta, 2007**

ON A SIX POINT SCALE, WHERE 6 MEANS "VERY HEALTHFUL" AND 1 MEANS "NOT HEALTHFUL AT ALL," HOW HEALTHFUL WOULD YOU SAY THESE FOODS ARE? PLEASE SELECT ONE FOR EACH ITEM.

| Answer | Yogurt (n=693) | Orange juice (n=686) | Pasta (n=693) |
|---|---|---|---|
| 6 very healthful | 44% | 58% | 11% |
| 5 | 31% | 25% | 14% |
| 4 | 14% | 11% | 36% |
| 3 | 4% | 4% | 24% |
| 2 | 1% | 2% | 10% |
| 1 not healthful at all | * | * | 3% |
| Don't know | 5% | 0% | 1% |

"n" denotes number of respondents.
*Less than 0.5%.

SOURCE: Chung-Tung Jordan Lin, "A2. On a six point scale, where 6 means 'very healthful' and 1 means 'not healthful at all,' how healthful would you say these foods are? Please select one for each item," in *Experimental Study of Health Claims on Food Packages: Preliminary Topline Frequency Report*, U.S. Food and Drug Administration, Center for Food Safety and Applied Nutrition, May 2007, http://www.cfsan.fda.gov/~comm/crnutri4.html (accessed November 17, 2007)

## TABLE 11.5

**Consumer knowledge of foods that may reduce health risks, 2007**

HAVE YOU EVER HEARD OR READ THAT [EATING/DRINKING FOOD] REGULARLY MAY HELP LOWER THE RISK OF THE FOLLOWING HEALTH PROBLEMS? PLEASE SELECT AN ANSWER FOR EACH HEALTH PROBLEM.

| Health problem | Yogurt (n=693) | | |
|---|---|---|---|
| | Yes | No | Don't know |
| Hypertension or high blood pressure | 17% | 67% | 17% |
| Cancer | 12% | 70% | 18% |
| Osteoporosis or bone problem | 62% | 28% | 11% |
| Diabetes or high blood sugar | 15% | 68% | 17% |
| Heart disease | 24% | 61% | 15% |

| Health problem | Orange juice (n=686) | | |
|---|---|---|---|
| | Yes | No | Don't know |
| Hypertension or high blood pressure | 24% | 60% | 16% |
| Cancer | 35% | 50% | 15% |
| Osteoporosis or bone problem | 39% | 47% | 14% |
| Diabetes or high blood sugar | 16% | 68% | 16% |
| Heart disease | 46% | 40% | 14% |

| Health problem | Pasta (n=693) | | |
|---|---|---|---|
| | Yes | No | Don't know |
| Hypertension or high blood pressure | 7% | 80% | 13% |
| Cancer | 4% | 83% | 13% |
| Osteoporosis or bone problem | 7% | 80% | 13% |
| Diabetes or high blood sugar | 7% | 80% | 12% |
| Heart disease | 9% | 78% | 12% |

"n" denotes number of observations.

SOURCE: Chung-Tung Jordan Lin, "B2. Have you ever heard or read that [eating/drinking food] regularly may help lower the risk of the following health problems? Please select an answer for each health problem. [RANDOMIZE LIST; RECORD FIRST ITEM.]," in *Experimental Study of Health Claims on Food Packages: Preliminary Topline Frequency Report*, U.S. Food and Drug Administration, Center for Food Safety and Applied Nutrition, May 2007, http://www.cfsan.fda.gov/~comm/crnutri4.html (accessed November 17, 2007)

pasta), the nutrients these foods contained, and their possible health benefits.

Of the three foods, yogurt and orange juice were considered healthier than pasta, with orange juice garnering the highest percentage (58%) of "very healthful" ratings. (See Table 11.4.) Nearly two-thirds (62%) of respondents queried about the health benefits of yogurt named reducing the risk of osteoporosis, and reducing the risk of hypertension (high blood pressure) was identified as a health benefit of orange juice by one-quarter (24%) of respondents questioned about it. (See Table 11.5.) Just 9% of respondents asked about the health benefit of pasta associated it with heart disease.

Nearly all the respondents (99%) who linked yogurt to osteoporosis named calcium as the nutrient that might help reduce the risk; three out of four (74%) of those who associated orange juice with hypertension said potassium might help reduce the risk; and about 17% of those who associated pasta with heart disease said the fictitious compound lysoton might help reduce the risk. (See Table 11.6.)

Interestingly, just 15% of respondents said they were dieting to lose weight. (See Table 11.7.) Nineteen percent of respondents were following low-fat diets, 16% were on low-carbohydrate diets, 13% were adhering to low-sodium diets, and 11% each chose low-cholesterol and low-sugar diets. Because respondents were permitted to choose more than one diet, some may be adhering to more than one plan. For example, it is likely that many people who are on low-fat diets also aim to consume a diet that is low in cholesterol. Similarly, low-carbohydrate diets are generally also low-sugar diets.

### AMERICANS KNOW OBESITY IS HARMFUL

The American public is, however, certain that obesity is harmful. As Table 11.8 shows, in 2007 nearly five out of six (83%) Gallup Poll survey respondents said they felt obesity was "very harmful," and an additional 15%

## TABLE 11.6

### Consumer knowledge of nutrients that may help to reduce health risks, 2007

FOR EACH OF THE FOLLOWING NUTRIENTS, WOULD YOU SAY IT MIGHT HELP REDUCE THE RISK OF [HEALTH PROBLEM]? IF YOU HAVE NEVER HEARD OF A NUTRIENT, PLEASE SELECT THAT OPTION.

| Nutrient | Yes | No | Have not heard | Don't know |
|---|---|---|---|---|
| **Osteoporosis or bone problem (n=428)** | | | | |
| Calcium | 99% | * | 0% | 1% |
| Potassium | 54% | 14% | * | 32% |
| Vitamin A | 46% | 13% | * | 40% |
| Phosphorus | 44% | 11% | 4% | 41% |
| **Hypertension or high blood pressure (n=161)** | | | | |
| Calcium | 60% | 15% | * | 25% |
| Potassium | 74% | 6% | * | 19% |
| Vitamin C | 68% | 6% | * | 25% |
| Vitamin A | 62% | 7% | * | 30% |
| **Heart disease (n=65)** | | | | |
| Lysoton | 17% | 12% | 48% | 23% |
| Fiber | 89% | 5% | 0% | 6% |
| Calcium | 58% | 18% | 0% | 23% |
| Potassium | 74% | 8% | 0% | 18% |

"n" denotes number of observations.
*Less than 0.5%.

SOURCE: Chung-Tung Jordan Lin, "B5. For each of the following nutrients, would you say it might help reduce the risk of [health problem]? If you have never heard of a nutrient, please select that option. [RANDOMIZE LIST IN EACH FOOD.]," in *Experimental Study of Health Claims on Food Packages: Preliminary Topline Frequency Report*, U.S. Food and Drug Administration, Center for Food Safety and Applied Nutrition, May 2007, http://www.cfsan .fda.gov/~comm/crnutri4.html (accessed November 17, 2007)

## TABLE 11.7

### Consumer use of various diets, 2007

WHICH OF THESE DIET PLANS HAVE YOU YOURSELF BEEN ON DURING THE PAST 30 DAYS? SELECT ALL THAT APPLY.

| Diet | (n=1036) |
|---|---|
| Low fat diet | 19% |
| Low carb or carbohydrate diet | 16% |
| Low sodium diet | 13% |
| Low calorie diet | 11% |
| Low cholesterol diet | 11% |
| Low sugar diet | 15% |
| Weight loss diet | 15% |
| None of these | 55% |
| Don't know | 1% |
| Prefer not to answer | * |

"n" denotes number of respondents.
*Less than 0.5%.

SOURCE: Chung-Tung Jordan Lin, "C2. Which of these diet plans have you yourself been on during the past 30 days? Select all that apply," in *Experimental Study of Health Claims on Food Packages: Preliminary Topline Frequency Report*, U.S. Food and Drug Administration, Center for Food Safety and Applied Nutrition, May 2007, http://www.cfsan.fda.gov/ ~comm/crnutri4.html (accessed November 17, 2007)

considered it "somewhat harmful." Interestingly, survey respondents' own self-reported weight did not appear to influence their belief that being overweight is harmful. The vast majority of both respondents who considered themselves about the right weight (86%) and those who said they were overweight (79%) were aware that being significantly overweight is very harmful. (See Table 11.9.)

In fact, Americans equate the health risks associated with obesity to those of smoking. The overwhelming majority of Gallup Poll survey respondents said that being obese was "very harmful" (83%) or "somewhat harmful" (15%) to one's health. (See Table 11.10.) Comparable percentages of respondents deemed smoking "very harmful" (79%) or "somewhat harmful" (14%) to health.

### Americans Understand That Obesity Is a Public Health Problem

According to the press release "Despite Conflicting Studies about Obesity, Most Americans Think the Problem Remains Serious" (July 14, 2005, http://www.hsph.harvard .edu/news/press-releases/archives/2005-releases/press0714 2005.html), the Harvard School of Public Health conducted a June 2005 opinion poll and found that three-quarters of Americans at that time considered obesity an "extremely" (34%) or "very serious" (41%) public health problem in the United States. Most Americans believed that scientific experts have been accurately representing (58%) or even underestimating (22%) the health risks of obesity. A scant 15% of the survey respondents thought that the health risks were being overstated by scientific experts.

The poll found that about one-third (35%) of Americans were monitoring the fat and carbohydrate content in their daily diet. Like the Gallup Poll, this survey found that even though more than half (54%) of those surveyed considered themselves to be overweight, less than one-third (32%) were attempting to lose weight.

Americans are aware that being moderately overweight leads to serious health problems; however, they remain unconvinced that overweight and obesity lead to premature death. Half (51%) of the respondents in the Harvard poll thought that someone who is moderately overweight would be more likely than someone of healthy weight to die prematurely. Nearly three-quarters (73%) believed that a moderately overweight person would be more likely than someone of healthy weight to develop a chronic illness such as diabetes or hypertension.

In view of apparently conflicting reports of the supremacy of one diet over another, it is not surprising that the Harvard poll found differing sentiments about the credulity of obesity researchers and nutrition experts. About half (48%) of respondents said they had a "great deal" (14%) or a "good amount" (34%) of trust in the advice scientific experts offer about how to lose and control weight. Sixty-one percent of respondents said they paid "a lot" (13%) or a "fair amount" (48%) of attention to nutritional recommendations from medical and scientific experts about how to manage their weight.

TABLE 11.8

**Public opinion on the harm posed by obesity, July 2007**

IN GENERAL, HOW HARMFUL DO YOU FEEL OBESITY IS TO ADULTS WHO ARE SIGNIFICANTLY OVERWEIGHT—VERY HARMFUL, SOMEWHAT HARMFUL, NOT TOO HARMFUL, OR NOT AT ALL HARMFUL?

| | Very harmful | Somewhat harmful | Not too harmful | Not at all harmful | Depends (vol.) | No opinion |
|---|---|---|---|---|---|---|
| 2007 Jul 12–15 | 83% | 15 | * | * | * | 1 |

*Less than 0.5%.

SOURCE: "In general, how harmful do you feel obesity is to adults who are significantly overweight—very harmful, somewhat harmful, not too harmful, or not at all harmful?" in *Personal Weight Situation*, The Gallup Organization, 2007, http://www.gallup.com/poll/7264/Personal-Weight-Situation.aspx#1 (accessed November 14, 2007). Copyright © 2007 by The Gallup Organization. Reproduced by permission of The Gallup Organization.

---

**TABLE 11.9**

**Percent of adult poll respondents who believe overweight is harmful, by survey respondents' body weight, July 2007**

IN GENERAL, HOW HARMFUL DO YOU FEEL OBESITY IS TO ADULTS WHO ARE SIGNIFICANTLY OVERWEIGHT—VERY HARMFUL, SOMEWHAT HARMFUL, NOT TOO HARMFUL, OR NOT AT ALL HARMFUL?

| | Very harmful % | Somewhat harmful % | Not too harmful % | Not at all harmful % | Depends (vol.) % | No opinion % |
|---|---|---|---|---|---|---|
| Describe self as overweight | 79 | 20 | * | * | — | * |
| Describe self as "about right" as far as weight is concerned | 86 | 11 | 1 | 1 | 1 | 1 |

*Less than 0.5%.
Note: Too few Americans classify themselves as "underweight" to provide meaningful results.

SOURCE: Frank Newport, "In general, how harmful do you feel obesity is to adults who are significantly overweight—very harmful, somewhat harmful, not too harmful, or not at all harmful?" in *Americans Put Obesity on Par with Smoking in Terms of Harmful Effects*, Gallup News Service, The Gallup Organization, July 20, 2007, http://www.gallup.com/poll/28177/Americans-Put-Obesity-par-Smoking-Terms-Harmful-Effects.aspx (accessed November 15, 2007). Copyright © 2007 by The Gallup Organization. Reproduced by permission of The Gallup Organization.

---

**TABLE 11.10**

**Public opinion on how harmful obesity is to health, 2007**

HOW HARMFUL ARE THE FOLLOWING TO ONE'S HEALTH?

| | Very harmful % | Some-what harmful % | Not too harmful % | Not at all harmful % | Depends (vol.) % | No opinion % |
|---|---|---|---|---|---|---|
| Being obese | 83 | 15 | * | * | * | 1 |
| Smoking | 79 | 14 | 3 | 2 | 1 | * |

*Less than 0.5%.

SOURCE: Frank Newport, "How Harmful Are the Following to One's Health," in *Americans Put Obesity on Par with Smoking in Terms of Harmful Effects*, Gallup News Service, The Gallup Organization, July 20, 2007, http://www.gallup.com/poll/28177/Americans-Put-Obesity-par-Smoking-Terms-Harmful-Effects.aspx (accessed November 15, 2007). Copyright © 2007 by The Gallup Organization. Reproduced by permission of The Gallup Organization.

## Americans Sympathize with People Who are Obese

There is considerable sympathy for people who are obese. Nearly three-quarters (74%) of Gallup respondents expressed sympathy for people who are obese and under-standing of the difficulties associated with losing weight. (See Table 11.11.) Furthermore, as with the belief that obesity is harmful, sympathy for people who are obese seems independent of whether respondents' view themselves as about the right weight or overweight—79% of respondents who described themselves as overweight and 73% of those who said they were "about right" were sympathetic toward people who are obese. (See Table 11.12.)

Table 11.13 reveals that more women (78%) than men (70%) expressed sympathy for people who were obese. Table 11.14 shows that overall, there is more sympathy for people who are obese (74%) than for those who smoke (58%).

Even though the Gallup Poll results do not reveal why the public feels more sympathy for people who are obese than for smokers, it may be because rates of obesity are increasing among people of all ages, race, and ethnicity, whereas smoking has declined. Because increasing numbers of Americans are overweight or obese, perhaps they are more inclined to feel empathy for others who share the problem. In "Americans Put Obesity on Par with Smoking in Terms of Harmful Effects" (July 20, 2007, http://www.gallup

**TABLE 11.11**

**Public opinion on people who are obese, 2007**

WHICH OF THE FOLLOWING STATEMENTS BETTER DESCRIBES YOUR VIEW TOWARD PEOPLE WHO ARE OBESE—[ROTATED: YOU ARE UNSYMPATHETIC TOWARD PEOPLE WHO ARE OBESE BECAUSE THEY DO NOT LOSE WEIGHT EVEN THOUGH THEY KNOW BEING OVERWEIGHT IS HARMFUL TO THEIR HEALTH, (OR) YOU ARE SYMPATHETIC TOWARD PEOPLE WHO ARE OBESE BECAUSE YOU UNDERSTAND THAT IT IS DIFFICULT FOR THEM TO LOSE WEIGHT EVEN IF THEY WANT TO]?

| | Unsympathetic toward obese | Sympathetic toward obese | Both/mixed (vol.) | No opinion |
|---|---|---|---|---|
| 2007 Jul 12–15 | 21% | 74 | 4 | 1 |

SOURCE: "Which of the following statements better describes your view toward people who are obese—[ROTATED: you are unsympathetic toward people who are obese because they do not lose weight even though they know being overweight is harmful to their health, (or) you are sympathetic toward people who are obese because you understand that it is difficult for them to lose weight even if they want to]?" in *Personal Weight Situation*, The Gallup Organization, 2007, http://www.gallup.com/poll/7264/Personal-Weight-Situation.aspx#1 (accessed November 14, 2007). Copyright © 2007 by The Gallup Organization. Reproduced by permission of The Gallup Organization.

**TABLE 11.12**

**Percent of adult poll respondents who are sympathetic toward people who are obese, by survey respondents' body weight, 2007**

WHICH OF THE FOLLOWING STATEMENTS BETTER DESCRIBES YOUR VIEW TOWARD PEOPLE WHO ARE OBESE—[ROTATED: YOU ARE UNSYMPATHETIC TOWARD PEOPLE WHO ARE OBESE BECAUSE THEY DO NOT LOSE WEIGHT EVEN THOUGH THEY KNOW BEING OVERWEIGHT IS HARMFUL TO THEIR HEALTH, (OR) YOU ARE SYMPATHETIC TOWARD PEOPLE WHO ARE OBESE BECAUSE YOU UNDERSTAND THAT IT IS DIFFICULT FOR THEM TO LOSE WEIGHT EVEN IF THEY WANT TO]?

| | Unsympathetic toward obese % | Sympathetic toward obese % | Both/Mixed (vol.) % | No opinion % |
|---|---|---|---|---|
| Describe self as overweight | 17 | 79 | 4 | 1 |
| Describe self as "about right" as far as weight is concerned | 22 | 73 | 4 | 1 |

SOURCE: Frank Newport, "Which of the following statements better describes your view toward people who are obese—[ROTATED: you are unsympathetic toward people who are obese because they do not lose weight even though they know being overweight is harmful to their health, (or) you are sympathetic toward people who are obese because you understand that it is difficult for them to lose weight even if they want to]?" in *Americans Put Obesity on par with Smoking in Terms of Harmful Effects*, The Gallup Organization, July 20, 2007, http://www.gallup.com/poll/28177/Americans-Put-Obesity-par-Smoking-Terms-Harmful-Effects.aspx (accessed November 15, 2007). Copyright © 2007 by The Gallup Organization. Reproduced by permission of The Gallup Organization.

**TABLE 11.13**

**Percent of adults who are sympathetic toward people who are obese, by survey respondents' gender, 2007**

WHICH OF THE FOLLOWING STATEMENTS BETTER DESCRIBES YOUR VIEW TOWARD PEOPLE WHO ARE OBESE—[ROTATED: YOU ARE UNSYMPATHETIC TOWARD PEOPLE WHO ARE OBESE BECAUSE THEY DO NOT LOSE WEIGHT EVEN THOUGH THEY KNOW BEING OVERWEIGHT IS HARMFUL TO THEIR HEALTH, (OR) YOU ARE SYMPATHETIC TOWARD PEOPLE WHO ARE OBESE BECAUSE YOU UNDERSTAND THAT IT IS DIFFICULT FOR THEM TO LOSE WEIGHT EVEN IF THEY WANT TO]?

| | Unsympathetic toward obese % | Sympathetic toward obese % | Both/Mixed (vol.) % | No opinion % |
|---|---|---|---|---|
| Men | 25 | 70 | 3 | 1 |
| Women | 17 | 78 | 4 | 1 |

SOURCE: Frank Newport, "Which of the following statements better describes your view toward people who are obese—[ROTATED: you are unsympathetic toward people who are obese because they do not lose weight even though they know being overweight is harmful to their health, (or) you are sympathetic toward people who are obese because you understand that it is difficult for them to lose weight even if they want to]?" in *Americans Put Obesity on par with Smoking in Terms of Harmful Effects*, The Gallup Organization, July 20, 2007, http://www.gallup.com/poll/28177/Americans-Put-Obesity-par-Smoking-Terms-Harmful-Effects.aspx (accessed November 15, 2007). Copyright © 2007 by The Gallup Organization. Reproduced by permission of The Gallup Organization.

**TABLE 11.14**

**Poll respondents' sympathy for smokers versus people who are obese, 2007**

WHICH OF THE FOLLOWING STATEMENTS BETTER DESCRIBES YOUR VIEW TOWARD PEOPLE WHO SMOKE—[ROTATED: YOU ARE UNSYMPATHETIC TOWARD SMOKERS BECAUSE THEY CONTINUE TO SMOKE EVEN WHEN THEY KNOW IT'S HARMFUL TO THEIR HEALTH AND THE HEALTH OF THOSE AROUND THEM, OR YOU ARE SYMPATHETIC TOWARD SMOKERS BECAUSE THEY ARE ADDICTED, AND YOU UNDERSTAND THAT IT IS DIFFICULT TO STOP EVEN IF THEY WANT TO]?

| | Unsympathetic % | Sympathetic % | Both/Mixed (vol.) % | No opinion % |
|---|---|---|---|---|
| Being obese | 21 | 74 | 4 | 1 |
| Smoking | 37 | 58 | 4 | 2 |

SOURCE: Frank Newport, "Which of the following statements better describes your view toward people who smoke—[ROTATED: you are unsympathetic toward smokers because they continue to smoke even when they know it's harmful to their health and the health of those around them, or you are sympathetic toward smokers because they are addicted, and you understand that it is difficult to stop even if they want to]? " in *Americans Put Obesity on par with Smoking in Terms of Harmful Effects*, The Gallup Organization, July 20, 2007, http://www.gallup.com/poll/28177/Americans-Put-Obesity-par-Smoking-Terms-Harmful-Effects.aspx (accessed November 15, 2007). Copyright © 2007 by The Gallup Organization. Reproduced by permission of The Gallup Organization.

.com/poll/28177/Americans-Put-Obesity-par-Smoking-Terms-Harmful-Effects.aspx), Frank Newport of the Gallup Organization opines that "the public is more willing to believe that obesity is a condition more under one's voluntary control than is the case for smoking."

## MANY AMERICANS WANT TO LOSE WEIGHT

The Gallup Poll reports that trying to lose weight is a common activity among the U.S. adult population—more Americans than ever before say they are seriously trying to lose weight. Table 11.15 shows that in November 2006, 28% of Americans said they were seriously trying to lose weight, up from 17% about fifty years earlier, in 1955. A February 2006 Gallup Poll found that Americans wanted to lose "a lot of weight" (18%) or "a little weight" (38%), whereas 39% wanted to maintain their weight and just 4% were seeking to gain weight. (See Table 11.16.)

## TABLE 11.15

**Percent of adult poll respondents who are seriously trying to lose weight, by gender, selected years, 1951–2006**

AT THIS TIME ARE YOU SERIOUSLY TRYING TO LOSE WEIGHT?

| | Yes % | No % | No opinion % |
|---|---|---|---|
| **National adults** | | | |
| 2006 Nov 9–12 | 28 | 71 | 1 |
| 2005 Nov 7–10 | 27 | 73 | * |
| 2004 Nov 7–10 | 29 | 71 | * |
| 2003 Nov 3–5 | 28 | 72 | * |
| 2002 Nov 11–14 | 24 | 75 | 1 |
| 2001 Jul 19–22 | 25 | 75 | * |
| 1999 Jul 22–25 | 20 | 80 | * |
| 1996 Feb 23–25 | 26 | 74 | * |
| 1990 Oct 18–21 | 18 | 82 | * |
| 1955 | 17 | 83 | * |
| 1953 | 25 | 75 | * |
| 1951 | 19 | 81 | * |
| **Men** | | | |
| 2006 Nov 9–12 | 24 | 75 | 1 |
| 2005 Nov 7–10 | 23 | 77 | * |
| 2004 Nov 7–10 | 23 | 77 | — |
| 2003 Nov 3–5 | 21 | 79 | * |
| 2002 Nov 11–14 | 19 | 80 | 1 |
| 2001 Jul 19–22 | 17 | 82 | 1 |
| 1999 Jul 22–25 | 16 | 84 | * |
| 1996 Feb 23–25 | 22 | 78 | * |
| 1990 Oct 18–21 | 11 | 88 | 1 |
| **Women** | | | |
| 2006 Nov 9–12 | 32 | 67 | * |
| 2005 Nov 7–10 | 30 | 70 | * |
| 2004 Nov 7–10 | 34 | 66 | * |
| 2003 Nov 3–5 | 35 | 65 | * |
| 2002 Nov 11–14 | 30 | 70 | * |
| 2001 Jul 19–22 | 32 | 68 | 0 |
| 1999 Jul 22–25 | 24 | 76 | * |
| 1996 Feb 23–25 | 30 | 70 | * |
| 1990 Oct 18–21 | 24 | 76 | * |

*Less than 0.5%.

SOURCE: "At this time are you seriously trying to lose weight?" in *Personal Weight Situation*, The Gallup Organization, 2007, http://www.gallup.com/poll/7264/Personal-Weight-Situation.aspx#1 (accessed November 14, 2007). Copyright © 2007 by The Gallup Organization. Reproduced by permission of The Gallup Organization.

## TABLE 11.16

**Percent of adult poll respondents wishing to lose, gain, or maintain their weight, February 2006**

WOULD YOU LIKE TO LOSE A LOT OF WEIGHT, OR ONLY A LITTLE WEIGHT?

[Combined responses]

| | Lose a lot of weight | Lose only a little weight | Stay at present | Put on weight | No opinion |
|---|---|---|---|---|---|
| 2006 Feb 9–12 | 18% | 38 | 39 | 4 | 1 |

SOURCE: "Would you like to [ROTATED: lose weight, stay at your present weight, or put on weight]? Would you like to lose a lot of weight, or only a little weight?" in *Personal Weight Situation*, The Gallup Organization, July 2007, http://www.gallup.com/poll/7264/Personal-Weight-Situation.aspx#1 (accessed November 14, 2007). Copyright © 2007 by The Gallup Organization. Reproduced by permission of The Gallup Organization.

## TABLE 11.17

**Percent of adult poll respondents choosing diet or exercise to lose weight, February 2006**

IF YOU HAD TO CHOOSE, WOULD YOU BE MORE LIKELY TO TRY TO LOSE WEIGHT—[ROTATED: BY DIETING, (OR MORE LIKELY TO TRY TO LOSE WEIGHT) BY EXERCISING]?

[Based on adults who would like to lose weight]

| | Dieting | Exercising | No opinion |
|---|---|---|---|
| 2006 Feb 9–12 | 36% | 61 | 3 |

SOURCE: "If you had to choose, would you be more likely to try to lose weight—[ROTATED: by dieting, (or more likely to try to lose weight) by exercising]?" in *Personal Weight Situation*, The Gallup Organization, July 2007, http://www.gallup.com/poll/7264/Personal-Weight-Situation.aspx#1 (accessed November 14, 2007). Copyright © 2007 by The Gallup Organization. Reproduced by permission of The Gallup Organization.

## TABLE 11.18

**Percent of adult poll respondents wanting to lose weight who would consider bariatric surgery, February 2006**

AS YOU MAY KNOW, SOME PEOPLE HAVE HAD SURGERY TO REDUCE THE SIZE OF THEIR STOMACH AS A MEANS OF LOSING WEIGHT. WHICH OF THE FOLLOWING BEST DESCRIBES YOU—[ROTATED: YOU WOULD DEFINITELY LIKE TO HAVE THIS SURGERY, YOU MIGHT BE INTERESTED IN HAVING IT BUT DON'T KNOW FOR SURE, YOU WOULD ONLY CONSIDER HAVING IT AS A LAST RESORT, (OR) YOU WOULD NEVER HAVE THIS SURGERY]?

[Based on adults who would like to lose weight]

| | Definitely like to have surgery | Might be interested | Only as last resort | Never have this surgery | No opinion |
|---|---|---|---|---|---|
| 2006 Feb 9–12 | 2% | 1 | 20 | 75 | 2 |

SOURCE: "As you may know, some people have had surgery to reduce the size of their stomach as a means of losing weight. Which of the following best describes you—[ROTATED: you would definitely like to have this surgery, you might be interested in having it but don't know for sure, you would only consider having it as a last resort, (or) you would never have this surgery]?" in *Personal Weight Situation*, The Gallup Organization, July 2007, http://www.gallup.com/poll/7264/Personal-Weight-Situation.aspx#1 (accessed November 14, 2007). Copyright © 2007 by The Gallup Organization. Reproduced by permission of The Gallup Organization.

Women were much more likely than men to be making serious efforts to lose weight. In November 2006 less than one-quarter (24%) of men said they were making serious attempts to lose weight, compared to 32% of women. (See Table 11.15.)

Significantly more survey respondents who wanted to lose weight indicated that they would be more likely to try exercising (61%) rather than dieting (36%) to lose weight. (See Table 11.17.) Three-quarters of the survey respondents who indicated that they would like to lose weight said they would never have bariatric surgery, and another 20% said they would consider surgery only as a last resort. Despite the increasing popularity of bariatric surgery, just 3% of those surveyed said they would definitely like to have surgery (2%) or might be interested it (1%). (See Table 11.18.)

## AMERICANS' CHANGING SHAPES AND SIZES

The results of a national size survey that gathered measurements from more than ten thousand people across the United States confirmed that Americans are not only getting heavier but also are changing in proportion. The "SizeUSA" project is an anthropometric research study (it studies human body measurements and makes comparisons of these measurements). Using a three-dimensional body scanner, researchers compiled measurements and analyzed them by gender, age group, and four ethnicities, as well as by geography, annual household income, marital status, education, and employment status.

The survey was performed to assist apparel manufacturers in producing clothing that will offer a better fit to more consumers. In "Sizing up America: Signs of Expansion from Head to Toe" (*New York Times*, March 1, 2004), Kate Zernike reports that the last such national survey of Americans was performed in 1941 by the U.S. Department of Agriculture (USDA). The USDA survey described the average American woman as a size 8, with a 35-inch bust, a 27-inch waist, and a 37.5-inch hip circumference. The 2003 "SizeUSA" survey found that the average white woman's bust, waist, and hip measurements in inches were 38-32-41 for women aged eighteen to twenty-five, and 41-34-43 for women aged thirty-six to forty-five. On average, African-American women measured 43-37-46, Hispanic women 42.5-36-44, and an "other" category, composed primarily of Asian-American women, measured 41-35-43. Based on the "SizeUSA" survey, the average American woman wears a size twelve or fourteen, rather than a size eight.

American men have also increased in size. The size 40 regular, which measures 40 inches at the chest with a 34-inch waist, a 40-inch hip, and a 15.5-inch collar, once considered the average, would be too small for many American men. The 2003 "SizeUSA" survey found that white men aged eighteen to twenty-five measured 41-35-41, and older white men aged thirty-six to forty-five measured 44-38-42. African-American men measured an average of 43-37-42, Hispanic men 44-38-42, and an "other" category, composed primarily of Asian-American men, measured 42-37-41.

Interestingly, measurements did not vary significantly by geography, education, or even income. The most significant variations in body shape were attributed to race, ethnicity, and age. For example, 11% of white women were described as having protruding stomachs, compared to 3% of Hispanic women and 4% of African-American women. More Hispanic women (20%) were described as having "full waists" than white (10%) or African-American (15%) women. Nearly one-quarter (24%) of African-American men were described as having a "prominent seat," compared to 9% of white men and 8% of Hispanic men.

The study concluded that along with expanding waists, American men over age forty-five were the most likely to have increased abdominal girth—"pot bellies"—and women older than thirty-six were the most likely to have big hips. Nearly 20% of men were described as "portly" and another 19% had "lower front waists," meaning the researchers had to look behind the overhanging belly to find the waist.

## AMERICANS' ATTITUDES ABOUT OVERWEIGHT

Despite escalating media coverage of overweight and obesity, and their associated health risks, many Americans do not appear to be overly concerned about overweight and obesity—their own or others'. They demonstrate little support for policy initiatives intended to prevent and combat obesity, and persist in the belief that obesity results from individual personal failings rather than a combination of genetic and environmental factors.

Taeku Lee and J. Eric Oliver examine in *Public Opinion and the Politics of America's Obesity Epidemic* (May 2002, http://ksgnotes1.harvard.edu/Research/wpaper.nsf/rwp/RWP02-017/$File/rwp02_017_lee.pdf) the prevailing sentiments about weight-related issues. Lee and Oliver sought to characterize Americans' attitudes about obesity to determine how attitudes and beliefs affect support for obesity-related policy changes. They assert that the concept of "moral failure" is at the root of public opinions that hold obesity as a personal choice and responsibility. They posit that obesity violates the valued American trait of self-reliance. Characterizing people who are obese as lazy, undisciplined, and lacking self-control enables the public to hold them responsible for their condition and may be used as justification for bias and discrimination. Lee and Oliver also posit that when obesity is understood as resulting from a lack of individual motivation, there will be little support for policies such as government regulations, civil protections, or taxes to prevent and decrease it.

Lee and Oliver observe that because obesity in the United States is a relatively recent phenomenon, public opinions about it are still forming, and most proposed policy changes—including taxes on sugary or high-fat snack foods, strengthening civil protections for individuals who are obese, and increasing the availability of public land for exercise—are unsupported or are actively opposed by a majority of Americans. In contrast, growing support exists for measures that regulate food advertising to children and that provide more nutritious school lunches. Lee and Oliver attribute the lack of enthusiasm for policy changes to low levels of awareness of the severity and scope of the problem and to deeply held negative stereotypes about people who are overweight and obese. They assert that as Americans learn that the rapid rise in obesity during the past two decades did not result from moral failure, they will be more inclined to advocate policies aimed at preventing and reducing obesity.

## ARE AMERICANS GETTING ENOUGH EXERCISE?

In "Only One-Third of Americans Are Frequent Exercisers" (November 22, 2006, http://www.gallup.com/poll/25546/Only-OneThird-Americans-Frequent-Exercisers.aspx), Frank Newport indicates that Americans' self-reports of the frequency and intensity of the exercise they engage in has remained constant since 2000 and continues to fall short of basic health recommendations. He notes that 65% of Americans surveyed reported that they did not participate in vigorous or moderate exercise five days a week or more. (See Figure 11.3.) Nearly half (45%) of the survey respondents reported that they never engage in vigorous exercise for at least twenty minutes. Just 12% of Americans say they exercise vigorously at least five days of the week. The average American engages in vigorous exercise less than two days per week.

More Americans engage in moderate exercise. Most survey respondents (84%) claimed that they occasionally obtain some type of moderate exercise. The average American engages in moderate exercise three days a week, and 29% get moderate exercise at least five days a week. Newport reveals that more respondents who engage in regular moderate or vigorous exercise also report excellent physical health, but cautions that the data do not determine causation—whether good health allows people to exercise or exercise creates good

**FIGURE 11.4**

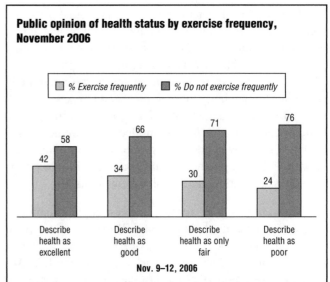

Public opinion of health status by exercise frequency, November 2006

SOURCE: Frank Newport, "Gallup Exercise Summary by Physical Health Ratings," in *Only One-Third of Americans Are Frequent Exercisers*, The Gallup Organization, November 22, 2006, http://www.gallup.com/poll/25546/Only-OneThird-Americans-Frequent-Exercisers.aspx (accessed November 19, 2007). Copyright © 2007 by The Gallup Organization. Reproduced by permission of The Gallup Organization.

health—or whether poor health prevents people from exercising. (See Figure 11.4.)

## CHILDHOOD OBESITY

Public health professionals and practitioners are researching and evaluating interventions aimed at preventing children and teens from becoming overweight as well as programs to help them lose weight. Children's and teen's attitudes about food, exercise, and weight influence the success of prevention and treatment programs.

In "Adolescents' Attitudes about Obesity and What They Want in Obesity Prevention Programs" (*Journal of School Nursing*, vol. 23, no. 4, 2007), Louise F. Wilson seeks to characterize adolescents' attitudes about overweight and obesity and to identify the features and attributes they value in prevention programs. She used a written questionnaire to survey middle school students to determine the program characteristics students felt would be most effective. Wilson finds that adolescents would be more likely to participate in, and adhere to, programs encouraging them to consume more water, fruits, and vegetables, eat less junk food, and exercise more. They expressed unwillingness to forgo soda, videogames, computer activities, or watching television to improve their health.

### Obesity Is among the Top Ten Concerns for U.S. Children

In March 2007 the C. S. Mott Children's Hospital and Knowledge Networks, Inc., conducted the National Poll on Children's Health (May 2, 2007, http://www.med

**FIGURE 11.3**

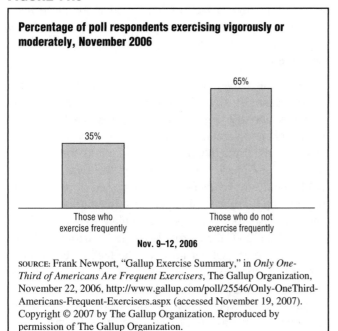

Percentage of poll respondents exercising vigorously or moderately, November 2006

SOURCE: Frank Newport, "Gallup Exercise Summary," in *Only One-Third of Americans Are Frequent Exercisers*, The Gallup Organization, November 22, 2006, http://www.gallup.com/poll/25546/Only-OneThird-Americans-Frequent-Exercisers.aspx (accessed November 19, 2007). Copyright © 2007 by The Gallup Organization. Reproduced by permission of The Gallup Organization.

.umich.edu/mott/research/chearhealthconcernpoll.html), a national online survey with a random sample of 2,076 adults. Of seventeen different health concerns for children, obesity was ranked number three, following smoking and drug abuse.

The poll found that Hispanics were more likely to express concern about obesity (42%), than whites (31%) or African-Americans (36%). The researchers opine that greater concern among Hispanics and African-Americans may reflect overall higher prevalence of obesity among Hispanic and African-American children and teens. Respondents with higher educational attainment (bachelor's degree or higher) rated childhood obesity as the number-one health issue for children, with 40% viewing it as a significant problem. In contrast, respondents with less than a high school education ranked childhood obesity tenth, with just one-quarter describing it as a major problem.

## Parents May Not Accurately Gauge Children's Weight

Parents play a pivotal role in terms of preventing childhood obesity by shaping their children's early eating and physical activity habits. Anjali Jain et al. observe in "Why Don't Low-Income Mothers Worry about Their Preschoolers Being Overweight?" (*Pediatrics* vol. 107, no. 5, May 2001) that mothers of overweight preschoolers frequently appear unaware of, or unconcerned about, their children's weight. To explore mothers' perceptions of overweight in children, why children become overweight, and barriers that prevent effective treatment of childhood obesity, the investigators conducted group interviews with low-income mothers of preschool children (aged twenty-four to sixty months old) who were overweight and determined to be at risk for obesity.

Jain et al. find that unlike health professionals, who assess children's weight status by plotting height and weight on standard growth charts, mothers were more likely to express concern about children's overweight when their children were teased by peers or unable to participate in physical activities. The mothers did not consider their children overweight if the children were active, had a good appetite, and ate a healthy diet. Instead of describing their children as overweight, mothers described them as "thick," "strong," "big-boned," or "solid." The interviewed mothers also believed that an inherited tendency to be overweight, in terms of inherited metabolism or body type, practically guaranteed that the child would become overweight regardless of environmental factors. Given this perception, it is not surprising that the mothers believed they were unable to affect a child's biological predisposition to be overweight.

In "Perception versus Reality: An Exploration of Children's Measured Body Mass in Relation to Caregivers' Estimates" (*Journal of Health Psychology*, vol. 12, no. 6,

2007), Anna Akerman, Marsha E. Williams, and John Meunier compare parents' reports of their children's height and weight against the measurements the researchers obtained. The researchers find that their measurements varied from the parents' perceptions of their children's body status. Parents of overweight children consistently underreported their children's BMI, and parents of underweight children overestimated their children's BMI. Akerman, Williams, and Meunier believe that parents have a "positive bias in cognition" that enables them to selectively interpret and correct for their children's deviations from a healthy body weight. In turn, this creates an alternative reality for them, one in which undesirable imperfections in their children do not exist.

## Americans Blame Parents, Schools, and the Food Industry for Children's Weight Gain

In "Poll Shows Growing Concern about Role of Advertising in Child Obesity" (*Wall Street Journal*, August 20, 2007), Beckey Bright reports on the *Wall Street Journal*/Harris Interactive August 2007 poll that surveyed attitudes about childhood obesity among 2,503 adults. The survey found that most Americans (84%) viewed childhood obesity as a major problem and 78% of parents with children under age twelve saw it as an "issue of growing concern." Eighty-three percent believed that parents have the greatest impact in terms of reducing childhood obesity.

Most of the survey respondents blamed the lack of exercise as a cause of children's overweight and felt that encouraging more physical activity will help solve the problem. The overwhelming majority (94%) felt schools should promote regular exercise. Nearly the same proportion (89%) favored parental efforts to limit time spent using computers, playing videogames, and watching television to encourage children to spend more time being physically active.

Most respondents also felt that children's diets must change. They believe schools and parents should restrict children's access to snack foods, sugary soft drinks, and fast food—88% said schools must do more to ensure that healthy foods are available and 83% felt parents must be more vigilant about their children's diets.

More than three-quarters (78%) of respondents cited food advertising targeting children as a "major contributor" to the problem, up from 65% the previous year. Despite the recent move of major food industry companies to improve the nutritional value of many of their offerings and engage in more responsible marketing and advertising practices, 60% of respondents favored government regulation of food industry advertising aimed at children.

There was strong support for measures the food industry might take to address the problem. Ninety-one percent of respondents expressed support for "using child-friendly

FIGURE 11.5

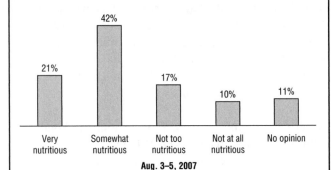

**Parents' opinions about the nutrition value of their children's school lunches, August 2007**

GENERALLY SPEAKING, WOULD YOU DESCRIBE THE LUNCHES SERVED IN THE SCHOOL THAT YOUR OLDEST CHILD ATTENDS AS—VERY NUTRITIOUS, SOMEWHAT NUTRITIOUS, NOT THAT NUTRITIOUS, OR NOT NUTRITIOUS AT ALL?

(Asked of those who are parents of children in grades K–12)

Aug. 3–5, 2007

SOURCE: Lydia Saad, "Generally speaking, would you describe the lunches served in the school that your oldest child attends as—very nutritious, somewhat nutritious, not that nutritious, or not nutritious at all?" in *Parents Indicate School Cafeterias Could Do Better*, The Gallup Organization, August 15, 2007, http://www.gallup.com/poll/28402/Parents-Indicate-School-Cafeterias-Could-Better.aspx#1 (accessed November 19, 2007). Copyright © 2007 by The Gallup Organization. Reproduced by permission of The Gallup Organization.

FIGURE 11.6

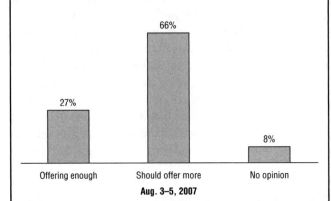

**Parents' opinions about whether schools offer enough healthy foods, August 2007**

THINKING ABOUT THE AMOUNT OF FOODS SUCH AS FRESH FRUITS, VEGETABLES, AND WHOLE GRAIN BREADS OFFERED IN YOUR CHILD'S SCHOOL, DO YOU THINK THE SCHOOL IS OFFERING ENOUGH OF THESE KINDS OF FOOD, OR SHOULD IT OFFER MORE?

(Asked of those who are parents of children in grades K–12)

Aug. 3–5, 2007

SOURCE: Lydia Saad, "Thinking about the amount of foods such as fresh fruits, vegetables, and whole grain breads offered in your child's school, do you think the school is offering enough of these kinds of food, or should it offer more?" in *Parents Indicate School Cafeterias Could Do Better*, The Gallup Organization, August 15, 2007, http://www.gallup.com/poll/28402/Parents-Indicate-School-Cafeterias-Could-Better.aspx#1 (accessed November 19, 2007). Copyright © 2007 by The Gallup Organization. Reproduced by permission of The Gallup Organization.

characters to promote healthier foods like fruits and vegetables," and 73% favored "limiting advertising to children to healthier foods that are lower in calories, fat and/or sugar." About two-thirds (64%) said that "no longer using popular characters from television shows and movies to market products to children" would help.

## Parents Feel Schools Should Offer Healthier Food Choices

An August 2007 Gallup Poll found that nearly two-thirds of parents described their children's school lunches as "very nutritious" (21%) or "somewhat nutritious" (42%), whereas more than one-quarter rated them "not too nutritious" (17%) or "not at all nutritious" (10%). (See Figure 11.5.) Two-thirds (66%) of the parents surveyed felt their children's schools should offer more healthy foods such as whole grain breads, fresh fruits, and vegetables. (See Figure 11.6.)

Despite their expressed dissatisfaction with schools' food choices and offerings, parents do not assign schools disproportionate responsibility for childhood obesity. Just 9% blame school lunches a "great deal" for the problem, whereas 22% think school lunches contribute a "moderate amount" to the problem. More than a third (37%) of respondents assigned "not much" blame to school lunches, and 30% said they were not at all to blame. (See Figure 11.7.)

FIGURE 11.7

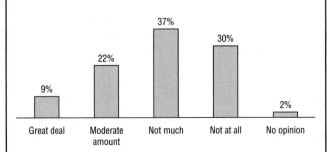

**Parents' opinions about whether school lunches are to blame for childhood obesity, August 2007**

HOW MUCH DO YOU THINK SCHOOL LUNCHES ARE TO BLAME FOR THE PROBLEM OF CHILDHOOD OBESITY—A GREAT DEAL, A MODERATE AMOUNT, NOT MUCH, OR NOT AT ALL?

(Asked of those who are parents of children in grades K–12)

Aug. 3–5, 2007

SOURCE: Lydia Saad, "How much do you think school lunches are to blame for the problem of childhood obesity—a great deal, a moderate amount, not much, or not at all?" in *Parents Indicate School Cafeterias Could Do Better*, The Gallup Organization, August 15, 2007, http://www.gallup.com/poll/28402/Parents-Indicate-School-Cafeterias-Could-Better.aspx#1 (accessed November 19, 2007). Copyright © 2007 by The Gallup Organization. Reproduced by permission of The Gallup Organization.

## The Cost of Healthy Foods Plays a Role in Children's Weight Gain

Roland Sturm and Ashlesha Datar of the Rand Corporation confirm in "Body Mass Index in Elementary School Children, Metropolitan Area Food Prices and Food Outlet Density" (*Public Health*, September 2, 2005) that the cost of fresh fruits and vegetables is closely associated with weight gain among children and has a better predictive value for overweight than whether they live near or frequent fast-food outlets. Sturm and Datar examined the weight gain of 6,918 children of varying socioeconomic backgrounds from fifty-nine U.S. metropolitan areas as the children progressed from kindergarten to third grade. The investigators did not analyze the children's diets but instead compared their weight gain to the price of different types of foods and the number of food outlets in their area.

Sturm and Datar find that young children who lived in communities where fruits and vegetables were expensive were more likely to gain excessive amounts of weight than those who lived in areas where produce costs less. On average, children in the study gained twenty-nine pounds. In Mobile, Alabama, the region with the highest relative price for produce, children gained about 50% more excess weight as measured by BMI than children nationally. In contrast, children in Visalia, California, the area with the lowest relative cost for fruits and vegetables, experienced excess weight gain that was about half the national average. The researchers opine that providing free fruits and vegetables to schoolchildren would improve their diet. The USDA launched such a program in about one hundred schools, Sturm and Datar observe, and it was met with enthusiastic support from parents and teachers. It is not yet known, however, whether the program has influenced participating children's weight.

# IMPORTANT NAMES AND ADDRESSES

**Academy for Eating Disorders**
60 Revere Dr., Ste. 500
Northbrook, IL 60062-1577
(847) 498-4274
FAX: (847) 480-9282
E-mail: info@aedweb.org
URL: http://www.aedweb.org/

**American Academy of Sleep Medicine**
One Westbrook Corporate Center, Ste. 920
Westchester, IL 60154
(708) 492-0930
FAX: (708) 492-0943
URL: http://www.aasmnet.org/

**American Cancer Society**
1599 Clifton Rd. NE
Atlanta, GA 30329
(404) 329-5705
FAX: (404) 325-9341
URL: http://www.cancer.org/

**American Diabetes Association**
1701 N. Beauregard St.
Alexandria, VA 22311
1-800-342-2383
E-mail: AskADA@diabetes.org
URL: http://www.diabetes.org/

**American Dietetic Association**
120 S. Riverside Plaza, Ste. 2000
Chicago, IL 60606-6995
1-800-877-1600
URL: http://www.eatright.org/

**American Heart Association**
7272 Greenville Ave.
Dallas, TX 75231
1-800-242-8721
URL: http://www.americanheart.org/

**American Obesity Association**
8630 Fenton St., Ste. 814
Silver Spring, MD 20910
(301) 563-6526

FAX: (301) 563-9595
URL: http://www.obesity.org/

**American Society of Bariatric Physicians**
2821 S. Parker Rd., Ste. 625
Aurora, CO 80014
(303) 770-2526
FAX: (303) 779-4834
URL: http://www.asbp.org/

**American Society for Metabolic and Bariatric Surgery**
100 SW Seventy-fifth St., Ste. 201
Gainesville, FL 32607
(352) 331-4900
FAX: (352) 331-4975
E-mail: info@asbs.org
URL: http://www.asbs.org/

**Arthritis Foundation**
PO Box 7669
Atlanta, GA 30357-0669
1-800-283-7800
URL: http://www.arthritis.org/

**Atkins Nutritionals, Inc.**
2002 Orville Dr. N, Ste. A
Ronkonkoma, NY 11779-7661
(212) 457-9345
URL: http://www.atkins.com/

**Center for Science in the Public Interest**
1875 Connecticut Ave. NW, Ste. 300
Washington, DC 20009
(202) 332-9110
FAX: (202) 265-4954
E-mail: cspi@cspinet.org
URL: http://www.cspinet.org/

**Centers for Disease Control and Prevention**
1600 Clifton Rd. NE
Atlanta, GA 30333
(404) 498-1515
1-800-311-3435
URL: http://www.cdc.gov/

**Council on Size and Weight Discrimination**
PO Box 305
Mt. Marion, NY 12456
(845) 679-1209
FAX: (845) 679-1206
E-mail: info@cswd.org
URL: http://www.cswd.org/

**Eating Disorders Coalition**
611 Pennsylvania Ave. SE,
Ste. 423
Washington, DC 20003-4303
(202) 543-9570
URL: http://www.eatingdisorderscoalition
.org/

**Federal Trade Commission**
600 Pennsylvania Ave. NW
Washington, DC 20580
1-877-382-4357
URL: http://www.ftc.gov/

**International Food Information Council**
1100 Connecticut Ave. NW, Ste. 430
Washington, DC 20036
(202) 296-6540
FAX: (202) 296-6547
E-mail: foodinfo@ific.org
URL: http://www.ific.org/

**International Size Acceptance Association**
PO Box 82126
Austin, TX 78758
URL: http://www.size-acceptance.org/

**National Association to Advance Fat Acceptance**
PO Box 22510
Oakland, CA 94609
(916) 558-6880
URL: http://www.naafa.org/

**National Association of Anorexia Nervosa and Associated Disorders**
PO Box 7
Highland Park, IL 60035
(847) 831-3438
FAX: (847) 433-4632
URL: http://www.anad.org/

**National Association of Cognitive-Behavioral Therapists**
102 Gilson Ave.
Weirton, WV 26062
(304) 723-3982
1-800-853-1135
FAX: (304) 723-3982
E-mail: nacbt@nacbt.org
URL: http://www.nacbt.org/

**National Center for Health Statistics**
3311 Toledo Rd.
Hyattsville, MD 20782
1-800-232-4636
E-mail: nchsquery@cdc.gov
URL: http://www.cdc.gov/nchs/

**National Center on Sleep Disorders Research**
National Heart, Lung, and Blood Institute, NIH
6701 Rockledge Dr.
Bethesda, MD 20892
(301) 435-0199
FAX: (301) 480-3451
URL: http://www.nhlbi.nih.gov/about/ncsdr/

**National Diabetes Information Clearinghouse**
1 Information Way
Bethesda, MD 20892-3560
1-800-860-8747
FAX: (703) 738-4929
E-mail: ndic@info.niddk.nih.gov
URL: http://diabetes.niddk.nih.gov/

**National Digestive Diseases Information Clearinghouse**
2 Information Way
Bethesda, MD 20892-3570
1-800-891-5389
FAX: (703) 738-4929
E-mail: nddic@info.niddk.nih.gov
URL: http://digestive.niddk.nih.gov/about/

**National Eating Disorders Association**
603 Stewart St., Ste. 803
Seattle, WA 98101
(206) 382-3587
1-800-931-2237
E-mail: info@NationalEatingDisorders.org
URL: http://www.nationaleatingdisorders.org/

**National Heart, Lung, and Blood Institute Health Information Center**
PO Box 30105
Bethesda, MD 20824-0105
(301) 592-8573
FAX: (240) 629-3246
E-mail: nhlbiinfo@nhlbi.nih.gov
URL: http://www.nhlbi.nih.gov/

**National Institute of Diabetes and Digestive and Kidney Diseases**
Bldg. 31, Rm. 9A04
31 Center Dr., MSC 2560
Bethesda, MD 20892-2560
(301) 496-3583
URL: http://www.niddk.nih.gov/

**National Mental Health Association**
2000 N. Beauregard St., Sixth Fl.
Alexandria, VA 22311
(703) 684-7722
1-800-969-6642

FAX: (703) 684-5968
URL: http://www.nmha.org/

**National Women's Health Information Center**
8270 Willow Oaks Corporate Dr.
Fairfax, VA 22031
1-800-994-9662
URL: http://www.4woman.gov/

**Rudd Center for Food Policy and Obesity**
309 Edwards St.
Yale University
New Haven, CT 06520-8369
(203) 432-6700
URL: http://www.yaleruddcenter.org/home.aspx

**TOPS Club, Inc.**
4575 S. Fifth St.
Milwaukee, WI 53207
(414) 482-4620
E-mail: topsinteractive@tops.org
URL: http://www.tops.org/

**Weight-Control Information Network**
1 WIN Way
Bethesda, MD 20892-3665
1-877-946-4627
FAX: (202) 828-1028
E-mail: win@info.niddk.nih.gov
URL: http://win.niddk.nih.gov/index.htm

**Weight Watchers International, Inc.**
175 Crossways Park West
Woodbury, NY 11797-2055
(516) 390-1400
URL: http://www.weightwatchers.com/index.aspx

# RESOURCES

The Centers for Disease Control and Prevention (CDC) tracks nationwide health trends, including overweight and obesity, and reports its findings in several periodicals, especially its *Health, United States* and *Morbidity and Mortality Weekly Reports*. The *National Vital Statistics Reports*, which is issued by the CDC's National Center for Health Statistics (NCHS), gives detailed information on U.S. births, birth weights, and death data and trends. The NCHS also compiles and analyzes demographic data—the heights and weights of a representative sample of the U.S. population—to develop standards for desirable weights. The National Health Interview Surveys, the National Health Examination Surveys, the National Health and Nutrition Examination Surveys, and the Behavioral Risk Factor Surveillance System offer ongoing information about the lifestyles, health behaviors, and health risks of Americans. Working with other agencies and professional organizations, the CDC produced *Healthy People 2010* (2007, http://hp2010.nhlbihin.net/), which serves as a blueprint for improving the health status of Americans.

The U.S. Department of Agriculture provides nutrition guidelines for Americans, and the Federal Trade Commission (FTC) has launched initiatives to educate consumers and the media about false and deceptive weight-loss advertising. The FTC is one of about fifty members of the Partnership for Healthy Weight Management, a coalition of scientific, academic, health-care, government, commercial, and public-interest representatives, that aims to increase public awareness of the obesity epidemic and to promote responsible marketing of weight-loss products and programs.

The relationship between birth weight and future health risks has been examined by many researchers, and the studies cited in this text were reported in the *American Journal of Obstetrics and Gynecology*, the *British Medical Journal, Circulation*, the *International Journal of Cancer*, and *Pediatrics*. Data from the CDC Pregnancy Nutrition Surveillance System show that very overweight women benefit from reduced weight gain during pregnancy to help reduce the risk for high-birth-weight infants.

The World Health Organization and the National Institutes of Health provide definitions, epidemiological data, and research findings about a comprehensive range of public health issues, including diet, nutrition, overweight, and obesity. The Central Intelligence Agency's *World Factbook* provides longevity estimates. The National Heart, Lung, and Blood Institute conducts research about obesity and overweight. Weight-control information and updated weight-for-height tables that incorporate height, weight, and body mass index are published by the National Institute of Diabetes and Digestive and Kidney Diseases (the part of the National Institutes of Health that is primarily responsible for obesity- and nutrition-related research). The National Institute of Mental Health offers information about eating disorders as well as the mental health issues related to obesity.

The origins, causes, and consequences of the obesity epidemic have been described in numerous professional and consumer publications, including *Ageing Research Reviews, American Journal of Clinical Nutrition, American Journal of Health Promotion, American Journal of Lifestyle Medicine, American Journal of Preventive Medicine, American Journal of Psychiatry, American Journal of Public Health, American Psychologist, Annual Review of Nutrition, Annals of Internal Medicine, Annals of the New York Academy of Sciences, Archives of Disease in Childhood, Archives of Family Medicine, Archives of Internal Medicine, Archives of Pediatrics and Adolescent Medicine, British Journal of Diabetes and Vascular Disease, British Journal of Gynecology, Cancer Epidemiology Biomarkers and Prevention, Chemistry and Industry, Consumer Reports, Current Opinion in Lipidology, Diabetes, Diabetes Care, Disease Management, Eating Behaviors,*

*Health Affairs, Health Psychology, International Journal of Eating Disorders, International Journal of Obesity, Journal of the American Medical Association, Journal of Clinical Endocrinology and Metabolism, Journal of Clinical Oncology, Journal of General Internal Medicine, Journal of Occupational and Environmental Medicine, Journal of School Health, Journal of School Nursing, Lancet, Medicine and Science in Sports and Exercise, Metabolism: Clinical and Experimental, National Law Journal, New England Journal of Medicine, Nutrition Journal, Obesity, Obesity Research, Obesity Reviews, Preventing Chronic Diseases, Proceedings of the National Academy of Sciences, Psychological Medicine, Public Health, Science,* and *Surgery for Obesity Related Diseases.*

Several excellent books and publications provided valuable insight into the obesity epidemic. Peter N. Stearns, in *Fat History: Bodies and Beauty in the Modern West* (2002), and Laura Fraser, in *Losing It: False Hopes and Fat Profits in the Diet Industry* (1997), offer detailed histories of magical cures and weight-loss fads. Other titles referenced in this edition include books by Kelly D. Brownell and Katherine Battle Horgen, *Food Fight: The Inside Story of the Food Industry, America's Obesity Crisis, and What We Can Do about It* (2004), and Greg Critser, *Fat Land: How Americans Became the Fattest People in the World* (2003). *Diabesity: The Obesity-Diabetes Epidemic That Threatens America—And What We Must Do to Stop It* (2005) by Francine R. Kaufman, who served as the president of the American Diabetes Association, contends that the diabesity epidemic "imperils human existence as we now know it."

Medical and public-health societies, along with advocacy organizations, professional associations, and foundations, offer a wealth of information about the relationship between weight, health, and disease. Sources cited in this edition include the American Dietetic Association, the American Heart Association, the American Medical Association, the American Obesity Association, the Center for Consumer Freedom, the Center for Science in the Public Interest, the International Size Acceptance Association, the National Academy of Sciences, the National Association to Advance Fat Acceptance, the National Eating Disorders Association, the Pharmacy Benefit Management Institute, and the Public Health Advocacy Institute.

The Gallup Organization makes available valuable poll and survey data about Americans' attitudes about overweight, obesity, physical activity, diet, and nutrition. Finally, many professional associations, voluntary medical organizations, and foundations dedicated to research, education, and advocacy about eating disorders, overweight, and obesity provided the up-to-date information included in this edition.

# INDEX

*Page references in italics refer to photographs. References with the letter* t *following them indicate the presence of a table. The letter* f *indicates a figure. If more than one table or figure appears on a particular page, the exact item number for the table or figure being referenced is provided.*

## J

Jameson, Gardener, 81
Jazzercise, Inc., 137
Jet fuel, 136–137
Joint injury, 35–37
Junk food, 64–65, 131, 132f

## K

Kaiser weight-loss program, 124
Kempner, Walter, 83
Knee problems, 36–37

## L

Labeling, 130(t8.3), 131, 134
Lanzalotta, Stephen, 83
Law and Obesity Project, PHAI, 131
Lawsuits, 131, 132–133, 137
Laxative use, 47f
Leading causes of death, 19, 21t–22t
Legislation
    Child Nutrition Act, 67
    "Fat and Short Law" (San Francisco), 137
    food industry protection, 133
    Healthy Students Act (proposed), 132
    Medicare Prescription Drug,
        Improvement, and Modernization
        Act, 117
    school programs, 132
    Stop Obesity in Schools Act
        (proposed), 132
*Letter on Corpulence, Addressed to the
    Public* (Banting), 79
Life expectancy, 19–20, 20t, 151–152
Linn, Robert, 82
Longevity, 151–152
Lost productivity, 113, 118
Low birth weight infants, 3–5, 3t
Low-calorie diets, 87, 87(t5.5), 88t, 89t,
    90t, 91t
Low-carbohydrate diets, 79–83, 87–90,
    92–95, 141–142, 143
Low-fat diets, 82, 90–95, 94t, 143
Low-income population, 169

## M

Macronutrients, 84t
Maintenance. *See* Weight maintenance
Mammograms, 136
Manz, Esther, 80
Marinetti, Filippo Tommaso, 80
Marketing. *See* Advertising
Maternal weight, 41–42, 42t
Mazel, Judy, 82
McDonald's Corporation, 134
Mead Johnson Company, 80
Measurement
    body fat, 7–9, 8(t1.4)
    health, 19
Meat consumption, 10–11

Media
    adolescents' television viewing habits,
        69(f4.22)
    eating disorders, 51–54, 76
    junk-food marketing to children and
        adolescents, 15, 64–65
    nutrition and health education, 65
    prevention programs, 168
Medicaid, 116–117
Medical insurance, 5, 23, 116–118, 132, 169
Medical research funding, 118, 119t–122t
Medicare, 23, 116–117
Medicare Prescription Drug, Improvement,
    and Modernization Act, 117
Medications. *See* Prescription drugs
Melanocortin 4 receptor (MC4R)
    deficiency, 27
Men Shoot for 9 program, 162–163
Mental health. *See* Psychological issues
Menus, 160t–161t
Metabolic syndrome, 42–43, 70, 73
Metabolism, 23, 27, 29
Metrecal, 80
Metropolitan Life Insurance Company
    (MetLife), 5
Mexican American cuisine, 91(t5.10)
Mild overweight, 110
Milk consumption, 11, 60, 62(f4.8)
Monogenic causes of obesity, 27
Montignac, Michel, 83
"Moral failure," obesity as, 17
Morbidity
    birth weight, 3–5
    obesity during pregnancy, 41
    overweight and obesity, 2, 19–20, 29–30
    *See also* Specific diseases
Mortality
    cancer, 40
    leading causes of death, 21t–22t
    overweight and obesity, 2, 19–20
    weight cycling, 109
Moving, 167
Mutations, gene, 27, 48
Myocardial infarction, 42–43
MyPyramid, 155–157, 157t, 158t, 159t, 162t
MyPyramid for Kids, 163f
Myths, diet and weight-loss, 141–144

## N

National Association to Advance Fat
    Acceptance (NAAFA), 136, 140
National Conference of State Legislatures, 132
National Health and Nutrition Examination
    Surveys, 9–10
National Health Examination Surveys, 9–10
Natural weight-loss products, 143
Negative-calorie foods myth, 142
Neotame, 81
New Jersey, 67

New Mexico, 112–113
Nidetch, Jean, 80
No diet or exercise required claims,
    147–149, 149(t9.3)
No diet or exercise required schemes,
    147–149
No restrictions on eating claims, 149
Noakes, Manny, 83
Noninsulin-dependent diabetes, 33, 35
Nonnutritive sweeteners, 81
Nonprescription diet aids, 123
Nonprescription weight-loss aids, 104–105
Nutrigenetics, 146
Nutrition. *See* Diet and nutrition; Nutrition
    education
Nutrition education
    *Dietary Guidelines for Americans*
        (USDA), 84–85
    effectiveness, 164–166
    5 a Day for Better Health program,
        157–159, 162–163
    funding, 154–155
    health-care providers, 77
    media efforts, 65
    MyPyramid, 155–157, 157f, 159t, 162t
Nutritionists, 123

## O

Obesity
    by age, 23f
    attitudes toward, 179
    body mass index classification, 8(t1.3)
    cancer, 39–40
    causes, 10–15
    children, 180–183
    definition, 7–10
    diabetes, 33, 35
    discrimination, 134–140
    disease model, 21–23
    fatty liver disease, 38–39
    gallbladder disease, 37–38
    genetics, 24, 27–29
    health consequences, 28t
    health consequences for children and
        adolescents, 69–70, 73
    health-care costs, 111–113, 118
    hospital costs of childhood and
        adolescent obesity, 114–116
    infant weight gain as prediction of, 4–5
    morbidity, influence on, 19–20
    osteoarthritis and joint injury, 35–37
    prevalence, 1, 9–10, 9f, 10f
    prevention programs for children and
        women, 168–169
    psychological issues, 45–46
    public opinion, 174–177
    public opinion on health problems of
        family members, 174(t11.3)
    by sex, age, race/ethnicity, and poverty
        level, 24t–27t

# Q

Quick weight loss claims, 150, 151*t*

# R

Race/ethnicity
- birth weight, 3*t*
- childhood obesity, concern about, 181
- eating disorders, 144
- exercise, 97–98, 98*t*–99*t*, 101*f*
- hypertension, 34*t*–35*t*
- overweight and obesity prevalence, 1, 9, 24*t*–27*t*
- overweight prevalence among adolescents, 59*f*
- prevention programs, 154
- serum cholesterol levels, 31*t*–33*t*
- size survey, 179

Rebound weight gain, 103–104
Recipes, 14
Recovery from eating disorders, 50
Red Flag Campaign, 151
Relapse prevention, 146
Research
- caloric restriction, 151–152
- funding, 118, 119*t*–122*t*
- weight management, 146

Respect Fitness Health Initiative, 139
Restaurants, 13–15
*The Rice Diet Cookbook* (Rosati), 83
Rimonabant, 103
Risk factors
- binge-eating disorder, 46
- eating disorders, 76
- obesity, 27–29
- overweight and obesity among children and adolescents, 69–70, 73
- weight loss, 109–110

Roeper, Richard, 53
Rosati, Kitty Gurkin, 83
Roux-en-Y gastric bypass surgery, 105–106

# S

Saccharin, 81
Sample menus, 160*t*–161*t*
San Bushmen, 105
School attendance, 73
School food choices
- competitive foods, 65–67, 66*f*, 67*t*, 133–134
- junk food, 64–65, 132*f*
- public opinion, 181, 182, 182*f*
- vending machines, 131

Screening
- gynecological cancer, 136
- overweight children and adolescents, 74

Sears, Barry, 82
Sedentary lifestyle. *See* Culture
Self-assessment of diet, 171, 172(*f*11.1)
Self-help weight loss programs, 107

Serum cholesterol levels, 30, 31*t*–33*t*
Shoppers, discrimination against, 137
Simple carbohydrates, 88
Single gene mutations, 27
Sitting time, worksite, 167
Size acceptance movement, 139–140
Size inflation, 125
Size survey, 179
Skinfold thickness measurements, 7
Sleep apnea and sleep disorders, 40
"Small Steps" campaign, 168
Smoking, 176–177, 177(*t*11.14)
Snacking, 14, 165–166
Social consensus theory, 139
Societal norms, 29
Socioeconomic status. *See* Economic issues
Sodium content, 155*t*
Soft drinks, 64–65, 67, 123, 131
Solovay, Sondra, 137
*The South Beach Diet* (Agatston), 82–83
Southern cuisine, 90(*t*5.8)
Standing, 167
Starchy foods, 143
States
- obesity and chronic disease prevention funding, 165*f*
- obesity lawsuits, 133
- obesity prevalence, 1
- prevention funding, 163–164
- public policy, 131–132
- school food legislation, 132, 132*f*

Statistical information
- adolescents' television viewing habits, 69(*f*4.22)
- birth weight by race/ethnicity, 3*t*
- blood pressure levels for boys ages 1–17 years, 71*t*
- blood pressure levels for girls ages 1–17 years, 72*t*
- body mass index based on self-reported height and weight, 173(*t*11.1)
- body mass index chart, 6*t*
- body mass index classification of overweight and obesity, 8(*t*1.3)
- body mass index for boys, ages 2–20, 57*f*
- calorie counting and low-fat food choices, 63(*f*4.12)
- calorie requirements, 156*t*
- cholesterol levels in high-risk children and adolescents, 74*t*
- chronic health conditions causing limitation of activity, 38*f*
- consumer assessment of healthfulness of certain foods, 174(*t*11.4)
- consumer knowledge of foods that may reduce health risks, 174(*t*11.5)
- consumer knowledge of nutrients that may help reduce health risks, 175(*t*11.6)
- consumer use of various diets, 175(*t*11.7)

diabetes, 37*f*
diet pills, powders or liquids, use of, among adolescents, 63
eating out, 12*t*–13*t*
economic and health burden of chronic disease, 112*t*
fasting among adolescents, 63(*f*4.13)
food energy and macronutrients per capita per day, 84*t*
food energy contributed from major food groups, 85*t*
food expenditure trends, 17*t*–18*t*
food intake pattern calorie levels, 158*t*
fruit and vegetable consumption, 164*f*
fruit and vegetable consumption among adolescents, 62(*f*4.7)
gestational weight gain, recommended, 42*t*
high fructose corn syrup consumption, 16*t*
hospital stays, 114*t*–116*t*
hypertension, 34*t*–35*t*
interpretation of body mass index, by age, 58*f*
leading causes of death, 21*t*–22*t*
life expectancy, 20*t*
medical research funding, 119*t*–122*t*
milk consumption among adolescents, 62(*f*4.8)
MyPyramid food intake patterns, 159*t*
obesity prevalence, 9*f*
obesity prevalence, by age and sex, 10*f*
obesity trends, 2*f*
overweight among high school students, by state, 60*f*–61*f*
overweight and obesity, by age, 23*f*
overweight and obesity by sex, age, race/ethnicity, and poverty level, 24*t*–27*t*
overweight prevalence among adolescents, by race/ethnicity, 59*f*
percentage of adults engaged in regular physical activity, by age and gender, 101(*f*6.2)
percentage of adults engaged in regular physical activity, by race/ethnicity, 101(*f*6.3)
percentage of students trying to lose weight, 62(*f*4.10)
percentage of students who exercise to lose or keep from gaining weight, 63(*f*4.11)
physical activity levels of adults, 98*t*–99*t*
physical education programs, 67–70
prevalence of overweight children and adolescents, 56*f*
proportion of adults reporting no leisure-time physical activity, 100*f*
public opinion on current weight, 173(*t*11.1)
public opinion on exercise, 180*f*
public opinion on healthfulness of diet, 172(*f*11.1)
public opinion on obese people, 177*t*

Women
    discrimination against obese
        shoppers, 137
    eating disorders, 51–54
    health-care discrimination, 136
    obesity prevention programs, 168–169
    reproductive health, 40–42

Workers' compensation claims,
    119–120
Worksite prevention programs,
    166–167
World Health Organization (WHO),
    127–129, 130–131, 155, 168
Wyoming, 133

## Y
"Youth Risk Behavior Surveillance,"
    59–60

## Z
*The Zone: A Dietary Road Map* (Sears),
    82